AF608357

CISM COURSES AND LECTURES

The series presents lecture notes, monographs, edited works and proceedings in the field of Mechanics, Engineering, Computer Science and Applied Mathematics.
Purpose of the series is to make known in the international scientific and technical community results obtained in some of the activities organized by CISM, the International Centre for Mechanical Sciences.

INTERNATIONAL CENTRE FOR MECHANICAL SCIENCES

COURSES AND LECTURES - No. 428

INTERFACIAL PHENOMENA AND THE MARANGONI EFFECT

EDITED BY

MANUEL G. VELARDE
INSTITUTO PLURIDISCIPLINAR, MADRID

RADYADOUR Kh. ZEYTOUNIAN
UNIVERSITY OF LILLE

Springer-Verlag Wien GmbH

This volume contains 90 illustrations

Originally published by Springer-Verlag Wien New York in 2002

SPIN 10903230

In order to make this volume available as economically and as rapidly as possible the authors' typescripts have been reproduced in their original forms. This method unfortunately has its typographical limitations but it is hoped that they in no way distract the reader.

ISBN 978-3-211-83696-5 ISBN 978-3-7091-2550-2 (eBook)
DOI 10.1007/978-3-7091-2550-2

FOREWORD

This book contains material delivered at a Summer Course held at CISM (Udine) in July 2000, mostly dealing with nonequilibrium capillarity, i.e., phenomena induced by surface tension gradients which derive from temperature or surfactant concentration variations. Exception is the text of Prof. L. M. Pismen dealing with the contact angle/line problem whose serious and not merely heuristic understanding was long due.

Marangoni (1865), provided a wealth of detailed information on the effects of variations of the potential energy of liquid surfaces and, in particular, flow arising from variations in temperature and surfactant composition. Among the phenomena involving Marangoni flows, we note that associated with the name of Bénard (1900), which refers to the formation of a cellular structure in a thin liquid layer heated from below.

One aspect of this science is seen today to bear on important phenomena associated with the processing of modern materials. The role of the basic effect in technology was probably first demonstrated by chemical engineers in the field of liquid-liquid extraction. Indeed, phenomena attributable to Marangoni flows have been reported in innumerable instances relevant to modern technologies, such as in hot salt corrosion in aeroturbine blades; the drying of solvent-containing paints; the drying of silicon wafers used in electronics; in materials processing, particularly in metallic systems which have been suspected to demonstrate Marangoni flows: e.g. in pyro-metallurgy, in which interest lies in interfacial mass transfer and in the continuous casting of lubrificating films; in the complex mixing hydrodynamics of furnaces containing melts; in the erosion of the walls of ceramic crucibles containing liquid metals, in melt and crystal processes in microgravity conditions, associated with space research programmes. These instances reflect both a scientific and industrial interest.

The codirectors of the Course, and coeditors, wish to express their appreciation to Professors L. M. Pismen, D. T. Papageorgiu, V. Ya. Shkadov and K. C. Mills for their acceptance of both duties, lecturing and writing lecture notes. They also acknowledge the CISM Scientific Council, with special attention to Prof. A. Barrero and Rector Prof. Sandor Kaliszky for their encouragement and support. Last but not least, they are grateful to the Secretarial staff of CISM for their efficient handling of administrative matters beyond the call of duty. This Course was also part of the training activity of the EU Network ICOPAC HPRN-CT2000-00136.

M. G. Velarde
R. Kh. Zeytounian

CONTENTS

STATIC AND DYNAMIC THREE-PHASE CONTACT LINES

L.M. Pismen

Technion-Israel Institute of Technology, Haifa, Israel

Abstract. The lectures review the statics and dynamics of the gas-liquid-solid contact line, with the emphasis on the role of intermolecular forces and mesoscopic dynamics in the immediate vicinity of the three-phase boundary. We discuss paradoxes of the existing hydrodynamic theories and ways to resoluve them by taking account of intermoleculr forces, activated slip in the first molecular layer, diffuse character of the gas-liquid interface and interphase transport.

Introduction

The unifying line is that all phenomena involving three-phase contact lines are mesoscopic by their nature. This means that their macroscopic properties, which interest us when we compute large scale hydrodynamic flow, are intimately dependent on interactions on the microscopic level. As a consequence, purely hydrodynamic description turns out to be inadequate, and has to be complemented by mesoscopic models of the fluid in the vicinity of a two-phase (gas-liquid or fluid-solid) or three-phase (gas-liquid-solid) interface, where properties are different from those of the bulk fluid.

Eventually, matching of molecular dynamics with continuous hydrodynamic description may be necessary to take a precise account of microscopic properties of interfaces. Our aim is, however, more modest: to gain qualitative understanding. We shall restrict therefore to the simplest but most universal kind of interactions - van der Waals forces, and follow their influence on static and dynamic properties of interfaces and contact lines. We shall review two kinds of models of interfacial regions, assuming either discontinuous or diffuse interphase boundary. The description will remain continuous, even when we go down to molecular-scale distances.

1 Equilibrium surface tension

1.1 Thermodynamics of two-phase systems

Before approaching the problem of dynamics of contact line, we shall briefly review the equilibrium properties of gas-liquid interfaces and their dependence on the proximity to solid surfaces. We shall consider the simplest one-component system: a liquid in equilibrium with its vapor. Thermodynamic equilibrium in a two-phase system implies equilibrium of the interphase boundary, which tends to minimize its area. The thermodynamic quantity that expresses additional energy carried by the interface is *surface tension*, defined as the derivative of the Helmholtz or Gibbs free energy with respect to interfacial area Σ:

$$\gamma = \left(\frac{\partial F}{\partial \Sigma}\right)_{V,T} = \left(\frac{\partial \Phi}{\partial \Sigma}\right)_{p,T} \tag{1}$$

Surface tension originates in intermolecular forces of cohesion. These forces are short-range but *nonlocal.* The equilibrium density should minimize the free energy. Near the interphase boundary, the density varies in space, and the interfacial energy can be computed as an excess energy of this inhomogeneous layer.

Computation of this extra energy appears to be a formidable problem, and necessarily involves a sequence of assumptions and approximations. Since strong interactions cannot be treated perturbatively, it is usually assumed that interactions are weak and long-range. This, however, hardly can be justified, unless for weakly non-ideal gases. Fortunately, strong interactions in a dense fluid nearly balance each other, so that the remaining net force acting on each molecule is weak and can be expressed as a superposition of pair interactions with molecules in its vicinity. Another strong assumption is that interactions with molecules at discrete and, generally, correlated positions are smeared out in space, and replaced by the action of a continuous effective medium. This is a *mean field* that does not take into account fluctuations. In a uniform medium, these assumptions allow to derive the van der Waals equation of state starting from the statistical theory of hard spheres with attractive interactions. Our treatment of inhomogeneous fluids will not go beyond this level. Such a theory, of course, cannot be quantitative, but what we need is a simple working model that could be extended to non-equilibrium situations.

It is convenient to use as thermodynamic variables temperature T and density $\rho = N/V$; N is the number of particles and V is volume. Further on, we shall restrict to isothermal processes, while density will be allowed to change in space. In this variables. the Helmholtz free energy is expressed as $F = Nf(\rho, T)$, and pressure p and chemical potential μ are defined as

$$p = -\left(\frac{\partial F}{\partial V}\right)_{N,T} = \rho^2 \frac{\partial f}{\partial \rho}, \quad \mu = \left(\frac{\partial F}{\partial N}\right)_{V,T} = \frac{\partial(\rho f)}{\partial \rho}. \tag{2}$$

Allowing for variable density, the expression for the local Helmholtz free energy per molecule $f[\rho(\boldsymbol{x})]$ derived under the above assumptions reads

$$f[\rho(\boldsymbol{x})] = f_0 - \frac{T}{2} \int \left(\mathrm{e}^{-U(r)/T} - 1\right) \rho(\boldsymbol{x} + \boldsymbol{r})\, \mathrm{d}^3\boldsymbol{r}. \tag{3}$$

Here U is the pair interaction potential dependent on the distance $r = |\boldsymbol{r}|$; we use the units with the Boltzmann constant scaled to unity; $f_0 = T \ln \rho$ is a reference free energy of an "ideal" state with interactions switched off; the factor $\frac{1}{2}$ compensates counting twice the interacting molecules in each pair. A suitable interaction potential is the Lennard–Jones potential

$$U = -A_l r^{-6} \left[1 - \tfrac{1}{2}(d/r)^6\right], \tag{4}$$

where d is the separation at the energy minimum and A_l is a parameter defining the interaction strength. The short-range term can be replaced by hard- core repulsion:

$$U = \begin{cases} -A_l r^{-6} & (r > d) \\ \infty & (r < d) \end{cases}, \tag{5}$$

where d is the nominal hard-core molecule diameter. Both potentials are shown together in Fig. 1.

When density is constant, the integral in Eq. (3) with hard-core interaction potential gives the free energy of a homogeneous van der Waals fluid. In the hard-core region $r < d$ the exponential

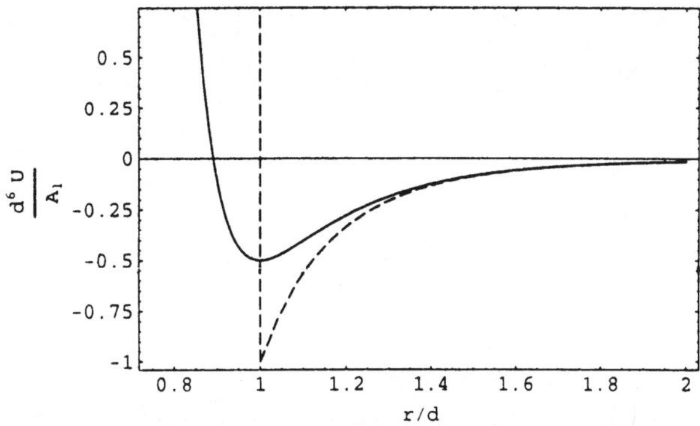

Figure 1. The Lennard–Jones (solid line) and hard core (dashed line) potentials.

term in Eq. (3) vanishes, while at $r > d$ the it is expanded in Taylor series. The result is

$$\overline{f}(\rho, T) = T \ln \rho + \rho(bT - a) \approx T \ln \frac{\rho}{1 - b\rho} - a\rho, \tag{6}$$

where $b = \frac{2}{3}\pi d^3$ is the excluded volume and

$$a = -2\pi \int_d^\infty U(r) r^2 \, \mathrm{d}r = \frac{2\pi A_l}{3d^3}. \tag{7}$$

The approximate expression in (6), applicable at $b\rho \ll 1$, yields by Eq. (2) the van der Waals equation of state $(p + a\rho^2)(1 - b\rho) = \rho T$.

Separating the homogeneous part, Eq. (3) can be rewritten as

$$f(\boldsymbol{x}) = \overline{f}[\rho(\boldsymbol{x})] + \tfrac{1}{2} \int_{r>d} U(r)[\rho(\boldsymbol{x} + \boldsymbol{r}) - \rho(\boldsymbol{x})] \, \mathrm{d}^3\boldsymbol{r}. \tag{8}$$

The two terms in the above expression give, respectively, the free energy of a homogeneous state and a "distortion energy" due to changes of density in space. The equilibrium density is defined by the minimum of $F = \int \rho f \mathrm{d}^3\boldsymbol{x}$ subject to the constraint of particle number conservation. This condition is enforced by introducing a Lagrange multiplier – chemical potential μ. Thus, the integral to be minimized is

$$\begin{aligned} \mathcal{F} = F - \mu N = \int \rho(\boldsymbol{x})[\overline{f}(\rho) - \mu] \mathrm{d}^3\boldsymbol{x} \\ + \tfrac{1}{2} \int \rho(\boldsymbol{x}) \, \mathrm{d}^3\boldsymbol{x} \int_{r>d} U(r)[\rho(\boldsymbol{x} + \boldsymbol{r}) - \rho(\boldsymbol{x})] \, \mathrm{d}^3\boldsymbol{r}. \end{aligned} \tag{9}$$

Since we are dealing with a two-phase system, the integrand of the first term should have two minima ρ_l, ρ_v corresponding to two stable uniform equilibrium states of higher and lower density (liquid and vapor). A flat interface separating the two phases is at equilibrium when both

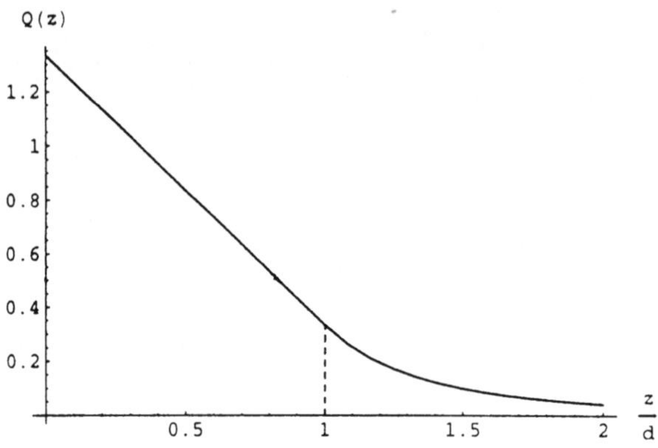

Figure 2. The surface energy density in the liquid as a function of depth.

stable equilibria are at Maxwell construction, i.e. the minima are of equal depth. This can be achieved at a certain value of chemical potential. The latter serves as a bias parameter that shift the equilibrium in favor of the dense (dilute) phase when it increases (decreases). The interfacial energy is contributed both by deviations from the equilibrium density levels in the transitional region and by the distortion energy localized there. Thus, the surface tension can be computed, once the density distribution in the transitional layer is known.

1.2 Surface tension of sharp and diffuse interface

Computation of density can be avoided if one assumes that it changes abruptly across the interphase boundary, i.e. jumps between both equilibrium values ρ_l, ρ_v over a molecular-scale distance d. This is the *sharp interface* approximation. The interfacial energy is contributed then only by the second term in Eq. (9). We consider a flat boundary at $z = 0$ and neglect the vapor density compared to the liquid density, i.e. set $\rho = \rho_l = \text{const}$ at $z < 0$ and $\rho = 0$ at $z > 0$. Then a short computation using the hard- core interaction potential (5) yields the interfacial energy per unit area

$$\gamma = \pi A_l \rho_l^2 \int_{-\infty}^{0} Q(z)\mathrm{d}z = \frac{\pi A_l \rho_l^2}{2d^2}. \tag{10}$$

where the energy density in each layer is computed using the radial integration variable $q = r^2$ with the lower integration limit $q_0 = (\zeta - z)^2$ at $\zeta - z < d$, $q_0 = d^2$ at $0 < \zeta - z \leq d$:

$$Q(z) = \int_0^{\infty} \mathrm{d}\zeta \int_{q_0}^{\infty} q^{-3}\,\mathrm{d}q = \begin{cases} \frac{1}{6}|z|^{-3} & (|z| > d) \\ \frac{1}{2}d^{-3}\left(\frac{4}{3} - |z|/d\right) & (|z| < d) \end{cases} \tag{11}$$

Although any continuum computation in a molecular scale region is questionable, the above expression serves us well, as it smoothly connects the $|z|^{-3}$ dependence of the energy density in the bulk region with a linear dependence near the interface (see Fig. 2) and avoids a nonphysical divergence that would arise if the change in excluded volume was not taken into account.

The opposite limiting case is that of a *diffuse* boundary with the density changing only slightly over distances comparable with the characteristic interaction length. Then one can expand

$$\rho(\boldsymbol{x}+\boldsymbol{r}) = \rho(\boldsymbol{x}) + \boldsymbol{r}\cdot\nabla\rho(\boldsymbol{x}) + \tfrac{1}{2}\boldsymbol{r}\boldsymbol{r} : \nabla\nabla\rho(\boldsymbol{x}) + \ldots \tag{12}$$

Using this in Eq. (9) we see that the contribution of the linear term to the nonlocal integral vanishes when the system is isotropic and, as a consequence, the interaction term is spherically symmetrical, and the lowest order contribution is due to the quadratic term:

$$F_{\mathrm{nl}}(\boldsymbol{x}) = -\tfrac{1}{2}K\int \rho(\boldsymbol{x})\nabla^2\rho(\boldsymbol{x})\,\mathrm{d}^3\boldsymbol{x} = \tfrac{1}{2}K\int |\nabla\rho(\boldsymbol{x})|^2\,\mathrm{d}^3\boldsymbol{x}, \tag{13}$$

where

$$K = -4\pi\int_d^\infty U(r)\,r^4\,\mathrm{d}r = \frac{4\pi A_l}{d}. \tag{14}$$

Thus, Eq. (9) is replaced by

$$\mathcal{F} = \int \left[\rho\bar{f}(\rho) - \mu\rho + \tfrac{1}{2}K|\nabla\rho(\boldsymbol{x})|^2\right]\mathrm{d}^3\boldsymbol{x}. \tag{15}$$

The corresponding Euler–Lagrange equation is

$$K\nabla^2\rho - g(\rho) + \mu = 0, \tag{16}$$

which has the form of a nonlinear reaction-diffusion equation containing a nonlinear function $g(\rho) = \mathrm{d}(\rho\bar{f}(\rho))/\mathrm{d}\rho$.

Coexistence between two semi-infinite phases separated by a planar boundary is described by the one-dimensional version of Eq. (16):

$$K\rho_{zz} - g(\rho) + \mu = 0. \tag{17}$$

The two static solutions are approached at $z \to \pm\infty$. The interfacial energy is computed most easily by using ρ as an independent, and the distortion energy density $W = \frac{1}{2}K\rho_z^2$ as a dependent variable. Then Eq. (17) is rewritten as

$$W'(\rho) - g(\rho) + \mu = 0. \tag{18}$$

Since W vanishes in the bulk fluid, this equation has to be integrated with the boundary conditions $W(\rho_l) = W(\rho_v) = 0$. The problem is, thus, overdetermined, and the value of chemical potential μ at which the phases coexist has to be determined from the solvability condition

$$\mu = \frac{\rho_l\bar{f}(\rho_l) - \rho_v\bar{f}(\rho_v)}{\rho_l - \rho_v}. \tag{19}$$

This is the well-known "equal area" condition (Fig. 3). One can also prove that pressure $p = \rho^2 f'(\rho)$ is constant in both coexisting phases. Indeed, differentiating Eq. (18) yields $W''(\rho) = \rho^{-1}p'(\rho)$, and the pressure difference between the two phases is computed as

$$p(\rho_l) - p(\rho_v) = \int_{\rho_v}^{\rho_l} \rho W''(\rho)\mathrm{d}\rho = -\int_{\rho_v}^{\rho_l} W'(\rho)\mathrm{d}\rho = 0.$$

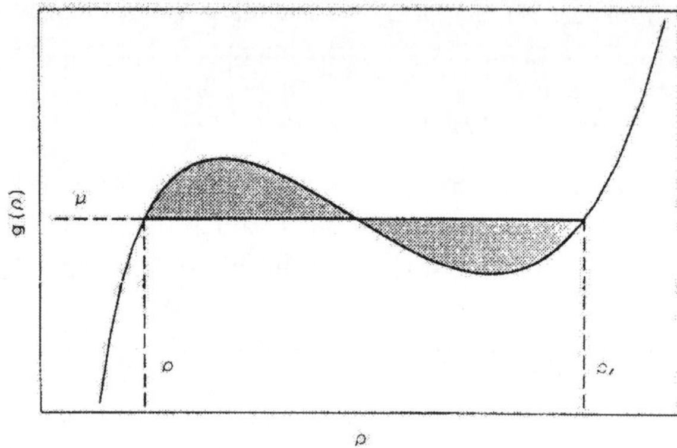

Figure 3. Illustration of the "equal area" condition.

Integrating Eq. (17) proves that the distortion energy equals the "potential" energy at any point:

$$\tfrac{1}{2}K\rho_z^2 = G(\rho) - G(\rho_v), \tag{20}$$

where $G(\rho) = \rho\overline{f}(\rho) - \mu\rho$. Using this "virial theorem", we compute the surface tension

$$\gamma = K\int_{-\infty}^{\infty} \rho_z^2 dz = \int_{\rho_v}^{\rho_l} \sqrt{2K[G(\rho) - G(\rho_v)]}\, \mathrm{d}\rho. \tag{21}$$

The integral can be computed numerically using here $\overline{f}(\rho)$ given by the second expression in Eq. (6). The result (Fig. 4) is more informative than the sharp-interface computation, as it also gives the dependence of surface tension on temperature. The surface tension decreases with growing temperature and vanishes at the critical point $a/(bT_c) = \frac{27}{8}$, $b\rho_c = \frac{1}{3}$. The low-temperature limit $\gamma \propto \rho_l^2\sqrt{aK}/b \propto \rho_l^2 A_l/d^2$ is qualitatively the same as in the sharp interface model, and the interface thickness reduces to $\sqrt{K/a} \propto d$ away from the critical point. The density profile and surface energy can be computed analytically in the vicinity of a critical point [at $T_c - T = O(\epsilon^2)$ and $\rho - \rho_c = O(\epsilon)$] where $g(\rho)$ can be approximated by a cubic polynomial.

1.3 Thermodynamics of curved interfaces

The above computation has been carried out for a flat interface, but we know that surface tension affects both equilibrium and dynamics only when the interface is curved. One may ask whether the computed value of γ is indeed the same surface tension that is responsible for the pressure drop across a curved interface. The answer is positive but qualified: yes, approximately, provided the curvature radius is much larger than the characteristic thickness of the interface.

A simple way to verify this is to compute extra energy of a weakly deformed interface, which is proportional to its area. If the position of the interface is $z = h(\boldsymbol{\xi})$ and $\nabla h \ll 1$, where $\boldsymbol{\xi}$ denotes the 2D vector in the support plane and ∇ is the 2D gradient, the additional surface

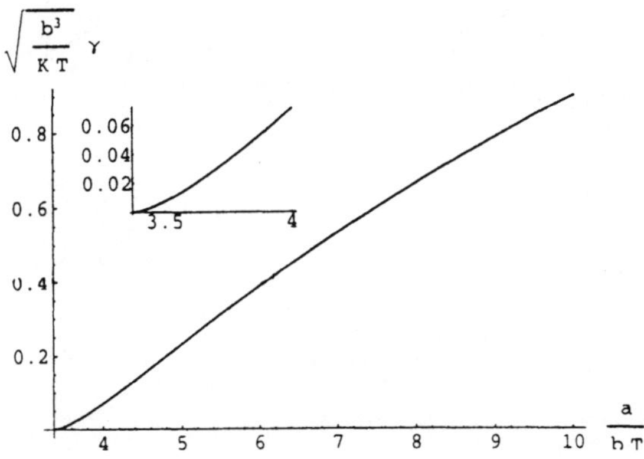

Figure 4. The dependence of the dimensionless surface tension on the dimensionless temperature. The inset shows the vicinity of the critical point.

energy is, to the lowest order,

$$F_\kappa = \gamma \int \left(\sqrt{1 + |\nabla h|^2} - 1 \right) \mathrm{d}\Sigma \approx \frac{\gamma}{2} \int |\nabla h|^2 \mathrm{d}\Sigma. \tag{22}$$

The force exerted by the deformed interface is obtained by varying this expression with respect to the interfacial position. The interface is at equilibrium when this force is compensated by a pressure increment computed as

$$p = \delta F_\kappa / \delta h = -\gamma \nabla^2 h. \tag{23}$$

The Laplacian $\nabla^2 h$ expresses the curvature κ of a weakly curved interface. This computation, in fact, does not contain any physical measure of the "weakness" of the deformation and can be extended, using some more sophisticated differential geometry, to arbitrary smooth surfaces, yielding the pressure increment equal to the mean Gaussian curvature multiplied by surface tension.

A physically meaningful proof taking account of the molecular origins of surface tension should reveal limitations of Eq. (23). The derivation is most transparent in the diffuse interface theory. We return to Eq. (16) and rewrite it in a coordinate frame aligned with a weakly deformed interface. The nominal position of a diffuse interface can be defined in a number of ways; the most natural choice is fix it as the locus of the unstable zero of the function $g(\rho)$. The spatial position of the interface is then defined as a function $\boldsymbol{x}(\boldsymbol{\xi})$ of surface coordinates $\boldsymbol{\xi}$. The latter can be chosen at will; for example, if we have a film with a weakly non-planar interface on a planar support, it is convenient to parametrize the interface as above by the coordinates $\boldsymbol{\xi}$ in the supporting plane, so that the interface position is defined by a $(2+1)$-vector $\boldsymbol{x} = \{\boldsymbol{\xi}, h(\boldsymbol{\xi})\}$. Given the interface, one can find unit tangent vectors $\boldsymbol{t}_\alpha$ along the surface coordinates ξ_α, the normal $\boldsymbol{n} = \boldsymbol{t}_1 \times \boldsymbol{t}_2$ and the curvature tensor $\boldsymbol{\kappa}$, defined through the derivatives of the normal with respect to the surface coordinates: $\partial \boldsymbol{n} / \partial \xi_\alpha = \kappa_{\alpha\beta} \boldsymbol{t}^\beta$.

Next, we define the coordinate axis ζ directed along $\boldsymbol{n}$ with the origin on the nominal interface. To fix the signs, we assume that the dense phase prevails at $\zeta < 0$ and the normal is directed

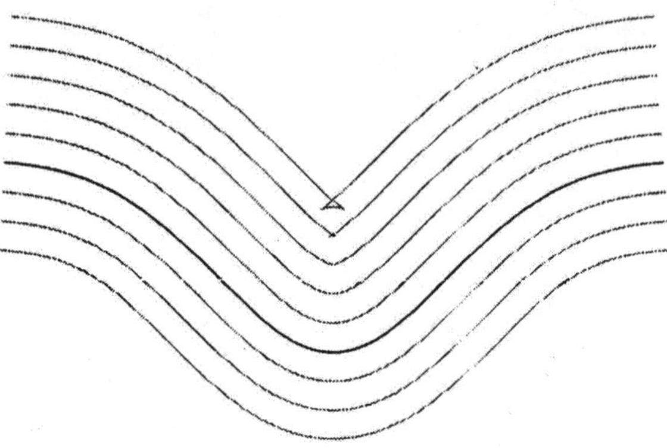

Figure 5. Construction of the aligned coordinate frame.

towards the dilute phase. The coordinate surfaces ζ = const are obtained by shifting the interface along the normal by a constant increment. Evidently, this shift causes the length to increase on convex, and to decrease on concave sections (Fig. 5). We can see that the aligned frame is not well defined far from the interface due to a singularity developing on the concave side at distance about the smallest value of the local curvature radius, i.e. the smallest inverse eigenvalue of the curvature tensor. The aligned frame is well defined only sufficiently close to the interface, so we have to assume $\zeta = O(\epsilon) \ll 1$ when the curvature radius is of $O(1)$. Then the length element in the aligned frame is

$$\mathrm{d}s^2 = \mathrm{d}\zeta^2 + \mathrm{d}\boldsymbol{\xi} \cdot (\boldsymbol{I} + \zeta\boldsymbol{\kappa}) \cdot \mathrm{d}\boldsymbol{\xi} + O(\zeta^2), \tag{24}$$

where $\boldsymbol{I}$ is the identity matrix. The metric tensor read from this expression can be used to write the Laplacian in the aligned frame. The aligned frame can be used for computations, provided the density changes from ρ_v to ρ_l within an $O(\epsilon)$ distance from the nominal interface. This means that the derivative $\partial/\partial\zeta$ should be scaled as ϵ^{-1}. Accordingly, the Laplacian is expressed as

$$\nabla^2 = \epsilon^{-2}\partial^2/\partial\zeta^2 + \epsilon^{-1}\kappa\,\partial/\partial\zeta + O(1), \tag{25}$$

where $\kappa = \mathrm{Tr}\,\boldsymbol{\kappa}$ is the mean Gaussian curvature. The derivatives in the transverse directions appear only in the next order, which we will not need.

The coordinate ζ has to be rescaled by the factor ϵ to accommodate the rapid change of ρ across the interface. The additional small term in the Laplacian causes a week distortion of the density profile, while the chemical potential μ has to be shifted by an $O(\epsilon)$ increment to keep the curved interface stationary. To find this shift, we expand

$$\rho = \rho_0(\zeta) + \epsilon\rho_1(\zeta) + \ldots, \quad \mu = \mu_0 + \epsilon\mu_1 + \ldots. \tag{26}$$

Using Eqs. (25), (26) in Eq. (16), we recover in the zero order Eq. (17) with ρ, μ, z replaced by ρ_0, μ_0, ζ. This equation is verified as before by a stationary diffuse interface solution satisfying the asymptotic condition $\rho_0(\zeta) \to \rho_v, \rho_l$ at $\zeta \to \pm\infty$.

The first-order equation reads

$$K\rho_1''(\zeta) - g'(\rho_0)\rho_1 + \Psi(\zeta) = 0. \tag{27}$$

It contains the inhomogeneity

$$\Psi(\zeta) = \kappa K \rho_0'(\zeta) + \mu_1. \tag{28}$$

The shift of the chemical potential μ_1 is determined by the solvability condition of Eq.(27), which requires the inhomogeneity $\Psi(\zeta)$ to be orthogonal to the translational Goldstone mode $\rho_0'(\zeta)$ of the homogeneous part:

$$\mu_1 = \frac{K\kappa}{\rho_l - \rho_v} \int_{-\infty}^{\infty} [\rho_0'(\zeta)]^2 \mathrm{d}\zeta = \frac{\gamma\kappa}{\rho_l - \rho_v}. \tag{29}$$

This relation is equivalent to the Gibbs–Thomson law relating the equilibrium pressure or temperature with the radius of a meniscus. We will be more interested, however, in fluid-mechanical consequences of this relation, since the difference of chemical potential between interface patches with different curvature acts as a driving force of the fluid flow along the interface.

For a sharp interface model, the additional interfacial energy of a curved interface can be obtained by repeating the computation in the beginning of the preceding subsection with distances recomputed with the help of the metric tensor of the coordinate frame aligned with the interface. One can also consider directly a spherical interface with the radius equal to the inverse mean curvature. The result is the same as in Eq. (23) or (29), and is applicable when κ^{-1} far exceeds the molecular diameter d.

2 Equilibrium near a three-phase junction

2.1 Static contact angles

Proximity of a solid surface should modify the free energy in the same way as proximity of the liquid-vapor interface. The additional term in the free energy integral (9) is

$$F_s = \int \rho(\boldsymbol{x})\,\mathrm{d}^3\boldsymbol{x} \int_s U_s(r)\rho_s(\boldsymbol{x}+\boldsymbol{r})\,\mathrm{d}^3\boldsymbol{r}. \tag{30}$$

where U_s is the attractive part of the fluid-solid interaction potential and $\int_s$ means that the integration is carried over the volume occupied by the solid; all other integrals here and in (9) are now restricted to the volume occupied by the fluid.

The sharp-interface computation in Section 1.2 can be repeated almost without change for a boundary between a solid with the density ρ_s at $z < 0$ and fluid at $z > 0$. If the solid-fluid hard-core interaction potential is expressed as $U_s = A_s/r^6$, we compute the energy of the liquid-solid interface as

$$\gamma_l = \frac{\pi\rho_l}{d^2}\left(\frac{A_l\rho_l}{2} - A_s\rho_s\right), \tag{31}$$

The energy of the vapor-solid interface γ_v is expressed in the same way, with ρ_l replaced by ρ_v. These expressions can be used in the Young–Laplace formula that derives the contact angle θ from the balance condition for a three-phase contact line:

$$\gamma_v - \gamma_l = \gamma\cos\theta, \tag{32}$$

If ρ_v is negligible, using Eqs. (10), (31) in Eq. (32) yields

$$\cos\theta = \frac{2A_s\rho_s}{A_l\rho_l} - 1. \tag{33}$$

The liquid wets the solid completely ($\theta \to 0$) when $A_s\rho_s \geq A_l\rho_l$. A similar computation can be carried out in the diffuse interface theory, but is somewhat more problematic, since one has to define a boundary condition for the fluid density on the solid surface.

Take note that the surface tensions have been computed here and in Section 1.2 for a semi-infinite system where vapor-liquid and fluid-solid interfaces are well separated. Therefore the "standard" contact angle defined by Eq. (33) is apt to change near the contact line where the interfaces converge and surface energies are no longer defined by Eqs. (10), (31). Thus, the "standard" contact angle in no way can be identified with the "true" contact angle at which the interfaces meet each other. In fact, one can argue whether the "true" angle can be reasonably defined at all, since the notion of a well-defined interface itself breaks down at molecular distances. Equation (33) becomes applicable away from the contact line, where the layer is thick and the interfaces are well separated. However, the actual inclination angle is apt to change at large distances due to external forces, such as gravity or dynamic pressure. Therefore, even though the Young–Laplace formula (32) is an exact thermodynamic relation (which can be related to the translational symmetry of a contact line in an infinite three-phase system), the "standard" angle may be in fact unobservable at either small or large distances from the solid.

2.2 Disjoining potential

The change of the chemical potential due to overlapping of the transitional layers near the vapor-liquid and fluid-solid interfaces is called *disjoining potential*. The original notion, introduced by Derjaguin, is *disjoining pressure* that causes separation of both interfaces, i.e. thickening of a wetting film. When the disjoining pressure is negative (sometimes called then *conjoining* pressure), the interfaces attract, causing the film rupture. Away from the critical point, the characteristic length defining the thickness of the overlap region is the molecular scale d. Since, however, van der Waals forces decay with distance relatively slowly (by a power, rather than exponential, law) and compete with weak macroscopic forces, the overlap is felt already in films of mesoscopic thickness where the continuum approach is still applicable. The overlap region extends to macroscopic scales in the vicinity of the critical point when the vapor-liquid interface becomes truly diffuse.

The shift of the free energy due to interactions with the solid is given by the liquid-solid interaction term (30) minus the lost part of the fluid-fluid interaction term in Eq. (9). We will compute it here for a liquid layer with a sharp interface at $z = h$ parallel to the solid surface at $z = h$ (the diffuse interface theory is discussed in Section 5.2). Setting $\rho = \rho_l = \text{const}$ at $0 < z < h$ and neglecting the gas density, the modified free energy integral (9) can be computed by integrating the van der Waals interactions laterally as in Eq. (11). The correction due to the interaction with the solid is computed as

$$F_{ls} = \pi\rho_l\left(\tfrac{1}{2}A_l\rho_l - A_s\rho_s\right)Q(z), \tag{34}$$

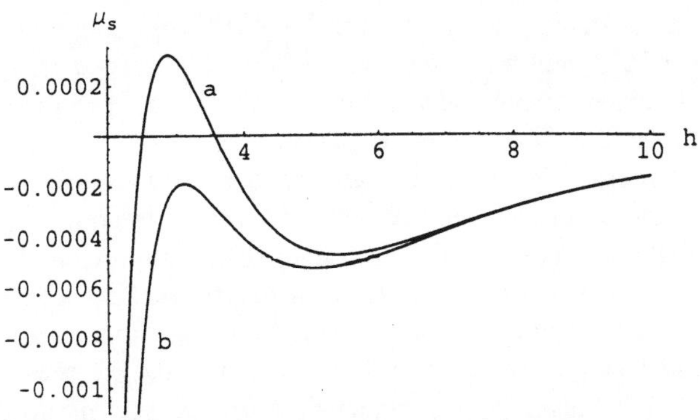

Figure 6. A non-monotonic function $\mu_s(h)$ resulting from combined action of polar and van der Waals forces. The two curves with a positive and negative maximum are drawn for $l = 1$ and $\chi/A = 0.12$ (a) and $\chi/A = 0.13$ (b).

where $Q(z)$ is given by Eq. (11). This yields the shift of the equilibrium chemical potential due to interactions with the solid surface:

$$\mu_s = \frac{\partial F_{ls}}{\partial \rho_l} = \frac{A}{6h^3 \rho_l} + \text{const at } h \geq d, \tag{35}$$

where A is the Hamaker constant:

$$A = \pi \rho_l \left(A_s \rho_s - A_l \rho_l \right) \tag{36}$$

(this differs from the standard definition by the factor π).

At $A > 0$, when the liquid wets the solid, the interface is "repelled" from the solid, so that the liquid layer thickens; this can be balanced by a negative shift of the chemical potential. The equilibrium thickness of the film grows with growing chemical potential, approaching in the limit $\mu_s \to 0$ macroscopic dimensions of a bulk phase. On the contrary, at $A < 0$ (partially wetting liquid) the gas-liquid and liquid-solid interfaces "attract" each other. Since the attraction increases with diminishing distance, this situation is unstable, and leads eventually to film rupture and formation of a contact line with a finite contact angle. It is, admittedly, rather confusing that "repelling" interaction stems from attraction between the fluid and the solid, and vice versa.

Of course, other kinds of fluid-solid interactions, besides van der Waals forces, exist in nature, and interactions with different spatial dependence may combine to give a non-monotonic function $\mu_s(h)$. For example, adding attractive *polar* interactions that decay exponentially with the distance yields (after omitting a constant term and rescaling)

$$\mu_s = \chi e^{-h/l} - \tfrac{1}{6} A h^{-3} \tag{37}$$

When the liquid is wetting ($A > 0$), the right-hand side has a negative minimum, provided $\chi > e^4 A/(8l)^3$. This means that in a certain range of μ_s, e.g. between the minimum of the curve

$\mu_s(h)$ in Fig. 6 and its negative maximum or zero, there are two stable solutions that correspond to equilibrium layers of different finite thickness. A thin film can coexist with the bulk state (a film of macroscopic thickness) when the maximum of this curve is positive.

We shall see that the disjoining potential plays a crucial role in the contact line motion. Before we go on to dynamics, it should be noted that even static problems, such as computing an equilibrium shape of a liquid droplet, with due account for intermolecular interactions near the contact line is very much non-trivial. The overall shape minimizing the free energy of the system is influenced by forces operating on widely separated scales – from molecular distances to the drop size, and, due to intermolecular interactions, the conditions near the three-phase line are very sensitive to surface inhomogeneities, both geometric and chemical. No wonder that the dynamics of the contact line is not yet well understood after decades of effort.

3 Classical hydrodynamic theory

3.1 Hydrodynamic setting

Hydrodynamic problems involving moving contact lines may arise in different contexts, such as:

- spreading of a droplet on a horizontal surface (Fig. 7a);
- pull-down of a meniscus on a moving wall (Fig. 7b);
- advancement of the leading edge of a film down an inclined pláne (Fig. 7c);
- condensation or evaporation on a partially wetted surface;
- climbing of a film under the action of Marangoni force.

In the latter case, there is persistent flow in the vicinity of the contact line even when the latter does not move.

One can expect the problems of this kind to be not simple technically, since they involve a free boundary whose shape has to be determined simultaneously with the computation of the flow field. We shall see, however, that the difficulty is not mere technical, but extends to the physics of the problem. Our aim is to understand this basic difficulty, and therefore we formulate the flow equations in the simplest possible way. First, we restrict to slowly moving fluids and neglect inertial effects. This is, in fact, not a serious restriction, since the Reynolds number relevant for the motion in the vicinity of the contact line should be based on the local film thickness, and goes down to zero as the contact line is approached. We shall also assume here that the fluid is incompressible. Thus, the hydrodynamic equations for the velocity field $\boldsymbol{u}(\boldsymbol{x})$ are the Stokes and continuity equations:

$$-\nabla(p+V)+\eta\nabla^2\boldsymbol{u}=0, \tag{38}$$

$$\nabla\cdot\boldsymbol{u}=0. \tag{39}$$

Here η is the dynamic viscosity and V is the potential of external body forces. The latter may include macroscopic forces, such as gravity, as well as intermolecular forces exerted on the fluid by a bounding solid wall or support. The classic boundary condition on the solid wall is no slip: $\boldsymbol{u}=0$. The conditions on the free boundary are given by the tangential and normal stress balances and the kinematic condition which states that the interface moves with the local normal velocity of the flow.

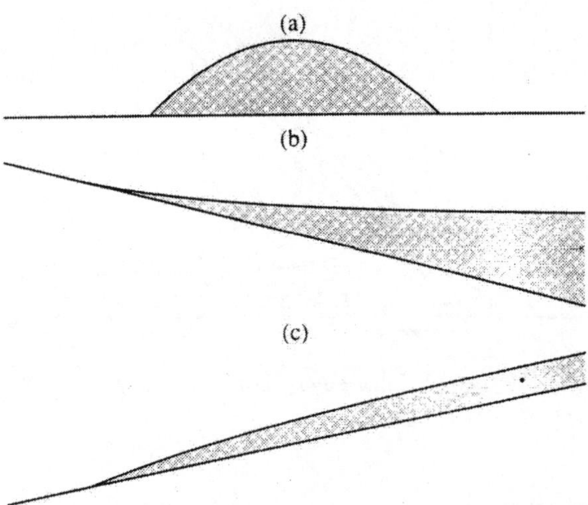

Figure 7. Some configurations involving moving contact lines: (a) spreading of a droplet; (b) pull-down of a meniscus; (c) flow down an inclined plane.

The main source of trouble lies in the no slip condition. At first glance, this condition seems to be altogether incompatible with motion of the contact line. If the material elements of one of the two immiscible phases adjacent to the solid wall do not move, how can they be replaced by material elements of the other phase? This is possible through *caterpillar* motion, as shown in Fig. 8. This flow kinematics makes the motion feasible. The trouble is, however, that there is no stagnation point on the three-phase contact, and the flow velocity vector must rotate along an arbitrarily small contour drawn about the contact line, while its absolute value remains finite. This gives rise to a velocity gradient and, hence, viscous stress growing inversely proportional to the distance to the contact line.

The singularity could be still tolerable if the total force needed to advance the contact line was finite, but it diverges logarithmically. In Scriven's idiom, "not even Heracles could sink a solid"! This is, indeed, a consequence of a formal solution of the Stokes equation obtained under the assumption that the interphase boundary is a straight line, as in Fig. 8 and the flow pattern is self-similar. This means that both radial and angular velocities in polar coordinates centered on the contact line are independent of the radial distance r and, as a result, the viscous stress diverges as r^{-1} at $r \to 0$. One can argue, of course, that this formulation of the problem is non-physical, as the interface must bend under the action of strong stresses developing near the contact line. On the other hand, a logarithmic singularity of the total force is very weak, and one could hope that it might disappear when the problem is solved in a mathematically correct way or, at least, when more subtle physical factors relevant in the vicinity of the contact line are taken into account. The singularity turned out, however, to be very resilient, and it remains troublesome even in more sophisticated versions of the theory.

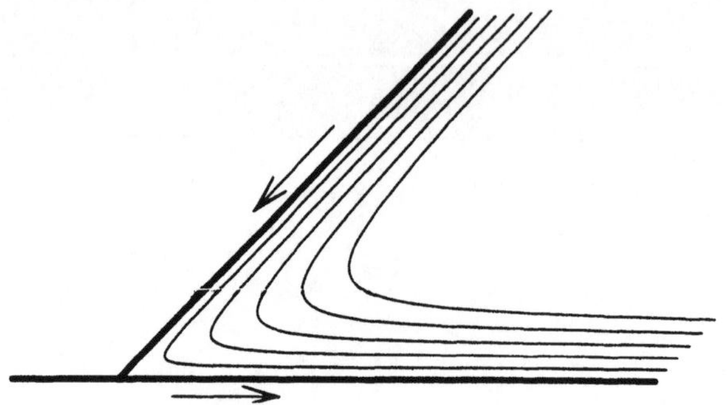

Figure 8. Flow pattern near a contact line.

3.2 Lubrication equations

The free boundary problem becomes analytically tractable when the *lubrication approximation* is used. This removes technical but not basic difficulties, and therefore we will use this approximation in further discussion. The lubrication approximation assumes the characteristic length scale in the "vertical" direction z (normal to the solid surface) to be much smaller than that in the "horizontal" (parallel) directions spanned by the 2D vector $\boldsymbol{x}$. The approximation is applicable in a liquid film with a large aspect ratio, when the interface is weakly inclined and curved. The scaling is consistent if one assumes $\partial_z = O(1)$, $\nabla = O(\sqrt{\delta}) \ll 1$, where ∇ is now the 2D gradient in the plane of the solid support. Then the continuity equation Eq. (39) requires that the vertical velocity v should be much smaller than the horizontal velocity (denoted by the 2D vector $\boldsymbol{u}$), i.e. $v \propto \sqrt{\delta}\boldsymbol{u}$.

The lubrication equations are derived by expanding both equations and boundary conditions in powers of the scale ratio δ and retaining the lowest order terms. First, we deduce from the vertical component of the Stokes equation, reduced in the leading order to $\partial(p+V)/\partial z = 0$ that $p+V$ is constant across the layer and therefore can be determined from the normal stress balance at the interface. We assume here that the only external force is gravity (acting in the dense phase only), and the gravity acceleration vector $\boldsymbol{g} = g\boldsymbol{\imath}$ (where g is acceleration of gravity and $\boldsymbol{\imath}$ is a unit vector along the slope) is directed at a small angle $\alpha\sqrt{\delta} \ll 1$ to the supporting plane. This can apply to pull-down of a meniscus on a moving support (Fig. 7b, $\alpha > 0$) or film spreading down an inclined plane (Fig. 7c, $\alpha < 0$), while $\alpha = 0$ for droplet spreading on a horizontal plane. The gravity potential is, to the leading order,

$$V = \rho_l g(h - \alpha\boldsymbol{\imath}\cdot\boldsymbol{x} - z). \tag{40}$$

If the viscous stress and pressure in the dilute phase are neglected, the leading order normal stress balance at the interface $h(\boldsymbol{x})$ reads

$$-p + \sqrt{\delta}\eta\partial_z(v + \boldsymbol{u}\cdot\nabla h) = \delta\gamma\nabla^2 h. \tag{41}$$

The viscous stress should be negligible compared to the other two terms; otherwise, it would be impossible to disentangle the horizontal and vertical components. Provided it holds, the surface tension is balanced by pressure, so that $p = -\delta\gamma\nabla^2 h$. The surface tension term can be formally made of $O(1)$ if one assumes $\gamma = \widehat{\gamma}/\delta = O(\delta^{-1})$. This assumption is consistent with the molecular theory of surface tension where the characteristic scale appearing in the computation of γ is the molecular scale d, which should be much smaller than any macroscopic length scale of the problem.

The horizontal component of the Stokes equation takes now the form

$$-\nabla W + \eta\partial_z^2 \boldsymbol{u} = 0. \tag{42}$$

where the driving potential is

$$W = -\widehat{\gamma}\nabla^2 h + \rho_l g(h - \alpha \boldsymbol{\imath} \cdot \boldsymbol{x}). \tag{43}$$

The estimate of $\boldsymbol{u}$ following from Eq. (42) is $\boldsymbol{u} = O(\sqrt{\delta})$. This equation should be viewed as an ODE defining parallel flow in the film cross-section. It is solved subject to the no slip condition $\boldsymbol{u}(\boldsymbol{x}, 0) = 0$ (in the Galilean frame of the solid support) together with the tangential stress balance condition at the free interface. The latter reduces in the leading order to $\partial_z \boldsymbol{u}(h) = 0$. The solution is elementary:

$$\boldsymbol{u} = -\eta^{-1} z \left(h - \tfrac{1}{2} z\right) \nabla W. \tag{44}$$

The kinematic condition at the free interface can be replaced now by the integral mass balance in the film cross-section, which has a general form

$$\partial_t h + \nabla \cdot \boldsymbol{j} = 0, \tag{45}$$

where the volumetric flux $\boldsymbol{j}$ is expressed, in view of Eq. (44), as

$$\boldsymbol{j} = \int_0^h \boldsymbol{u}\, \mathrm{d}z = -\tfrac{1}{3}\eta^{-1} h^3 \nabla W. \tag{46}$$

This has a form of a kinetic equation for the "order parameter" h with the "mobility coefficient"

$$k = \tfrac{1}{3}\eta^{-1} h^3 \tag{47}$$

dependent on h in a strongly nonlinear fashion. The driving potential W is of the "Cahn–Hilliard" type, appropriate for the case when the order parameter is conserved. The evolution equation (45) can be written in a variational form

$$\partial_t h = -\nabla \cdot k(h) \nabla \frac{\delta \mathcal{W}}{\delta h}, \tag{48}$$

$$\mathcal{W} = \int \left[\frac{\widehat{\gamma}}{2} |\nabla h|^2 + \rho_l g \left(\frac{h^2}{2} - \alpha h \boldsymbol{\imath} \cdot \boldsymbol{x}\right)\right] \mathrm{d}^2 \boldsymbol{x}. \tag{49}$$

The two terms in the integrand are the interfacial and gravitational energies.

3.3 Steady motion

Consider a contact line that moves relative to the solid support along the x axis with a constant speed $u_0 = O(\sqrt{\delta})$. The velocity is assumed to be positive when the bulk fluid layer, situated at $x > 0$, advances to the left. As long as the contact line is straight, the stationary lubrication equation is one-dimensional. We rewrite this equation in the comoving frame, i.e. the Galilean frame where the contact line is at rest. After transformation to a dimensionless form, only two parameters remain: the dimensionless velocity (capillary number) $U = 3\eta u_0 \delta/\gamma$ and the Bond number $G = g\rho_l \delta \lambda^2/\gamma$, where λ is a length scale, as yet indefinite. After integrating once and setting the integration constant to zero to satisfy the condition of zero net flux through the contact line, this equation becomes

$$Uh^{-2} + h'''(x) - G(h'(x) - \alpha) = 0. \tag{50}$$

The relevant length can be identified with the distance to the contact line, so that the gravity term becomes negligible as the contact line is approached. Removing this term and setting $h'(x) = \sqrt{y(h)}$ yields

$$\frac{2U}{h^2 y^{1/2}} + y''(h) = 0. \tag{51}$$

It is convenient to transform to the logarithmic scale $\xi = \ln h$. The resulting equation is

$$y''(\xi) - y'(\xi) + 2Uy^{-1/2} = 0. \tag{52}$$

At $|\xi| \gg 1$, i.e. both at very large and at very small distances, the first term can be neglected, and the remaining first-order equation is easily integrated. Reverting to the original variables, we have

$$h'(x) \asymp \left(3U \ln \left|\frac{h}{h_0}\right|\right)^{1/3}, \tag{53}$$

where h_0 is an indefinite integration constant. This asymptotics is valid only when h is *logarithmically* large. Take note that the asymptotic expression does not approach any constant slope that could be identified with either "true" (at $h \to 0$) or "apparent" (at $h \to \infty$) contact angle. The inclination angle grows indefinitely in both limits, although it is practically felt only at exceedingly large or small distances: the cubic root of the logarithm is a very weak dependence. At large h this can be corrected by bringing back the gravity term. At small h there is, however, no reasonable solution, as Eq. (53) cannot be fit to either finite or zero contact angle. Growing inclination angle indicates the breakdown of the lubrication approximation, rather than a physical singularity. It is related, however, to the viscous stress singularity that persists, as we have seen, also in the full hydrodynamic problem.

3.4 Phenomenological slip condition

Returning to Eq. (51), we can observe that it might be very easy to repair the weak logarithmic divergence by modifying the mobility coefficient (47) in such a way that it would be proportional to h^n with $n < 3$ at $h \to 0$. Then, following the logarithmic transformation, the last term in Eq. (52) becomes be proportional to h^{n-3} and vanishes at $h \to 0$, provided the contact angle remains finite. The remaining linear equation is solved by $y = C_0 + C_1 e^{\xi}$, i.e. describes at $\xi \to -\infty$ relaxation to an arbitrary "true" contact angle $\theta_0 = \sqrt{y(0)}$.

The desired modification can be effected by introducing a weak slip on the solid surface, i.e replacing the boundary condition $\boldsymbol{u}(0) = 0$ by $\boldsymbol{u} = b\partial_z \boldsymbol{u}$ at $z = 0$ with $b \ll 1$. Then we obtain instead of Eq. (47)

$$k = \tfrac{1}{3}\eta^{-1}h^2(h + 3b), \tag{54}$$

so that $k(h) \propto h^2$ at $h \ll b$.

Introducing the phenomenological slip coefficient does not eliminate the singularity on the contact line, but makes it integrable. The lubrication equations are readily solved numerically when the mobility coefficient (54) is used. The constant b becomes irrelevant at macroscopic distances, but the entire profile of the film will depend on its value. due to high sensitivity of the solution to conditions at the contact line.

The phenomenological model contains still another adjustable parameter: the "true" contact angle $\theta_0 = h'(0)$, which may also depend on the propagation speed U. It is usually assumed that $\theta_0 \propto U^{1/3}$. This is often viewed as an empirical relationship, although nobody can measure the "true" contact angle. What is actually measured in the experiment is the "apparent" dynamic contact angle observable at macroscopic distances. It follows from Eq. (53) that this angle is indeed proportional to $U^{1/3}$, so it is natural to assume that this estimate does not change at $h \to 0$. This is likely to be true in the framework of the phenomenological slip model where the lubrication equation well behaves in this limit.

Moreover, the one-third law can be deduced without solving any equations at all, as it follows from the scale invariance of Eq. (51). The only parameter of this equation can be removed by rescaling $h \to hU^{-1/3}$, $y \to yU^{2/3}$. The latter formula gives directly the dependence $h'(x) \propto U^{1/3}$. It is equivalent to Tanner's law of spreading that states that the radius R of a spreading drop changes with time as $t^{1/10}$. The derivation is simple. If gravity can be neglected and the spreading is slow, the droplet retains (except, perhaps, very close to the contact line) its equilibrium shape that can be approximated by a spherical cap. Then its conserved volume is proportional (at $\theta \ll 1$) to $R^3\theta \propto R^3(\mathrm{d}R/\mathrm{d}t)^{1/3}$. This gives rise to the evolution equation $\mathrm{d}R/\mathrm{d}t \propto R^{-9}$, integrated to $R \propto t^{1/10}$. Thus, Tanner's law emerges as a direct consequence of scaling invariance in this very crude model of spreading.

A physical justification of the slip condition should be sought for in the molecular theory, to be considered in the following Section.

4 Hydrodynamic theory with intermolecular forces

4.1 Asymptotics near the contact line

The disjoining pressure at the free interface $\rho_l\mu_s$ given by Eq. (35) can be added to Eq. (43), yielding the overall driving potential

$$W = -\hat{\gamma}\nabla^2 h - \tfrac{1}{6}Ah^{-3} + \rho_l g(h - \alpha \imath \cdot \boldsymbol{x}). \tag{55}$$

The length scale relevant in the microscopic boundary layer where intermolecular forces are important can be defined by balancing the van der Waals and surface tension terms in this expression. The respective van der Waals length $\lambda^2 = \frac{1}{6}|A|\delta/\gamma$ typically falls into a nanoscale range. The gravity parameter based on this length, $G = \frac{1}{6}\rho_l g\delta^2|A|/\gamma^2$ is quite different from the standard Bond number. This parameter, equal to the squared ratio of the van der Waals to gravity/capillary length, is very small, and gravity can be neglected in a thin film influenced by intermolecular forces. Assuming $A > 0$ and is taking λ as the length scale, we write thre lubrication equation to be solved in this region in the form

$$\frac{2U}{h^2 y^{1/2}} + \frac{\mathrm{d}}{\mathrm{d}h}\left[y'(h) + \frac{2}{h^3}\right] = 0. \tag{56}$$

The static equation at $U = 0$ is invariant to the rescaling that changes the slope

$$x \to x/C,\ h \to h/\sqrt{C},\ y \to Cy. \tag{57}$$

The factor C is arbitrary, and the rescaling symmetry exists because no forces capable to fix the slope are present in this approximation. The static solution, immediately following from Eq. (56), is $y = h^{-2} + C$, which can be further integrated to

$$h = \sqrt{Cx^2 - C^{-1}}. \tag{58}$$

This solution has a zero corresponding to a "contact line" located at $x = C$, and the "true" contact angle is $\pi/2$. The asymptotic slope $h'(x) = \sqrt{y} \asymp \sqrt{C}$ cannot be identified with the macroscopic equilibrium contact angle, but remains indefinite and is not related to any material properties.

One can be tempted to use the static solution (58) as a zero approximation for a solution at $U \ll 1$. This approach is unsuitable for two reasons. First, the static solution (58) is unstable, as any perturbation at the foot of the film would tend to spread further into a layer of minimal (molecular) thickness. Second, velocity never can be treated as a small perturbation, since it can be eliminated from Eq. (56) by applying the transformation (57) with the scaling factor $C = U^{2/3}$. This means that an arbitrarily small velocity causes a finite deviation from the static solution at sufficiently large distances, and a regular perturbation solution at $U \ll 1$ is meaningless.

A correct asymptotics of Eq. (56) at $h \to 0$ is obtained by observing that the surface tension is negligible in a thin and almost flat "precursor film", so that the viscous stress is balanced by disjoining pressure, contrary to a spurious balance between surface tension and disjoining pressure in the static solution (58). This yields

$$h'(x) \asymp \tfrac{1}{3}Uh^2,\quad h \asymp -3/x \text{ at } h \to 0,\ x \to -\infty, \tag{59}$$

valid up to a molecular cut-off of h. The asymptotics at large h is given by Eq. (53) as before.

One can try and integrate Eq. (56) starting from the asymptotics (59) at $h \to 0$ and adjusting initial data to arrive at the far field asymptotics (53). This computation, if successful, would also yield a unique value of the constant h_0. Unfortunately, integration by shooting method is very sensitive to the choice of an initial point. The result would be also sensitive to any additional physical factors that may influence the asymptotics (59), and we should look into it more closely before attempting the computation.

4.2 Kinetic slip condition

The asymptotics $h \propto x^{-1}$ is somewhat unpleasant, since, if one takes it literally, the total volume of the precursor film diverges. The divergence is removed by a molecular cut-off. It is also possible that a very long precursor film on an adversely inclined plane is balanced by gravity or that the film never reaches a steady state with a constant profile $h(x)$ along all its length. What is still more important, the physics of motion in a layer adjacent to the solid surface may be quite different from the bulk motion described by the Stokes equation.

It is commonly assumed in physical kinetics that flux is proportional to the gradient of the driving thermodynamic potential. Such a relation has the same form as Eq. (46), but the mobility coefficient would be usually independent of the transported variable (or "order parameter"). Thus, the total flux through a thin layer of thickness d adjacent to the solid support can be expressed simply as

$$j = -\frac{dD}{T} \nabla W, \tag{60}$$

where D is the surface diffusivity and ∇ is the two-dimensional gradient along the surface. This is equivalent to introducing slide velocity

$$\boldsymbol{u}_s = \frac{j}{d\rho_l} = -\frac{D}{\rho_l T} \nabla W. \tag{61}$$

Extrapolating continuous description of fluid motion to a molecular scale might be conceptually difficult but unavoidable as far as interfacial dynamics is concerned. Long-range intermolecular interactions, such as London–van der Waals forces, still operate on a mesoscopic scale where continuous theory is justified, but they should be bounded by an inner cut-off d of atomic dimensions. Thus, distinguishing the first molecular layer from the bulk fluid becomes necessary even in equilibrium theory. In dynamic theory, the transport in the first molecular layer can be described by Eq. (60), whereas the bulk fluid obeys hydrodynamic equations supplemented by the action of intermolecular forces. Equation (61) serves then as the boundary condition at the solid surface. Moreover, at the contact line, where the bulk fluid layer either terminates altogether or gives way to a monomolecular precursor film, the same slip condition defines the slip component of the flow pattern.

We assume now that the velocity profile $\boldsymbol{u}(z)$ verifies Eq. (42) with the usual no-slip boundary condition on the solid support $u(0) = 0$ replaced by the slip condition $\boldsymbol{u}(d) = \boldsymbol{u}_s$ at the molecular cut-off distance d, where $\boldsymbol{u}_s$ is given by Eq. (61). The solution in the bulk layer $d < z < h$ is

$$\boldsymbol{u} = -\eta^{-1} \left[b^2 + h(z-d) - \tfrac{1}{2}(z^2 - d^2)\right] \nabla W, \tag{62}$$

where $b = \sqrt{D\eta/\rho_l T}$ is the effective slip length. The mobility coefficient k is obtained by integrating Eq. (62) across the layer. Including also the constant slip velocity $\boldsymbol{u}_s$ in the slip layer $0 < z < d$, we have

$$k(h) = \eta^{-1} \left[b^2 h + \tfrac{1}{3}(h-d)^3\right]. \tag{63}$$

Since both b and d are measurable on the molecular scale, this expression does not differ in a macroscopically thick layer from the standard lubrication mobility given by Eq. (47), and the correction becomes significant only in theimmediate vicinity of the contact line. The three microscopic scales d, b, and λ are about of the same order of magnitude. The natural choice for d is the nominal molecular diameter, identified with the cut-off distance in the van der Waals theory. The standard value is about 0.2nm. The slip length is likely to be of the same order of magnitude. The approximate relation between viscosity η and self-diffusivity D_m in a liquid yields $D_m\eta \approx 10^2 T/3\pi d$. The surface diffusivity should be somewhat lower than the diffusivity in the bulk liquid, and with $D/D_m \approx 0.1$ we have $b \approx d$.

4.3 Intermediate asymptotics

The transformed dimensionless lubrication equation, including intermolecular interactions as well as gravity and using the molecular cut-off d as the length scale, takes now the form

$$\frac{hU}{\sqrt{y}\,k(h)} + \frac{1}{2}y''(h) - \frac{a}{h^4} - G\left(1 - \frac{\alpha}{\sqrt{y}}\right) = 0, \tag{64}$$

where the dimensionless Hamaker constant is $a = (\lambda/d)^2 = \frac{1}{6}A\delta/(\gamma d^2)$. If the fluid is not volatile, the motion at the point where the film thins down to the minimal thickness $h = d$ should be pure slip, while standard caterpillar motion is retained at observable macroscopic distances. The leading edge of an advancing wetting film is followed by a thin precursor film, where surface tension is negligible and the action of intermolecular forces is balanced by viscous dissipation, as explained in the preceding subsection. The boundary condition at the leading edge will be then the same Eq. (61) with u_s replaced by the edge propagation speed U and ∇W computed at $h = d$ with surface tension neglected:

$$y(d) = \left(d^4U/3\lambda^2b^2\right)^2. \tag{65}$$

In the intermediate region, where h far exceeds the microscopic scales d, b, λ but is still far less than the gravity length $\lambda G^{-1/2}$, the film profile is determined by the balance between viscous stress and surface tension. The asymptotics of the truncated Eq. (64) (with G set to zero) at $h \to \infty$ is given by Eq. (53). The indefinite constant h_0 can be obtained by integrating Eq. (64) (with gravity neglected) starting from the boundary condition (65) and adjusting another necessary boundary value to avoid runaway to $\pm\infty$. There is a unique heteroclinic trajectory approaching the asymptotics (53). It is very sensitive to the initial conditions as well as to the molecular-scale factors operating close to the contact line. The growth of the inclination angle is never saturated, as long as macroscopic factors (gravity or volume constraint) are not taken into account.

Equation (64) can be integrated using the shooting method starting from the boundary condition (65) and adjusting $y'(d)$ to arrive at the required asymptotics at $h \to \infty$. The solution in the intermediate region depends on the molecular parameters d, b as well as on the propagation speed U. The latter's impact is most interesting for our purpose. Examples of the computed dependence of the inclination angle $\theta = \sqrt{y(h)}$ on the local film thickness h using the boundary condition (65) at different values of U are given in Fig. 9.

The curve segments at $h \gg 1$ can be fit to the asymptotic formula (53) to obtain the integration constant h_0. Recall that the asymptotic formula can be used only when h is *logarithmically* large, and the convergence is very slow, as the expansion is in the cubic root of logarithm; therefore h_0 can be only obtained approximately from the computed profiles. The dependence of h_0 on U based on Fig. 9 is shown in Fig. 10. We see here a rather strong variation of the integration constant, so that deviations from Tanner's law $\theta \propto U^{1/3}$ should be considerable.

4.4 Pull-down of a meniscus

The simplest stationary arrangement including gravity is realized when an inclined plane, dry at $x \to -\infty$ slides in the direction of a wetting layer (Fig. 7b). Solving Eq. (64) with the same

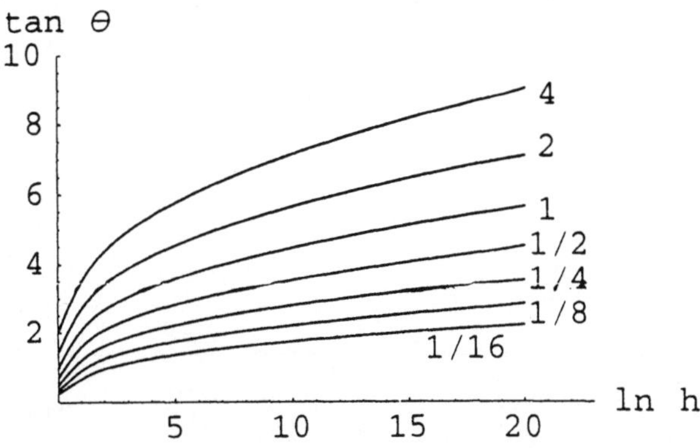

Figure 9. Dependence of the local surface inclination on the local film thickness at different values of the dimensionless propagation speed U. The numbers at the curves show the values of U. Other parameters used in the computations are $b = 1$, $a = 1/3$.

boundary condition (65) as before brings now to the asymptotics $y = \sqrt{\alpha}$ at $h \to \infty$ that corresponds to a horizontal free interface.

The curves $y(h)$ seen in Fig. 11a and Fig. 11c all depart from the intermediate asymptotic curve obtained for vanishing G as in the preceding subsection. However, due to extreme sensitivity of the shooting method to the choice of the missing initial value, one has to integrate from the outset the full equation rather than trying to start integration from some point on the intermediate asymptotic curve. One can see that the maximum inclination angle (which may be identified with the "visible" contact angle) grows as G increases. This increase is, however, not pronounced when the initial incline (identified with the "true" contact angle) is high. One can distinguish therefore between two possibilities: first, when the main dissipation is due to kinetic resistance in the first monomolecular layer that raises $y(d)$, and second when the viscous dissipation prevails and the inclination angle keeps growing in the region of bulk flow. Take note that even in the latter case the region where the inclination and curvature are high are close to the contact line when measured on a macroscopic scale.

Fig. 12 shows the dependence of the "visible" contact angle θ_m, defined as the maximum inclination angle and observed in the range where the gravity-dependent curves depart from the intermediate asymptotics, on the speed U. The angle drops close to zero at small flow velocities. Fig. 13 shows the actual shape of the meniscus obtained by integrating the equation $h'(x) = \sqrt{y(h)}$, $h(0) = d$. The dependence of the pull-down length Δ (computed as the difference between the actual position of the contact line and the point where the continuation of the asymptotic planar interface hits the solid surface) on the gravity length is shown in Fig. 14.

It comes, of course, as no surprise that introducing a molecular cut-off and applying a kinetic slip condition to the first molecular layer resolves the notorious singularities of hydrodynamic description. The hydrodynamic singularities are eliminated, however, only at molecular distances, and are still felt in sharp interface curvatures at microscopic distances identified here as the intermediate asymptotic region. The computations are eased considerably when non-physical divergence of both viscous stress and attractive Lennard–Jones potential beyond the cut-off limit

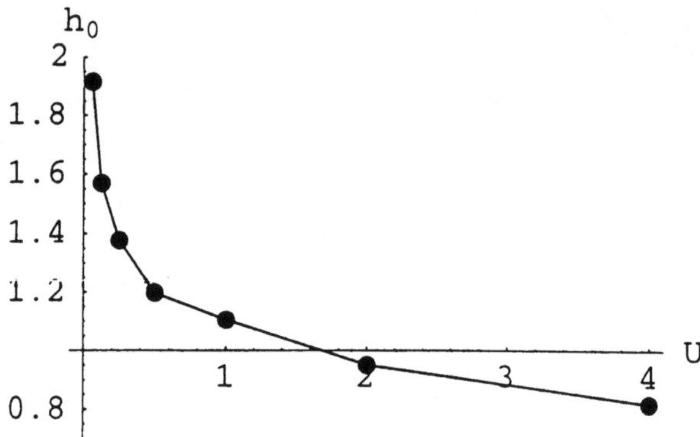

Figure 10. Dependence of h_0 on U

are eliminated. As a result, the stationary equations can be solved by shooting method with reasonable accuracy in a very wide range extending from molecular to macroscopic scales, and the "true" contact angle at the cut-off distance can be defined unequivocally.

The "true" angle (unobservable by available techniques) depends on the slip length as well as on the edge propagation speed, but not on gravity or asymptotic inclination angle. These macroscopic factors influence, however, the "visible" contact angle observed in the interval where the actual film profile departs from the intermediate asymptotic curve. Since the latter's location, though not shape, depends on the molecular-scale factors, as well as on the cut-off distance, the visible angle depends on both molecular and macroscopic factors. The universality of the intermediate asymptotics that allows to deduce Tanner's law $\theta \propto U^{1/3}$ from scaling only (Section 3.4) is impaired by the dependence of the integration constant h_0 on both molecular factors and velocity. Thus, the lack of simple recipes for predicting the value of dynamic contact angle is deeply rooted in the mesoscopic character of the contact line.

5 Diffuse interface theory

5.1 Basic equations

A difuse interface model uses for description of a two-phase system an "order parameter" which changes continuously across the interphase boundary. This variable is often called a "phase field", as its allows to define which of the alternative phases prevails at each location. We have already used this approach in Section 1.2 to compute equilibrium surface tension, and noticed that in a one-component fluid the only required order parameter is density. We shall now extend this model to non-equilibrium situations involving fluid flow.

A general diffuse interface (phase field) model coupled to hydrodynamics includes the following ingredients:

- a dynamic equation of the phase field variable(s) derived from an appropriate energy functional;

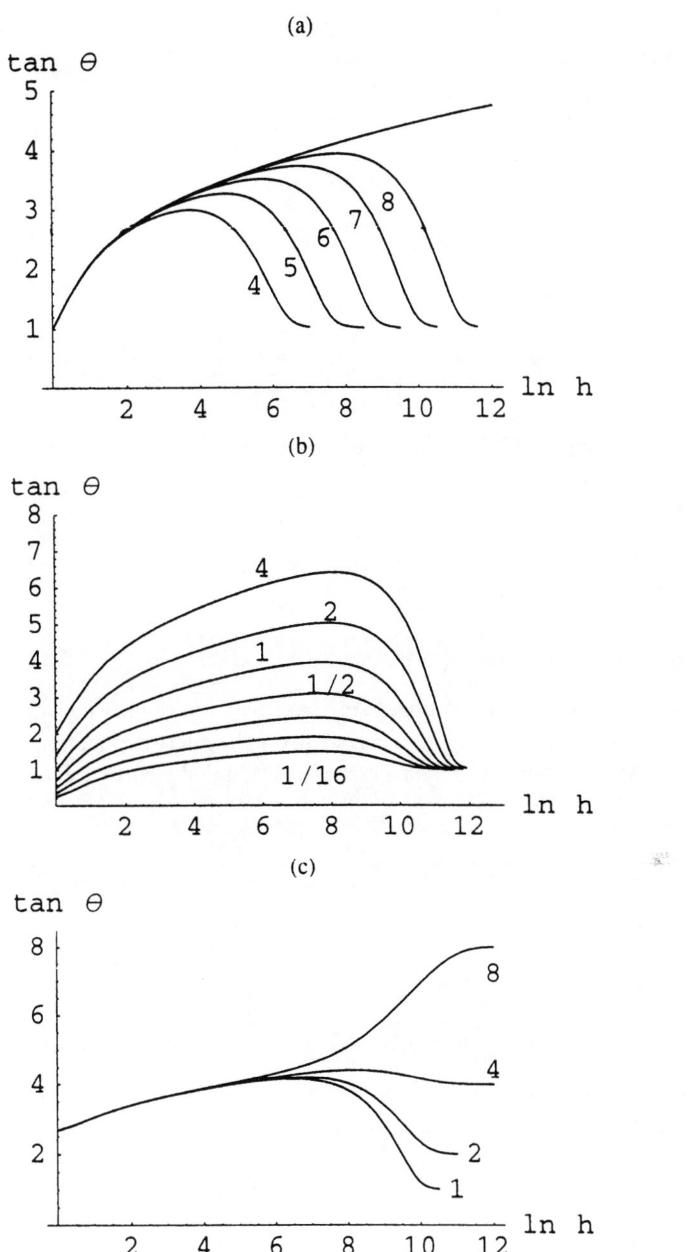

Figure 11. Dependence of the local surface inclination angle θ on the film thickness (a) at $U = 1, \alpha = 1$ and different values of the Bond number G; (b) at $G = 10^{-8}$ and different values of U; (c) at $U = 1$, $G = 10^{-8}$ and different values of the asymptotic inclination angle α. The numbers at the curves show the values, respectively, of $-\log G$, U and α. Other parameters used in all computations are $b = 1, a = 1/3$.

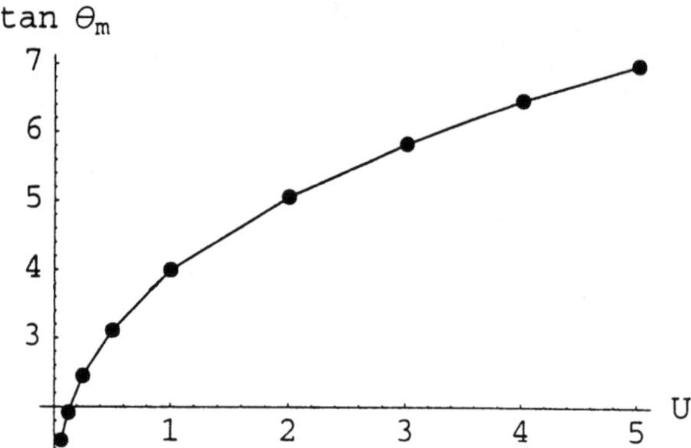

Figure 12. Dependence of the visible contact angle θ_m on U.

- a constituent relation defining the dependence of pressure or chemical potential on the phase variable(s);
- the continuity equation;
- the equation for a flow field $\mathbf{u}(\mathbf{x}, t)$.

In a one-component system, the appropriate phase field variable is density ρ, and the equation for the static density distribution (16) is derived from the energy functional (15).

The density field is coupled to hydrodynamics through the capillary tensor

$$\mathbf{T} = \mathcal{L}\mathbf{I} - \nabla\rho \otimes \partial\mathcal{L}/\partial\nabla\rho, \tag{66}$$

where $\mathbf{I}$ is the unity tensor. Neglecting the inertial effects, the flow is described by the generalized Stokes equation

$$\nabla \cdot (\mathbf{T} + \mathbf{S}) + \mathbf{F} = 0, \tag{67}$$

where $\mathbf{F} = -\nabla V$ is an external force and $\mathbf{S}$ is the viscous stress tensor with the components

$$S_{jk} = \eta(\partial_j v_k + \partial_k v_j) + (\zeta - \tfrac{2}{3}\eta)\delta_{jk}\nabla \cdot \mathbf{v}, \tag{68}$$

where η, ζ are dynamic viscosities (generally, dependent on ρ), and v_j are components of the velocity field $\mathbf{v}$. The system of equations is closed by the continuity equation

$$\rho_t + \nabla \cdot (\rho\mathbf{v}) = 0, \tag{69}$$

The Stokes equation (67) is rewritten using Eq. (66) as

$$-\nabla V - \rho\nabla\mu + \nabla \cdot (\eta\nabla\mathbf{v}) + \nabla[(\zeta + \tfrac{1}{3}\eta)\nabla \cdot \mathbf{v}] = 0. \tag{70}$$

Thus, local chemical potential dependent on proximity of the fluid-solid and liquid-vapor interface and sensitive to the latter's curvature, as discussed in Sections 1, 2, serves as a driving force of fluid motion. As long as the fluid is incompressible, the forms of the Stokes equation with $\rho\nabla\mu$ or ∇p are totally equivalent; Eq. (70) is, however, the correct form in the transitional interfacial layer where incompressibility is violated.

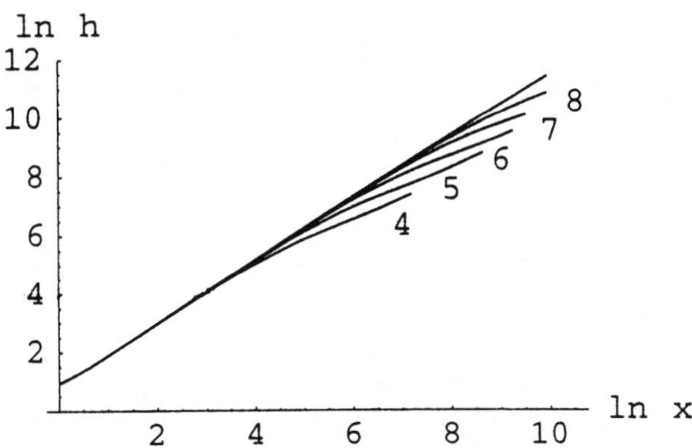

Figure 13. The shape of the meniscus for different values of b. The numbers at the curves show the values of $-\log G$.

5.2 Equilibrium density profile and chemical potential in a thin layer

Before approaching our main task of the analysis of motion in the vicinity of a three-phase boundary, it is necessary to clarify how the equilibrium density profile and chemical potential are modified in the proximity of a solid surface. We consider now the fluid occupying the semi-infinite domain $z > 0$ with the density changing in z direction only. This computation requires boundary conditions for the density to be set on the solid surface. This is rather difficult to do in the framework of a diffuse interface theory in a consistent way, since a sharp solid interface implies a discontinuity, and the question is sensitive to the character of interactions at short distances, as well as to such subtle factors as packing near the boundary and periodic crystalline field. A formal way to derive a boundary condition is to allow a non-vanishing variation of the density at the solid boundary $\delta\rho(0)$ when the energy functional (15) is varied. This leaves, after integrating by parts, the boundary term $K\rho'(0)\delta\rho(0)$. This is added to the variation of the fluid-solid interaction energy $\gamma_l'(\rho)\delta\rho(0)$, where the dependence on the fluid density near the wall is given by Eq. (31) with ρ_l replaced by $\rho(0)$. Setting the coefficient at $\delta\rho(0)$ to zero yields the boundary condition

$$\frac{\pi}{d^2}\left(A_l\rho(0) - A_s\rho_s\right) - K\rho'(0) = 0. \tag{71}$$

The same boundary condition can be obtained by adding to Eq. (15) the shift of chemical potential due to liquid-solid interactions given by Eq. (35), and assuming that the density near the solid surface coincides with the equilibrium density corresponding to the shifted value of chemical potential at the molecular distance d from the wall. Both computations are in somewhat inconsistent, since both Eq. (31) and Eq. (35) has been obtained under assumption of constant density, but the error must be small when the liquid density changes only slightly near the solid wall. This is indeed true when the strength of liquid-liquid and liquid-solid interaction is almost on par, so that the Hamaker constant is small relative to the absolute value of the constituent terms. We take note that this is also the condition of the contact angle, as defined by Eq. (33), to be small when $A < 0$. Then the boundary condition (71) is satisfied at $\rho'(0) \ll 1$ and ρ close to ρ_l.

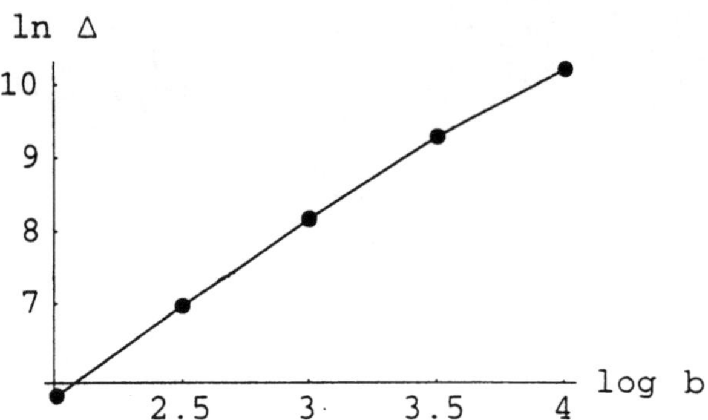

Figure 14. The dependence of the pull-down length Δ on $\log G$.

Another way to justify the above derivation is to assume that the solid–fluid interaction is short-range compared to the thickness of the diffuse vapor-liquid interface. Then it is likely prevail locally in the vicinity of a solid wall, so that both algebraic terms in Eq. (71) are large, and it can be replaced by a simpler Dirichlet boundary condition $\rho(0) = \rho_{ls}$. Formally, the fluid-solid surface tension coefficients can be computed, once the density profile satisfying Eq. (16) with the set boundary condition is known, and the "standard" contact angle found using Eq. (32). The range $\rho_v < \rho_{ls} < \rho_l$ corresponds then to partial wetting. The contact angle is zero (complete wetting) at $\rho_{ls} \geq \rho_l$. This "standard" angle has nothing to do with a "true" contact angle at the solid surface. The latter is not defined at all in the diffuse interface theory, since different isodensity levels behave in a qualitatively different way as the solid surface is approached. The only level that hits the solid surface at the right angle is $\rho = \rho_{ls}$; the levels with $\rho < \rho_{ls}$ are asymtotically parallel, and those with $\rho > \rho_{ls}$ antiparallel to the surface.

Since our aim is a qualitative description of a system far from criticality involving the vapor phase with negligible density, we shall be not involved with the algebraically difficult van der Waals form of $g(\rho)$, but work with a simple cubic polynomial that gives $\rho_v = 0$ and allows to carry out computations analytically. Thus, we choose

$$g(\rho) = \rho(1 - 2\rho)(1 - \rho), \tag{72}$$

which is at Maxwell construction at $\mu = 0$. We shall also rescale K to unity. Then $\overline{f}(\rho) = \frac{1}{2}\rho(1 - \rho)^2$, and the equilibrium surface tension is computed as

$$\sigma = \int_0^1 \rho(1 - \rho)d\rho = \tfrac{1}{6} \tag{73}$$

This formula can be also obtained directly using the standard kink solution that approaches $\rho_v = 0$ at $z \to \infty$ and $\rho_l = 1$ at $z \to -\infty$:

$$\rho_0(z) = (1 + e^z)^{-1}. \tag{74}$$

This solution is distorted in the vicinity of a solid wall. In the following, we shall adopt a simple Dirichlet boundary condition $\rho(0) = 1 - a$ with $|a| \ll 1$. Solving Eq. (18) with a cubic $g(\rho)$ is elementary: the exact solution is expressed in elliptic functions. It is, however, more elucidating to find an approximate solution. We construct the solution by perturbing a standard kink solution $\rho_0(z-h)$ centered at $z = h$, i.e. Eq. (74) for the cubic $g(\rho)$. The actual solution is approximated to the zero order by the standard kink only when $\rho_0(-h)$ is sufficiently close to unity; thus, h must satisfy the condition $h > \ln(1/a)$. The density profile is expanded in the small parameter a:

$$\rho = \rho_0(z-h) + a\rho_1(z;h) + \dots. \tag{75}$$

For the time being, we assume $\mu = 0$. Then the first-order equation is

$$\rho_1''(z) + g'(\rho_0)\rho_1 = 0, \tag{76}$$

subject to the boundary condition

$$\rho_1(0) = -1 + a^{-1}[1 - \rho_0(-h)] \approx -1 + \psi, \tag{77}$$

where $\psi = a^{-1}e^{-h} \leq 1$.

Due to the exponential decay of interactions, the correction to the zero-order solution is actually of a higher order of magnitude everywhere except an $O(\ln a^{-1})$ vicinity of the wall, where ρ_0 is close to unity. On this interval, Eq. (76) can be replaced by the equation with constant coefficients

$$\rho_1''(z) - \rho_1(z) = 0, \tag{78}$$

The solution decaying at $z \to \infty$ is

$$\rho_1(z) = -e^{-z}(1-\psi). \tag{79}$$

At $a > 0,\ h > \ln(2/a)$, the combined function

$$\rho_a = \rho_0 + a\rho_1 = \left(1 + e^{z-h}\right)^{-1} - e^{-z}\left(a - e^{-h}\right) \tag{80}$$

reaches a maximum at $z = \frac{1}{2}\ln(ae^h - 1) > 0$ (Fig. 15). Such a solution describes a liquid layer sandwiched between the vapor and the solid. At smaller values of h, the maximum disappears, and the solution can be interpreted as a pure vapor phase thickening near the solid wall. The same solution applies at $a < 0$ when the density increases at the solid surface, whether it is approached from the liquid phase or directly from the vapor phase. The approximation breaks down at $h < \ln(1/a)$, which is, in fact, the minimal possible thickness of the dense layer in this model.

Non-monotonic density profiles are unstable. Since, however, the influence of the wall decays exponentially with the distance, the dynamics is practically frozen whenever the interphase boundary is separated from the wall by a layer thick compared to the characteristic width of the diffuse interface. A static solution with a fixed h exists only at a certain fixed value of μ, which can be determined using a solvability condition of the first-order equation as in Section 1.3. In a wider context, an appropriate solvability condition serves to obtain an evolution equation for the nominal position h of the interphase boundary. The technique of derivation of solvability conditions for a problem involving a semi-infinite region and exponentially decaying interactions is non-standard and therefore deserves some attention.

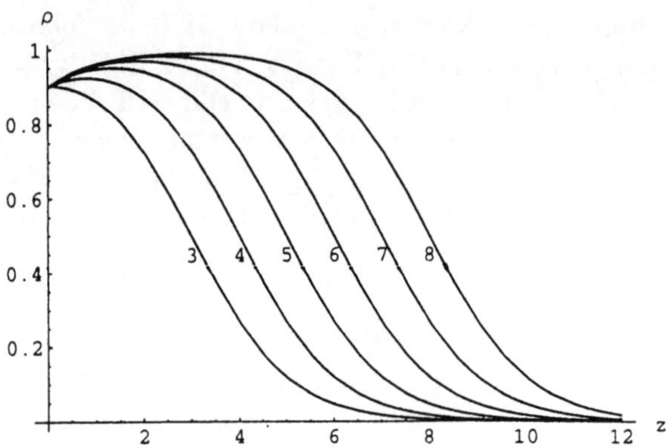

Figure 15. Stationary density profiles. Numbers indicate the values of the nominal thickness h.

An inhomogeneous first-order equation has the general form (27). When this equation is defined on the infinite axis, the solvability condition appears due to the presence of a Goldstone mode of the linear operator $\rho_0'(z)$ related to the translational symmetry of the kink. In the presence of a solid boundary, a difficulty arises, however, since the translational invariance is broken and no easily computable eigenfunction is available. In addition, the orders of magnitudes in the perturbative scheme should be estimated in a non-standard way in view of the exponential decay of interactions. The difficulties are overcome with the help of asymptotic matching technique. The solvability condition is computed, similar to Eq. (29), using the translational eigenfunction on the infinite axis, but the integration is not carried out over the entire axis (which now extends into the unphysical region $z < 0$), but starts at some location $z = z_0 > 0$ where ρ differs from the asymptotic value $\rho = 1$ by an $O(a)$ increment. This generates boundary terms in the solvability condition, which takes now the form

$$\int_{z_0}^{\infty} \frac{d\rho_0(z-h)}{dz} \Psi(z) dz = \left[\frac{d\rho_0(z-h)}{dz} \frac{d\rho_1(z)}{dz} - \frac{d^2\rho_0(z-h)}{dz^2} \rho_1(z) \right]_{z=z_0} \tag{81}$$

The boundary values of the first-order solution $\rho_1'(z)$ are obtained by solving the first-order equation (27) directly on the interval $0 \le z \le z_0$, where Eq. (27) can be replaced by the equation with constant coefficients (78) with the added inhomogeneity $\Psi(z)$. The solution of this equation is

$$\rho_1(z) = \rho_1^{(h)}(z) + \int_0^z \mathcal{G}(z-\zeta)\mathcal{H}(\zeta) d\zeta, \tag{82}$$

where $\rho_1^{(h)}$ is given by Eq. (79) and $\mathcal{G}(z-\zeta)$ is Green's function of the homogeneous operator in Eq. (27). The last term can be neglected, provided the lower limit of the integral in the left-hand side of Eq. (81) can be shifted to $-\infty$ without introducing a significant error. The matching is successful when Eq. (81) reduces to a form independent of z_0 in the leading order.

Now we apply this matching technique to computation of a constant value of chemical potential $\mu = \mu_c$ required to keep the kink at equilibrium (possibly, unstable) at a given location $z = h$. In this case, the inhomogeneity in Eq. (78) is just a constant $\Psi = \mu_c$, and the integral in the left-hand side of Eq. (81) is $\mu_c[\rho_0(\infty) - \rho_0(z_0)] = -\mu_c + O(a)$. Since this expression remains unchanged in the leading order when z_0 is shifted to $-\infty$, i.e. $\rho_0(z_0) = 1 - O(a)$ replaced by unity, it is sufficient to use in Eq. (81) the first term of Eq. (82) only. Retaining the leading term only, we obtain

$$\mu_c \equiv a^2 M(h) = 2a^2\psi(1-\psi) = 2e^{-h}\left(a - e^{-h}\right). \tag{83}$$

The first expression demonstrates that the computed chemical potentials in fact at most of $O(a^2)$, although the equation is nominally of the first order. The gained order of magnitude is due to the fast decay of interactions. Since the computed value is of a higher order, there is no need to correct the equilibrium profile computed in the preceding subsection to $O(a)$. For $a > 0$, the function $\mu_c(h)$ passes a maximum at the same value $h = \ln(2/a) = O(1)$ that marks the transition from monotonic to non-monotonic density profiles. Sustaining a static profile requires a bias in favor of the liquid state, and the value of μ_c at the maximum represents the critical value of chemical potential required to nucleate a thick liquid layer on the solid surface. For $a < 0$, μ_c in Eq. (83) is negative and increases monotonically with h; in this case, on the contrary, a bias in favor of the vapor phase is necessary to keep the interface stationary. The phase plane orbits at $\mu_c > 0$ and $\mu_c < 0$, as well as the plot of Eq. (83) are shown in Fig. 16.

5.3 Lubrication approximation

Two-dimensional motion can be rationally treated in the familiar "lubrication approximation", assuming the characteristic scale in the "vertical" direction (normal to the solid surface) to be much smaller than that in the "horizontal" (parallel) direction. When the interface is weakly inclined and curved, the density is weakly dependent on the coordinate x directed along the solid surface. The velocities v, u corresponding to weak disequilibrium of the phase field considered above will be consistently scaled if one assumes $\partial_z = O(1)$, $\partial_x = O(\sqrt{\delta})$, $u = O(\delta^{3/2})$, $v = O(\delta^2)$. It is further necessary for consistent scaling of the hydrodynamic equations that the "constant" part of the chemical potential μ, associated with interfacial curvature, disjoining potential, and external forces and weakly dependent on x, be of $O(\delta)$, while the "dynamic" part varying in the vertical direction and responsible for motion across isodensity levels, be of $O(\delta^2)$. We can assume therefore that $\mu + V$ is independent of z.

In two dimensions, the term ρ_{xx} is added to the inhomogeneity in the first-order equation (76). In this order, the vertical density profile can be represented by the standard kink solution $\rho_0(z - h(x,t))$, and the x dependence is due to slow variation of h in the "horizontal" direction. Thus,

$$\rho_{xx} = -\rho_0'(z-h)h_{xx} + \rho_0''(z-h)h_x^2. \tag{84}$$

The respective contribution to the solvability condition is, in the leading order,

$$-h_{xx}\int_{-\infty}^{\infty}[\rho_0'(z)]^2 dz = -\sigma h_{xx}, \tag{85}$$

while the contribution of the term containing h_x^2 vanishes in the leading order by symmetry.

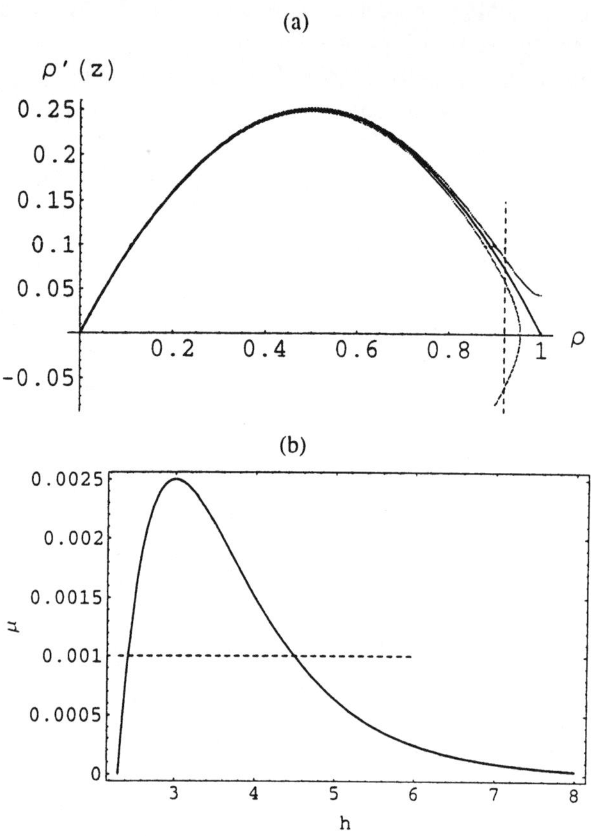

Figure 16. (a) Phase plane orbits at $\mu_c > 0$ (lower curve) and $\mu_c < 0$ (upper curve). (b) μ_c as a function of the layer thickness at $a > 0$. The dashed line in both pictures connects points corresponding to a monotonic and a non-monotonic profiles at the same value of μ_c.

Another possible contribution to the solvability condition may come from external forces. In the presence of gravity directed against the z axis, the equilibrium is achieved, according to Eq. (70), at $\mu = \mu_0 - a^2 Gz$ rather than $\mu = \mu_0 = \text{const}$. The rescaled acceleration of gravity is denoted as $a^2 G$, which presumes that it matches the other terms by the order of magnitude. The integral in Eq. (81) involving the variable part of μ is mostly accumulated in the diffuse interface region, so that we have in the leading order

$$-G \int_0^\infty z\rho_0'(z-h)dz \approx Gh. \tag{86}$$

Collecting Eqs. (85) and (86), we obtain the expression for the hydrostatic chemical potential

$$\mu = \delta \left[M(h) - \sigma h_{xx} + G(h-z) \right], \tag{87}$$

where $M(h)$ is defined by Eq. (83).

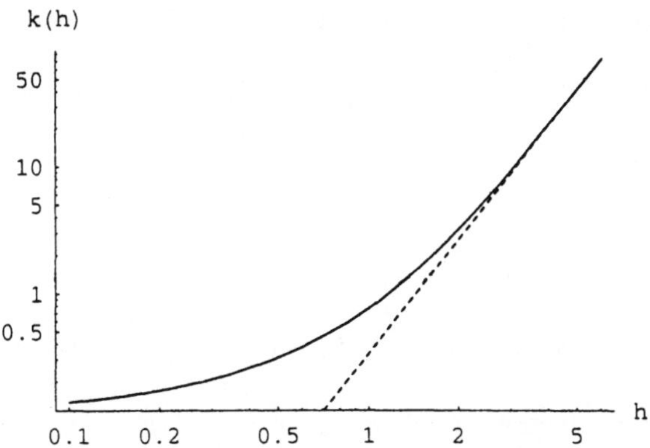

Figure 17. The function $\eta k(h)$, compared with the respective function for the sharp interface $\frac{1}{3}h^3$ (dashed line).

Equations of motion in lubrication approximation are obtained following the same routine as in Section 3.2. We shall restrict to one-dimensional motion along the x axis. The horizontal velocity u is determined from the horizontal component of the Stokes equation. Adding gravity as an external force, we write the leading order equation as

$$-\rho_0(z-h)W_x + (\eta u_z)_z = 0, \tag{88}$$

where the driving potential W is defined as

$$W = G\alpha x + M(h) - \sigma h_{xx} + G(h-z), \tag{89}$$

This expression follows from Eq. (87), with the addition of the gravity term acting when the supporting plane is weakly inclined. The inclination angle α must be of $O(\sqrt{\delta})$ to match by the order of magnitude the other terms in the equation. The density profile is given in the leading order by the standard kink solution (74) centered at the nominal interface position $h(x)$ slowly varying in the horizontal direction.

The solution of Eq. (88) satisfying the no-slip boundary condition on the solid boundary and the no stress condition at infinity has a general form

$$u(z) = \eta^{-1} W_x \Psi(z;h). \tag{90}$$

The function $\Psi(z;h)$ depends on an assigned dependence of viscosity on density, but the flux $u\rho_0$ in the dense layer (at z not much larger than h) is nearly the same for either $\eta =$ const or $\eta \propto \rho$, and is close to the standard lubrication solution $\Psi = -z(h - \frac{1}{2}z)$ valid for incompressible Poiseuille flow in a layer of thickness h with a free boundary as in Eq. (44).

The evolution equation of h is obtained by inserting Eqs. (74), (90) in the continuity equation (69) and integrating it from 0 to ∞. Using the relations

$$\int_0^\infty \rho_t dz = -h_t \int_0^\infty \rho_0'(z)dz = h_t + O(a), \int_0^\infty (\rho v)_z dz = 0,$$

we obtain the evolution equation in the general form [cf. (45)]

$$h_t = \partial_x \left[k(h) W_x \right]. \tag{91}$$

with the mobility coefficient

$$k(h) = -\eta^{-1} \int_0^\infty \rho_0(z-h) \Psi(z,h)\, dz. \tag{92}$$

The function $k(h)$, computed numerically and plotted in Fig. 11, differs only slightly from the respective function for the sharp interface given by Eq. (47) when h exceeds its minimal admissible value $h_0 = \ln(1/a)$. Taking into account small deviations from the standard kink solution near the wall adds only a higher-order correction.

Apart from a slightly modified volumetric rate, the specific contribution of the diffuse interface to Eq. (91) is carried by the function $M(h)$, which is dependent on the boundary conditions on the solid surface and expresses disjoining potential. The structure of Eq. (91) is identical to that of standard equations of motion of thin liquid films, which are recovered at large h when the disjoining potential becomes negligible. At small h, the disjoining potential is not singular. At the same time, the viscous stress singularity at the contact line is relaxed as the latter's location becomes indefinite.

Steady flow of a liquid film under the action of disjoining potential and gravity can be described by Eq. (91) rewritten in the frame moving with a speed U. We shall assume that the liquid layer thickens at $x \to \infty$, and assume U to be positive when the thick layer advances. Standard macroscopic arrangements fixing the asymptotic conditions at $x \to \infty$ are possible, e.g. $h \to \infty,\ h_x = -\alpha$ for a liquid wedge with the angle α or $h_x = 0,\ h = \sqrt{3U/\alpha G}$ for an asymptotically flat film on an inclined plane.

Admissible asymptotics at $x \to -\infty$ depends on the form of the function $M(h)$. If it is given by Eq. (83) with $a > 0$, the layer may attain asymptotically at $x \to -\infty$ the state of lowest energy $h = h_0 = \ln(1/a)$ (formally, this is possible at zero inclination α, although gravity effects are negligible in films of molecular thickness).

The starting point is Eq. (91) with the effective pressure given by Eq. (89). Removing extra parameters by rescaling and integrating once yields

$$h'''(x) - \left(M'(h) + G\right) h'(x) - \alpha G + \frac{U(h-h_0)}{k(h)} = 0, \tag{93}$$

where the integration constant has been introduced allowing for a precursor film with the thickness h_0 at $x \to -\infty$. A more convenient form of Eq. (93) is obtained using as the dependent variable $y = h_x^2$ and as the independent variable the nominal thickness h:

$$\frac{1}{2} y''(h) - \left(M'(h) + G\right) + \frac{1}{\sqrt{y}} \left(\frac{U(h-h_0)}{k(h)} - \alpha G \right) = 0. \tag{94}$$

Equation (94) is free from singularities which are usually caused by divergences of either viscous stress, or disjoining potential, or both, in a layer of vanishing thickness. It can be integrated numerically starting from the asymptotics at $x \to -\infty$. The asymptotics of Eq. (94) obtained

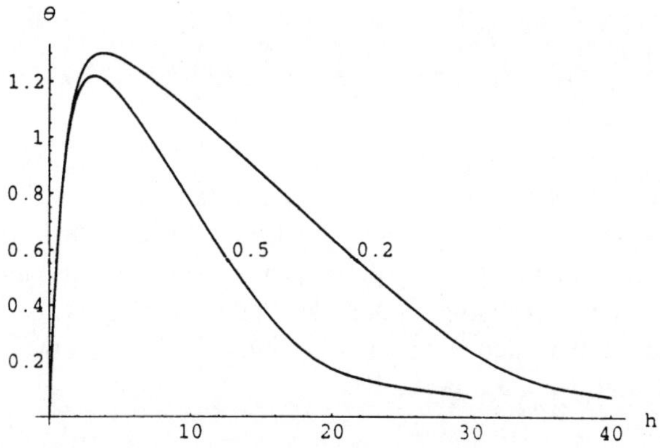

Figure 18. Dependence of the interface inclination angle θ on the nominal thickness h of a spreading dense layer for $3U = 0.5$ and $3U = 0.2$ (as indicated by numbers at the respective curves). The values of G found by shooting are, respectively, 0.035123081 and 0.0079817.

by expanding near $h = h_0$ is $y \asymp c^2(h - h_0)^2$, implying exponential decay to the "optimal" thickness $h - h_0 \propto e^{\kappa x}$, where the constant κ is a positive root of the characteristic equation

$$\kappa^3 - M'(h_0)\kappa + U/k(h_0) = 0. \tag{95}$$

Fixing, say, the value of U, one can use the shooting method to adjust the value of G satisfying the appropriate boundary condition at infinity, $\sqrt{y} = -\alpha$. A very fine adjustment of the parameter is needed to advance to moderate values of h. An example of a computed dependence of the interface inclination angle on the nominal thickness of a dense layer spreading on a horizontal support is shown in Fig. 18.

6 Spreading assisted by interphase transport

6.1 Interphase transport in diffuse interface theory

Equilibrium solutions with ρ varying along the z axis exist only at a particular constant value of μ, equal to zero in the adopted gauge. Any deviation of this value sets the interface into motion; the interface shift corresponds to evaporation or condensation retarded by viscous friction. The simplest case is steady propagation of the boundary between two semi-infinite phases. The stationary one-dimensional equations in the frame moving with the speed c of the steadily propagating interface are

$$(\rho v)_z = 0, \quad -\rho\mu_z + (\widehat{\eta}v_z)_z = 0 \tag{96}$$

where v is the single velocity component in this frame; external forces are omitted and $\widehat{\eta} = \zeta + \frac{4}{3}\eta$ is the renormalized viscosity, accounting also for the divergence term in Eq. (70). These equations are readily integrated yielding

$$j \equiv \rho v = \text{const}, \quad \mu = \mu_c + jR(z), \tag{97}$$

where

$$R(z) = \int \frac{1}{\rho}\frac{d}{dz}\left(\widehat{\eta}\frac{d\rho^{-1}}{dz}\right) dz. \tag{98}$$

The flux j is related to the propagation velocity c as $j = -c(\rho_l - \rho_v)$. The sign of c is chosen in such a way that it is positive when the dense (liquid) state advances. The constant μ_c, which may be fixed by external conditions, represents the driving force of the process.

It is reasonable to assume that the disequilibrium is weak, so that both μ_c and the constant flux j are multiplied by a book-keeping small parameter δ when Eq. (97) is used in Eq. (17). The perturbed equation can be expanded in a usual way, and the relation between the flux j and μ_c is obtained from the solvability condition (81):

$$\mu_c = -c \int_{-\infty}^{\infty} \rho_0'(z) R(z) dz. \tag{99}$$

The integral in the right-hand side can be interpreted as the effective friction factor. It depends on the basic density profile $\rho_0(z)$ as well as on the assumed dependence of the viscosity on density. If $\rho_0 = \rho_c + \widetilde{\rho}$ represents a weakly perturbed critical density, $R(z) = -\eta_c \rho_c^{-3} \widetilde{\rho}_z$, and the integral in Eq. (99) is proportional to surface tension. In the case of vanishing vapor density which interests us most, assuming η = const leads to a divergent integral. The divergence is not eliminated also when the viscosity is proportional to density. Taking, for example, $\widehat{\eta} = \nu\rho$, Eq. (98) is evaluated using the relation $\rho_z = -\rho(1-\rho)$ as $R(z) = -\nu \ln(\rho_0(z)/\rho_c)$. The weak divergence on the vapor side can be eliminated by assuming a small but finite vapor density ρ_v. Then evaluating the solvability condition yields

$$\begin{aligned}\mu_c &= -c \int_{-\infty}^{\infty} \rho_0'(z) R(z) dz \\ &= -c\nu \int_{\rho_v}^{\rho_l} \ln \frac{\rho_0(z)}{\rho_c} d\rho = -c\nu(1 + \ln \rho_c).\end{aligned} \tag{100}$$

where μ_c is the chemical potential at the location with a chosen density level ρ_c.

The dense layer advances ($c > 0$) at $\mu > 0$. This causes the chemical potential to drop at at locations with lower density ahead of the propagating interface. thereby effectively slowing down the advance of the dense layer. A sharp drop in the dilute layer. leading to a divergent friction factor (99), causes substantial deviations from the zero-order density profile.

6.2 Kinetic retardation of interphase transport

Taking into account "normal" viscous retardation only (with $\widehat{\eta} \approx \eta$) may exaggerate the actual phase transition rate, since transport through a sharp density gradient is in fact an activated process, except, perhaps, in an immediate vicinity of a critical point. When the interface is treated as a sharp discontinuity, this may be accounted for by introducing a finite evaporation rate (involving an appropriate activation energy) and a condensation "sticking coefficient". Both quantities are difficult to estimate quantitatively but, in principle, they insure a finite evaporation or condensation rate even under conditions when viscous retardation is absent.

In the framework of the phase field theory, kinetic retardation can be accounted for by replacing the stationary equation (16) or (17) by the respective gradient flow equation containing a large relaxation time τ. In one dimension, we have

$$\tau\rho_t = \rho_{zz} - g(\rho) + \mu. \tag{101}$$

On the infinite axis, this equation (with $\mu = \text{const}$) has a solution steadily propagating with a speed dependent on μ, and satisfying the stationary equation in the comoving frame

$$-\tau c\rho_z + \rho_{zz} - g(\rho) + \mu = 0. \tag{102}$$

In the case of weak disequilibrium, $\mu = O(\delta) \ll 1$, the propagation speed $c = O(\delta)$ is easily computed, as in the preceding subsection, using the solvability condition of Eq. (102) expanded in δ:

$$c = \frac{\mu(\rho_l - \rho_v)}{\tau\sigma} = \frac{6\mu}{\tau}, \tag{103}$$

where σ is defined by Eq. (21) and the numerical value is given for the cubic $g(\rho)$.

Equations (99) and (103) represent two opposite limits when, respectively, either viscous or kinetic retardation is prevalent. A rough estimate for the lower bound of the relaxation time is $\tau \propto l^2/D$, where l is the thickness of the diffuse interface and D is the diffusivity. The characteristic time of viscous retardation on the same length scale is $\tau_v \propto l^2/\nu$, where $\nu = \eta/\rho$. For common liquids, the Prandtl number $\Pr = \nu/D$ is large, and $\tau_v/\tau \propto D/\nu \ll 1$. Viscous retardation may be still felt at larger scales, complementing the kinetic retardation near the diffuse boundary. At $\Pr \gg 1$, the flow velocity is nearly constant throughout the transitional boundary region, and the propagation velocity defined by Eq. (103) can be viewed as the velocity of the slow drift of the interphase boundary due to the evaporation or condensation in the frame moving with the local velocity of the ambient fluid. At fixed propagation velocity, the increments due to the viscous and kinetic retardation are additive. In the dense layer, the former is negligible at $\Pr \gg 1$, although it becomes important in the dilute phase, as we have seen in the preceding subsections.

When Eq. (101) is coupled to hydrodynamics, ρ_t should be replaced by convective derivative, and the equation can be rewritten using the continuity equation (69) as

$$\tau\rho\nabla\cdot\mathbf{v} + \rho_{zz} - g(\rho) + \mu = 0. \tag{104}$$

This shows clearly that kinetic retardation is effective, as it should be, in the diffuse boundary region, where the fluid is compressible. The scaling of the lubrication approximation remains consistent only if the relaxation time τ in Eq. (101) is of $O(\delta)$. With this scaling, the speed of the vapor-liquid interface displacement is of $O(\delta^2)$, i.e. of the same order of magnitude as the vertical velocity.

6.3 Lubrication equations with interphase transport

The results of the computations that have been carried out so far in this section for an infinite fluid layer separated by a diffuse vapor-liquid interface can be applied to the spreading problem after minimal modification. In a bounded layer, the chemical potential in the liquid phase μ_c

driving the evaporation or condensation flux is determined by the combined action of surface tension and disjoining pressure. The disjoining potential can be computed with the help of the solvability condition, as in Section 5.2; the respective formulae remain in force, since the flux-related drop of the chemical potential occurs in the dilute phase only, and is negligible in the diffuse interface region, where the translational eigenfunction is localized.

The basic equation of the lubrication approximation, Eq. (91), is modified in the case of non-equilibrium spreading by an added evaporation or condensation term:

$$\frac{\partial h}{\partial t} = -\beta(W - W_0) + \partial_x \left[k(h)W_x\right]. \tag{105}$$

The evaporation or condensation flux is assumed here to be proportional to a shift of the chemical potential relative to some equilibrium level W_0. The value and even the order of magnitude of the coefficient β may vary widely, depending on the physical situation under consideration. This determines, in turn, the relative importance of the two terms on the right-hand side of Eq. (105). In the case of viscously retarded motion with finite ρ_v, the first term appears to be proportional to $\widehat{\eta}W$, although the term representing the horizontal transport through the liquid phase is of order $\widehat{\eta}\, \partial^2 W/\partial x^2$ and is negligible compared to the evaporation or condensation term in the lubrication limit, when the horizontal derivatives are small. In this case, flow across the isodensity levels associated with evaporation or condensation, driven by the deviation of the chemical potential from equilibrium, would be larger by $O(\delta^{-1})$ than hydrodynamic "horizontal" motion. This does not represent the most usual situations, where evaporation is hindered by large activation energies. In the present model a way to enter consistently this activation effect is to make the evaporation flux and the horizontal transport of the same order of magnitude. This can be done by imposing an $O(\delta^{-1})$ relaxation time τ. Although this connection between a molecular quantity τ and a macroscopic length scale δ may look artificial, this represents a distinguished limit where a balance between factors of different physical origin is attained. In the absence of horizontal flux (which would happen far from a solid boundary), one recovers the classical Thomson expression for the evaporation driven by the curvature of the liquid-vapor interface.

Mass transport across isodensity lines should become particularly important when the lubrication approximation breaks down. This should happen near the contact line in the case when two alternative fluid densities near the solid wall are possible. If, say, the boundary densities are $\rho_{sv} \ll 1$ and $\rho_{sl} = 1 - a$, $a \ll 1$, the three-phase "contact line" can be viewed as a sharp transition between $O(1)$ positive and negative values of the nominal thickness h, such that $e^{-|h|} \ll 1$ on either side. This can be treated as a shock of Eq. (91) or (93). The Hugoniot condition, which should ensure zero net flux through the shock, is the equality of chemical potentials on both sides. Unfortunately, this condition cannot be formulated precisely, since the sharp-interface limit of the surface tension term is inapplicable in the shock region. Moreover, our test computations of the profile of the dense layer using Eq. (94) with different boundary conditions imposed on the "shock" at $h = h_0$ showed that the spreading velocity is very sensitive to the conditions on the shock.

It remains therefore essential to solve the full system of density field and hydrodynamic equations in the shock region whenever a sharp transition between alternative surface densities is possible. The outer limit of the resolved shock structure should be matched with the lubrication equations (91), (93), or (105). The transport across isodensity lines in the shock region alleviates the viscous stress singularity remaining in the lubrication model. In its turn, the latter provides a gradual transition to the sharp-interface limit at large distances.

Equation (105) also appears in a natural way in sharp interface theory. It has been studied in the context of competition between layers of different thickness that arises when different kinds of intermolecular forces operate, as in Section 2.2. The film thickness may change then across a transition front, which will advance under combined action of evaporation/condensation and fluid flow driven by the difference of equilibrium chemical potential of the two alternative states.

Summary and perspectives

The main message of these lectures is the need to amend the classical hydrodynamic theory by direct inclusion of intermolecular interactions. This is necessary not only in the theory of contact line motion outlined here, but in all mesoscopic hydrodynamic problems, e.g. in fluid mechanics of microdevices, which attracts lately a lot of attention. The specific feature of the contact line problem is the connection between microscopic and macroscopic. The motion in the precursor film can and should be treated more precisely, on the statistical level with due account for fluctuations or directly through molecular dynamics simulations. A challenging problem is matching the microscopic theory with classical hydrodynamics applicable in macroscopic domains away from the immediate vicinity of the contact line.

Continuous models including intermolecular forces, in particular, the diffuse interface model provide a sound theoretical basis for studying equilibrium capillary phenomena in fluids. We have shown that these models can be extended in a natural way to study a thoroughly dynamical spreading process. The lubrication limit, where the contact angle is small, allows us to derive consistently an equation of motion for the liquid-vapor interface interacting with the solid surface. In the static limit, this equation yields back the equilibrium Young-Laplace theory.

Evaporation or condensation processes may play a significant role in the dynamics of contact lines. The driving force for the evaporation or condensation is the imbalance between the pressure drop across the interface and its equilibrium value. An advancing or receding contact angle differing from its equilibrium value makes the contact line a source or sink for evaporation or condensation. This interesting phenomenon might be checked experimentally by observing the accumulation of a non-volatile tracer diluted in the liquid phase that would be left by evaporation near a moving contact line.

Our analysis indicates that kinetic retardation of the interphase transport is essential for a well-balanced theory away from the critical point. The available simulations of the motion of a diffuse interface near a three-phase contact line, taking into account viscous retardation only and, in effect, assuming evaporation or condensation to be as easy as plain advection, may grossly overestimate the rate of interphase transport, but the latter remains essential even when its order of magnitude is reduced due to kinetic retardation.

In summary, the dynamics of contact line is a very rich and complex problem. The unexplored frontier of the research lies both in experiment required to resolve microscopic structure of dynamic three-phase boundary under controlled condition and in multiscale computations clarifying the role of molecular-scale inputs in macroscopic behavior.

Acknowledgments This work has been supported by the Israel Science Foundation and Minerva Center for Nonlinear Physics of Complex Systems.

Bibliographical notes

Section 1. The standard reference for general thermodynamic relations is

L.D. Landau, and E.M. Lifshitz, v. V, *Statistical Physics*, Part I, Pergamon Press, 1980.

A comprehensive review of intermolecular forces is found in

J.H. Israelachvili, *Intermolecular and Surface Forces*, Academic Press, New York, 1992.

For review of the molecular basis of capillarity, see

J.S. Rowlinson and B. Widom, *Molecular Theory of Capillarity*, Oxford University Press, 1982.

The diffuse interface theory goes back to van der Waals:

J.D. van der Waals, Z. f. Phys. Chem. **13** 657 (1894); English translation J.S. Rowlinson, J. Stat. Phys. **20** 197 (1979).

Section 2. The primary reference for the origins of disjoining potential is

B.V. Derjaguin, N.V. Churaev and V.M. Muller, *Surface Forces*, Consultants Bureau, New York, 1987.

Thermodynamics of the contact line is reviewed in detail by

P.G. de Gennes, Rev. Mod. Phys. **57**, 827 (1985).

For a more detailed treatment of intermolecular interactions near a static contact line, see

G.J. Merchant and J.B. Keller, Phys. Fluids A **4**, 477 (1992).
C. Bauer and S. Dietrich, Eur. Phys. J. B **10** 767 (1999).

For a recent treatment of effect of fluctuations, see

A. Hazareesing and M. Mézard, Phys. Rev. E **60** 1269 (1999).

Section 3. For early "classical" fluid-mechanical theory of a moving contact line, see

C. Huh and L.E. Scriven, J. Coll. Int. Sci. **35**, 85 (1971).
E.B. Dussan V and S.H. Davis, J. Fluid Mech. **65**, 71 (1974).

The phenomenological slip condition has been used in a number of fluid-mechanical computations:

L.M. Hocking, J. Fluid Mech. **79**, 209 (1977); **211**, 373 (1990).
H.P. Greenspan, J. Fluid Mech. **84**, 125 (1978).
P.J. Haley and M.J. Miksis, J. Fluid Mech. **223**, 57 (1991).

Various efforts to eliminate the viscous stress singularity are reviewed by

Yu.D. Shikhmurzaev, J. Fluid Mech. **334** 211 (1997).

Section 4. Various aspects of fluid-mechanical treatment of a moving contact line with account of van der Waals forces is found in

C.A. Miller and E. Ruckenstein, J. Coll. Interface Sci. **48**, 368 (1974).
H. Hervet and P.G. de Gennes, C. R. Acad. Sci. **299 II** 499 (1984).
P.G. de Gennes, X.Hue, and P.Levinson, J. Fluid Mech. **212**, 55 (1990).
L.M. Hocking, Phys. Fluids A **5**, 793 (1993).
L.M. Pismen, B.Y. Rubinstein, and I. Bazhlekov, Phys. Fluids **12**, 480 (2000).

For derivation and fluid-mechanical implications of the kinetic slip condition, see

E. Ruckenstein and C.S. Dunn, J. Coll. Interface Sci. **59**, 135 (1977).
L.M. Pismen and B.Y. Rubinstein, Langmuir, **17** 5265 (2001).

Section 5. Diffuse interface theory coupled with hydrodynamics is reviewed by

D.M. Anderson, G.B. McFadden, and A.A. Wheeler, Ann. Rev. Fluid. Mech. **30** 139 (1998).

For applications to the problem of contact line motion, see

P. Seppecher, Int. J. Engng Sci. **34** 977 (1996).
D. Jacqmin, J. Fluid Mech. **402**, 57 (2000).
L.M. Pismen and Y. Pomeau, Phys. Rev. E, **62** 2480 (2000).

HYDRODYNAMICS OF SURFACE TENSION DOMINATED FLOWS

D.T. Papageorgiu
University Heights, Newark, NJ, USA

1 Introduction

This series of lectures is concerned with the fluid dynamics of surface tension driven flows. Surface tension forces arise at the interface between two fluids (e.g. water and air) and can be of central importance in applications where a separating (and in many cases moving) interface is part of the process. Examples include, but are not limited to, the dynamics of liquid films, jets, and drops.

In our pursuit of a theoretical description of such problems, we are led to the following ubiquitous situation: In order to solve for the bulk flow field (governed by the Navier-Stokes equations, for instance), the position of the free surface and boundary conditions there, need to be known; the flow field, and other interfacial forces, act to move the interface to a new position which in turn affects the flow field. In general, this coupling is nonlinear and in many problems of practical interest no steady state exists. In fact, severe flow regimes can arise as in the breakup of Newtonian liquid jets where a topological singularity is encountered in finite time.

The mathematical problems are challenging and a combination of analytical and computational tools is often used in their solution. In this series of lectures I will attempt to give a flavor of such research for two different problems that we have been working on over the last few years. The first problem is concerned with the stability and breakup of liquid jets (or bridges). Aspects of linear stability will be described first (including discussion of absolute versus convective instability), followed by nonlinear asymptotic theories which lead to simpler evolution equations that are capable of describing the event of jet breakup. In fact, the model is in remarkable agreement with experiments and one of its mathematical predictions, namely the constant rate at which the minimum jet radius tends to zero near breakup, is used as an alternative rheological measurement of the ratio of surface tension coefficient to liquid viscosity (see McKinley and Tripathi (1999).)

The second problem is concerned with the motion of spherical bubbles through liquids containing soluble surfactants. Applications include thermocapillary motions and dropwise extrusion processes relying on enhanced interphase mass transfer. The mechanisms of the Marangoni effect in this particular situation will be described in detail later. Briefly, as the bubble moves through the fluid, surfactant adsorbs onto the front end, is convected towards the back end by the surface flow and in turn desorbs from the back end into the bulk. Different regimes arise depending on the ratio of the rate of convection to that of desorption into the bulk. It is typical in many systems for surfactant to build up at the back end. The surface tension, which is a decreasing function of surface surfactant concentration, becomes lower at the back end than at the front of the bubble. A Marangoni force acts in a direction opposing the flow (i.e. from regions of low to high surface tension) and consequently increases the drag on the bubble. Through our theoretical studies (some experiments will also be described) we will show that this increase in drag can be removed by choosing appropriate surfactants and increasing their concentration in the bulk. This finding opens the way to the possibility of controlling the size of wakes behind bubbles rising at order one Reynolds numbers and hence enhancing interphase mass transfer.

2 Basic equations of motion and boundary conditions.

A common feature of the problems to be presented, is the presence of a separating interface between two fluid phases. As discussed in the Introduction the fluid dynamics problem is complicated by the unknown dynamics of the interface; this motion must be determined as part of the solution, leading to very difficult nonlinear free boundary problems. It is useful, therefore, to seek models which are subsets of the full equations but which are, at the same time, of practical importance and derivable by rational asymptotic expansions, for instance.

Before proceeding to such types of analysis and computations in the sections that follow, we begin with a statement of the full problem with as much of the physics represented as possible. Our approach is to work with macroscopic models of the interface separating the fluid phases. This approach represents the interface by a sharp dynamic surface embedded in three-dimensional space, across which flow and concentration variables can jump in a manner specified by physical boundary conditions. The alternative microscopic approach seeks to describe the three-dimensional thin transition layer between the two phases using statistical or continuum mechanical methods. The reader is referred to Chapters 15-18 of the text by Edwards, Brenner and Wasan as well as the many references therein.

In what follows we present the equations and boundary conditions for the problems which are addressed here, namely axisymmetric jets (two-phase generalizations are straightforward) and spherically symmetric drops or bubbles. Before presentation of the mathematical models, we also give a brief description of the applications and what we hope too understand by the theoretical studies. For the sake of brevity, the elements of tensor algebra which are needed in deriving interfacial conditions will be stated and described as needed. For general descriptions the reader is referred to the texts by Aris [1], Edwards et al. [18] and McConnell [44].

2.1 AXISYMMETRIC JETS.

Applications of liquid jets

There are many processes and technological applications that take advantage of the tendency of capillary instability to break up a jet into droplets. In applications such as fuel injection and ink jet printing and spraying and atomization of liquids (see Hertz and Hermanrud [32] and Sweet [71]) a controlled and efficient breakup is desirable. A closely related class of applications involves the capillary breakup of liquid bridges. Applications include fiber spinning (Denn [16]), measurement of rheological properties such as surface tension, shear viscosity and extensional viscosity (see Tsamopoulos, Chen and Borkar [78], McKinley and Tripathi [45]), and studies of agglomeration of particles (see Ennis, Li, Tardos and Pfeffer [23]).

Mathematical Models

Consider the dynamics of a Newtonian liquid jet whose axis is aligned with the gravitational field. We will consider axisymmetric flows (experiments, theory and computations lend support for this assumption as long as the flow is not too fast and the fluid is Newtonian) and will use cylindrical coordinates (z, θ, r), with the jet axis in the z-direction as usual. The corresponding flow field is defined by

$$\mathbf{u}(\mathbf{x}, t) = (w, v, u), \qquad \mathbf{x} = z\mathbf{e}_z + \theta\mathbf{e}_\theta + r\mathbf{e}_r (= (z, \theta, r)), \tag{1}$$

where t is time, where the usual unit vectors of the cylindrical coordinate system have been used. The pressure inside the jet is p and that outside is the constant atmospheric value p_0. The jet is bounded by a free axisymmetric surface given by $r = S(z, t)$.

The bulk flow is given by the incompressible Navier-Stokes equations which follow. Note that eventhough we assume that the flow variables are independent of θ, we will allow for a non-zero azimuthal velocity $v(r, z, t)$. This can arise from a solid body rotation, for example and can be used to model the effects of centrifugal forces on capillary instability. The momentum and continuity equations (in component) form are:

$$\rho(w_t + uw_z + ww_z) = -p_z + \mu\left[\frac{1}{r}(rw_r)_r + w_{zz}\right] + \rho g, \tag{2}$$

$$\rho(v_t + uv_r + \frac{uv}{r} + wv_z) = \mu\left[\frac{\partial}{\partial r}\left(\frac{1}{r}(rv)_r\right) + v_{zz}\right], \tag{3}$$

$$\rho(u_t + uu_r - \frac{v^2}{r} + wu_z) = -p_r + \mu\left[\frac{\partial}{\partial r}\left(\frac{1}{r}(ru)_r\right) + u_{zz}\right], \tag{4}$$

$$\frac{1}{r}(ru)_r + w_z = 0. \tag{5}$$

In (2)-(4), the constants ρ, μ, g are the fluid density, viscosity and acceleration due to gravity, respectively. For the flows of interest here, namely the breakup of jets into droplets, the flows are axisymmetric with $v = 0$. This is assumed from now on.

The equations need to be solved subject to initial and boundary conditions. The complicating feature of the problem is the nonuniform free interface and we turn our attention next to the conditions to be satisfied there.

Interfacial boundary conditions

The interfacial boundary conditions arise from kinematic considerations (the interface is a material surface moving with the fluid - this is a basic premise of the macroscale theory), and from continuity of interfacial stresses. They can be listed as follows:

1. *Kinematic.* Physically, this condition states that fluid does not cross the interface.

2. *Tangential stress condition.* The stress at any point on the interface in a direction tangential to the interface jumps as we cross from one phase to the other by an amount equal to the force exerted by surface tension gradients (Marangoni forces).

3. *Normal stress condition.* The stress at any point on the interface in a direction normal to the interface as we cross from one phase to the other, jumps by an amount equal to the normal stress due to interfacial curvature (this is in turn given by the Young-Laplace equation and depends on the geometry); the jump in pressure at a point on the interface occurs as we cross into the phase which contains the center of curvature.

4. *Surfactant conservation equation.* This interfacial condition is a statement of solute balance on the interface. Several physicochemical effects can be modelled here including *diffusion-controlled* kinetics, *adsorption-controlled* kinetics, *insoluble* surfactants and surface solute production rate per unit area (for example by condensation or evaporation of solute across the interface).

The conditions outlined above are closely related. The surface tension coefficient, σ say, is a function of surface surfactant concentration. For insoluble surfactants we only need to track the movement of surfactant on the interface. This case will be derived in detail for liquid jets in this Section. For soluble surfactants, however, the bulk solute distribution (typically given by solution of a convection/diffusion equation) must be determined and an equation of state (an *adsorption isotherm*) is needed to express the surfactant concentration on the interface in terms of that in the adjacent sublayer concentration from the bulk side. A problem of this type is described in full for spherical bubble motion later.

Defining the region $0 \le r < S(z,t)$ by region 1 and that outside the jet by region 2, at any point on the surface $r = S(z,t)$ we assume that there is a unique outward pointing normal (into region 2) defined by $\mathbf{n}$ and a corresponding tangent vector $\mathbf{t}$. It is easy to see (using the definitions (1)) that

$$\mathbf{n} = \frac{1}{\sqrt{1+S_z^2}}(-S_z, 0, 1), \qquad \mathbf{t} = \frac{1}{\sqrt{1+S_z^2}}(1, 0, S_z). \tag{6}$$

The kinematic condition is most easily found by requiring that the total (material) derivative of the function $F = r - S(z,t)$; this gives

$$u = S_t + wS_z \qquad \text{on} \qquad r = S(z,t). \tag{7}$$

Next, we state the tangential and normal stress balance conditions in terms of the appropriate stress tensor defined as $\mathbf{T}$, and which has components

$$T_{ij} = -p\delta_{ij} + 2\mu e_{ij}, \tag{8}$$

where the summation notation is implied and the indices take values from 1 to 3. (For a discussion of the stress tensor as well as expressions for it in different coordinate systems, see Batchelor [2].) The tangential and normal stress balances take the following form:

$$[\mathbf{t}\cdot\mathbf{T}\cdot\mathbf{n}]_1^2 = -\mathbf{t}\cdot\nabla_s\sigma. \tag{9}$$

$$[\mathbf{n}\cdot\mathbf{T}\cdot\mathbf{n}]_1^2 = \sigma\nabla\cdot\mathbf{n} = \sigma\left(\frac{1}{R_1}+\frac{1}{R_2}\right), \tag{10}$$

where the jump notation $[\cdot]_1^2$ means the value of the expression inside the bracket from region 2 to region 1. The nonzero expressions for e_{ij} are given by (see [2]):

$$e_{11} = w_z, \quad e_{13} = e_{31} = \frac{1}{2}(u_z + w_r), \quad e_{33} = u_r. \tag{11}$$

Use of (11) and (6) in (9), (10) respectively, gives the following boundary conditions (written in full):

$$\left[\frac{2\mu}{1+S_z^2}\left(-S_z w_z + S_z u_r + \frac{1}{2}(u_z + w_r)\right)\right]_1^2 = -\mathbf{t}\cdot\nabla_s\sigma, \tag{12}$$

$$\left[-p + \frac{2\mu}{1+S_z^2}\left(S_z^2 w_z - S_z(u_z + w_r) + u_r\right)\right]_1^2 = \sigma\left[\frac{1}{S\sqrt{1+S_z^2}} - \frac{S_{zz}}{(1+S_z^2)^{3/2}}\right]. \tag{13}$$

Equations (7), (12) and (13) are the first three of the enumerated boundary conditions given above. We note that variations in surface tension are felt through the surface tension σ which is a function of the local surfactant concentration. We turn our attention to the equation that governs variations in solute concentration on the interface. We emphasize that we will treat the case of insoluble surfactants for this problem. Once the local surfactant concentration is known, the surface tension coefficient follows from the assumed equation of state. The theoretical results that follow in later chapters are applicable to different models for the isotherm.

Surface concentration boundary condition.

This boundary condition states that as the interface moves the insoluble surfactant will be redistributed by both interfacial stretching as well as by the local flow field. In a fixed frame of reference, the equation is derived by doing a mass balance over a surface element and a small time interval. Defining the surfactant concentration per unit area to be $\Gamma(z,\theta,t)$ the equation is found to be (see Wong, Rumschitzki and Maldarelli [82]):

$$\Gamma_t - \frac{\partial\mathbf{X}}{\partial t}\cdot\nabla_s\Gamma + \nabla_s\cdot(\Gamma\mathbf{u}_s) - D_s\nabla_s^2\Gamma + \Gamma\kappa\mathbf{u}\cdot\mathbf{n} = 0, \tag{14}$$

where subscripts s denote surface operators (these are derived explicitly later), $\mathbf{X}(z,\theta,t)$ is the position vector of the interface, $\mathbf{u}_s$ is the tangential surface velocity (i.e. the velocity in a tangent plane to a given point on the interface, D_s is a solute surface diffusion coefficient and κ is the curvature at a given point. The second term represents the effect of interfacial stretching on the local solute concentration, the third term is surface convection, the fourth term is surface diffusion and the last term is due to the normal motion of the interface.

In the remainder, we derive an explicit form of this equation in our cylindrical coordinate system relevant to liquid jets.

We will need several results from tensor analysis and rather than using general results (see [1], [18], [44]) for general curvilinear coordinate systems, we begin by specifying our system and working with the coordinates relevant to liquid jets - we will allow asymmetry, i.e. θ dependence, however. The appropriate general curvilinear coordinates here are $u^1 = z, u^2 = \theta$, and the surface can be defined parametrically by the position vector

$$\mathbf{x} = \mathbf{X}(z,\theta). \tag{15}$$

This defines a unique vector on the interface which for the present problem can be expressed as

$$\mathbf{X}(z,\theta) = S(z,\theta)\mathbf{e}_r + z\mathbf{e}_z. \tag{16}$$

Given a point on the interface $r = S(z,\theta)$, there is a tangent plane passing through it. The contravariant basis vectors for the tangent plane are given by

$$\mathbf{t}_1 = \frac{\partial \mathbf{X}}{\partial z}, \quad \mathbf{t}_2 = \frac{\partial \mathbf{X}}{\partial \theta},$$

which on use of the expression (16) and elementary vector calculus become

$$\mathbf{t}_1 = \mathbf{e}_z + S_z\mathbf{e}_r = (1,0,S_z), \qquad \mathbf{t}_2 = S\mathbf{e}_\theta + S_\theta\mathbf{e}_r = (0,S,S_\theta). \tag{17}$$

The metric for a surface element is given by

$$(ds)^2 = h_{11}(dz)^2 + 2h_{12}dzd\theta + h_22(d\theta)^2,$$

where

$$h_{11} = \mathbf{t}_1\cdot\mathbf{t}_1 = 1+S_z^2, \quad h_{12} = h_{21} = \mathbf{t}_1\cdot\mathbf{t}_2 = 0, \quad h_{22} = \mathbf{t}_2\cdot\mathbf{t}_2 = S^2.$$

The quantities h_{11}, h_{12}, h_{22} form the covariant components of the surface metric tensor which has determinant

$$h = h_{11}h_{22} - (h_{12})^2 = |\mathbf{t}_1\times\mathbf{t}_2|^2 = S^2(1+S_z^2). \tag{18}$$

The expression (18) features in the calculation of surface gradients. (An alternative derivation for the normal is also available through $\mathbf{n} = \frac{1}{\sqrt{h}}\mathbf{t}_1\times\mathbf{t}_2$.) It is useful to introduce the covariant basis vectors $\mathbf{t}^2$ and $\mathbf{t}^2$ in terms of the contravariant ones as follows:

$$\begin{aligned}\mathbf{t}^1 &= -\frac{1}{\sqrt{h}}\mathbf{n}\times\mathbf{t}_2 = \frac{1}{h}(h_{22}\mathbf{t}_1 - h_{12}\mathbf{t}_2) = \frac{1}{1+S_z^2}(1,0,S_z),\\ \mathbf{t}^2 &= \frac{1}{\sqrt{h}}\mathbf{n}\times\mathbf{t}_1 = \frac{1}{h}(-h_{12}\mathbf{t}_1 + h_{11}\mathbf{t}_2) = \frac{1}{S}(0,1,0).\end{aligned}$$

By definition the surface operator applied to σ is given by

$$\nabla_s\sigma = \mathbf{t}^1\frac{\partial\sigma}{\partial z} + \mathbf{t}^2\frac{\partial\sigma}{\partial\theta} = \frac{1}{1+S_z^2}\frac{\partial\sigma}{\partial z}(1,0,S_z), \tag{19}$$

with a similar expression holding for $\nabla_s\Gamma$. (Note that the last equality in (19) uses axisymmetry.) It is straightforward now to compute the Marangoni force $\mathbf{t}\cdot\nabla_s\sigma = \frac{1}{\sqrt{1+S_z^2}}\frac{\partial\sigma}{\partial z}$ that enters into the tangential stress balance equation (12).

Finally we need to calculate the surface divergence for vectors. This is needed both for the surface convection term and the diffusion term in equation (14) - the latter follows from the fact that $\nabla_s^2\Gamma = \nabla_s\cdot(\nabla_s\Gamma)$. We quote the following general result from vector calculus (see reference texts). If a vector field $\mathbf{q}(z,\theta)$ is decomposed into

$$\mathbf{q} = q_1(z,\theta)\mathbf{t}_1 + q_2(z,\theta)\mathbf{t}_2 + q_3(z,\theta)\mathbf{n}, \tag{20}$$

then we have the result

$$\nabla_s\cdot\mathbf{q} = \frac{1}{\sqrt{h}}\left(\frac{\partial}{\partial z}(\sqrt{h}q_1) + \frac{\partial}{\partial\theta}(\sqrt{h}q_2)\right) + \kappa q_3. \tag{21}$$

It remains to express the vector $\mathbf{u}_s$ in (14) as in (20) above. Recall that $\mathbf{u}_s$ is the tangential surface velocity. It is a vector in the tangent plane and can therefore be written as

$$\mathbf{u}_s = u_1\mathbf{t}_1 + u_2\mathbf{t}_2.$$

It follows, then,, that the velocity at the interface can be written as

$$\mathbf{u} = \mathbf{u}_s + u_n\mathbf{n}. \tag{22}$$

Also, we know that $\mathbf{u} = (w,0,u)$ for axisymmetric flows and so taking the dot product of (22) with $\mathbf{t}_1$, $\mathbf{t}_2$ and $\mathbf{n}$ respectively, gives

$$u_1 = \frac{1}{1+S_z^2}(w+uS_z), \quad u_2 = 0, \quad u_n = \frac{1}{\sqrt{1+S_z^2}}(-wS_z+u).$$

The remaining terms in (14) are easily computed and the final form of the equation for liquid jets is

$$\begin{aligned}\Gamma_t &- \frac{S_tS_z\Gamma_z}{1+S_z^2} + \frac{1}{S\sqrt{1+S_z^2}}\frac{\partial}{\partial z}\left(\frac{S\Gamma(w+uS_z)}{\sqrt{1+S_z^2}}\right)\\ &- \frac{D_s}{S\sqrt{1+S_z^2}}\frac{\partial}{\partial z}\left(\frac{S\Gamma_z}{\sqrt{1+S_z^2}}\right) + \frac{\Gamma(u-wS_z)}{(1+S_z^2)^2}\left(\frac{1+S_z^2}{S} - S_{zz}\right) = 0.\end{aligned} \tag{23}$$

Solution of equation (23) is required in order to evaluate the surface tension coefficient and its gradient. Different equations of state are given and discussed by [18]. Here we will use the Langmuir equation of state which can be written in the form

$$\sigma(\Gamma) = \sigma_0\left[1+\beta\ln(1-\tilde{\Gamma})\right], \tag{24}$$

where $\beta = \frac{RT\Gamma_\infty}{\sigma_0}$ wwith R the gas constant, T the temperature, Γ_∞ the maximum packing concentration, σ_0 the value of the surface tension for a clean interface and $\tilde{\Gamma} = \Gamma/\Gamma_\infty$. Note that different nondimensionalizations for Γ are possible, for instance by the average value that is initially present on the interface.

Nondimensionalization

For theoretical purposes it is more appropriate to address dimensionless versions of the equations. We need to consider, therefore, typical scales for lengths, velocities, pressure, time and solute concentration. A natural length scale is the undisturbed jet radius, R say whereas the clean surface tension value σ_0 and the fluid viscosity can be used to form a velocity scale σ_0/μ, a pressure scale σ_0/R and a time scale $\mu R/\sigma_0$. Note that in choosing a velocity scale we are using capillary scales; one can use an external velocity scale, U say, which could be the jet carrier speed. The eventual nonlinear analysis and jet breakup is equally applicable to either nondimensionalization, the latter being equivalent to the former in a Galilean frame moving with the constant background jet velocity. The linear stability analysis, however, is performed by perturbing about the uniform jet velocity U.

Defining the operator

$$\Delta \equiv \frac{\partial^2}{\partial r^2} + \frac{1}{r}\frac{\partial}{\partial r} + \frac{\partial^2}{\partial z^2},$$

the momentum and continuity equations for axisymmetric flows become

$$R_e(w_t + uw_z + ww_z) = -p_z + \Delta w + G, \tag{25}$$

$$R_e(u_t + uu_r + wu_z) = -p_r + \Delta u - \frac{u}{r^2}, \tag{26}$$

$$\frac{1}{r}(ru)_r + w_z = 0, \tag{27}$$

where the dimensionless groups are a Reynolds number (this is the inverse of the Ohnesorge number) and an inverse Froude number as well as a Peclet number P_e which appears later given by

$$R_e = \frac{\rho(\sigma_0/\mu)R}{\mu}, \qquad G = \frac{\rho g R^2}{\sigma - 0}, \qquad P_e = \frac{(\sigma_0/\mu)R}{D_s}. \tag{28}$$

The dimensionless boundary conditions are given next for thee case when the outer surrounding fluid (region 2) is passive and there is a constant atmospheric pressure there. We have on $r = S(z,t)$ (the same symbols are used for dimensionless variables as the previously given dimensional ones):

Tangential stress balance

$$\frac{2}{1+S_z^2}\left(S_z(u_r - w_z) + \frac{1}{2}(1-S_z^2)(u_z + w_r)\right) = -\frac{\beta}{\sqrt{1+S_z^2}}\frac{\Gamma_z}{1-\Gamma}. \tag{29}$$

Normal stress balance

$$(1+S_z^2)p - 2\left(w_z S_z^2 + u_r - S_z(u_z + w_r)\right) = \frac{\sigma(\Gamma)}{\sqrt{1+S_z^2}}\left(\frac{1+S_z^2}{S} - S_{zz}\right). \tag{30}$$

Kinematic

$$u = S_t + wS_z, \tag{31}$$

Conservation of surfactant

$$\Gamma_t - \frac{S_t S_z \Gamma_z}{1+S_z^2} + \frac{1}{S\sqrt{1+S_z^2}}\frac{\partial}{\partial z}\left(\frac{S\Gamma(w+uS_z)}{\sqrt{1+S_z^2}}\right)$$
$$- \frac{1}{P_e}\frac{1}{S\sqrt{1+S_z^2}}\frac{\partial}{\partial z}\left(\frac{S\Gamma_z}{\sqrt{1+S_z^2}}\right) + \frac{\Gamma(u-wS_z)}{(1+S_z^2)^2}\left(\frac{1+S_z^2}{S} - S_{zz}\right) = 0. \tag{32}$$

Finally, we have the dimensionless equation of state given earlier also

$$\sigma(\Gamma) = 1 + \beta \ln(1-\Gamma), \tag{33}$$

where β has already been defined and we are now using Γ for the dimesionless surfactant concentration.

Equations (25)-(27) along with the boundary conditions (29)-(32) must be solved subject to initial conditions at $t = 0$ for the velocity field (which should be solenoidal for consistency with (27)) and the surface surfactant concentration $\Gamma(z,0)$. As can be seen the problem is difficult due to the nonlinear coupling present. An additional difficulty which is of interest here, is the possibility of jet pinching which manifests itself as a finite-time singularity of the system; we will describe how results from the analysis of such events using asymptotic methods can be used in practical applications. Finally, note that if $\Gamma \equiv 0$, we have the case of clean interfaces with constant surface tension.

2.1.1 BUBBLE MOTION IN LIQUIDS CONTAINING SURFACTANTS

Background and applications

When a bubble (the discussion for drops is similar) moves through an unbounded fluid phase containing surfactants, there is a physico-chemical activity which acts to alter surface mobility which in turn affects features such as the drag on the particle. Such descriptions were first put forward in the pioneering works of Frumkin and Levich [27], [41]. In what follows we assume that surfactants are present in the liquid phase only and that they are soluble. As the bubble moves, the local surface flow sweeps surfactant to the back end by convection, and the surface concentration, Γ say, increases there. This increase causes surfactant to kinetically desorb into the trailing end sublayer, which in turn increases the sublayer concentration here (denoted by C_s) relative to the concentration in the bulk far from the bubble (denoted by C_∞). A diffusive flux of surfactant away from the trailing end is set up, therefore. The situation is the opposite at the leading end, where the depletion of the sublayer concentration there drives a flux of material from the bulk to the vicinity of the leading end. Eventually a steady state develops with the surfactant concentration at the back end increasing by $\Delta\Gamma$ above the equilibrium value Γ_0 (Γ_0 is given by the adsorption isotherm corresponding to the constant bulk value C_∞), and the desorption rate balances convection. In turn, the back sublayer concentration increases by ΔC above C_∞ with the diffusive flux away from the particle balancing desorption. At the front end the surface concentration decreases below Γ_0 and kinetic adsorption balances convection. In turn the sublayer concentration decreases below C_∞ with diffusion to the surface balancing adsorption there. Even though the average surface concentration scales with $\Gamma_0(C_\infty)$, the surface concentration is larger at the trailing than the leading end and as a result the surface tension σ is lower at the rear end relative to the front. This difference in surface tension causes a Marangoni stress which acts opposite to the direction of the surface flow (the surface flow could be caused by either buoyancy or thermocapillary forces, for example), and thus reduces interfacial mobility. This reduction has an adverse effect on bubble migration velocities for a fixed driving force and has been observed both theoretically and experimentally by many researchers. For a fairly complete list of references see Wang, Papageorgiou and Maldarelli [80], [81]. For experimental studies and a collection of results see Clift, Grace and Weber [7].

Past research has shown that even trace amounts of surfactants (which are inevitably present in most practical systems) cause considerable retarding effects and affect technological processes such as thermocapillary migrations in microgravity environments (see [70]), and reduction in interphase mass-transfer in dropwise extraction systems (see [22], [28], [33], [72]).

Thermocapillary migration is used to manage bubble motion in the absence of gravity. This is achieved by imposing a temperature gradient across a fluid phase containing bubbles. Thus, one end of the bubble is relatively warmer than the other and since surface tension is a decreasing function of temperature, the cooler pole has a higher surface tension than the warmer one. A Marangoni stress acts in the direction of the cooler pole along the interface and as a result this "tugging" force sets up a fluid streaming and consequently a pressure gradient, which propels the bubble towards the warmer fluid. It has been observed (see Kim and Subramanian [36], [37], and Nadim and Bohran [49], for example) that even trace amounts of surfactant can significantly retard such migrations. This can be understood by the retarding mechanism described above: Since the bubble moves towards the warmer fluid, surfactant will be swept by the flow towards the cooler end where it will act to reduce the surface tension and hence the Marangoni induced propulsion velocity is also reduced. Such inefficiencies could seriously hamper microgravity processes for making glass, superconducting materials and composites using miscibility gap solidification.

In the case of interphase mass transfer, e.g. a gas dissolution process such as CO_2 dissolving in water, and in the creeping flow regime (to fix matters), the associated Peclet numbers are large (e.g. the diffusion coefficient of the gas in the liquid is of order 10^{-5} cm^2s^{-1}; the velocities are of order 10 $cm\ s^{-1}$ for bubble radii of order 10^{-1} cm; the Peclet number P_e, the ratio of the product of velocity with the radius to the diffusion coefficient, is of order 10^5). Using boundary layer theory for high Peclet numbers, we can estimate the size of the mass transfer coefficient (the Sherwood number) for the transport of CO_2 assuming a constant value boundary condition at the interface, to be of order $P_e^{1/2}$ for a stress free surface. In the presence of surface active impurities, however, the bubble surface becomes partially or completely immobile, giving rise to a stagnant cap on the bubble surface where the no-slip condition is in effect. Under these conditions, boundary layer theory for a no-slip interface allows an estimate for the mass transfer transport coefficient of order $P_e^{1/3}$. This is asymptotically lower than the estimate $P_e^{1/2}$ for a clean surface. Such findings indicating reduction in mass transfer coefficients are reported by Raymond and Zieminski [58], where surface-active material was intentionally added to the water phase; in particular, the coefficients tend to values expected for a fluid/solid rather than a fluid/fluid interface.

The objective of our studies is to identify possible mechanisms and theoretically and experimentally study remobilization technologies whereby the induced Marangoni stress is reduced or even completely removed by the intentional addition of appropriate solutes. In this report, we describe theoretical aspects of our work.

Mathematical Model

Consider a spherical bubble of radius a which is moving through a viscous incompressible fluid which contains soluble surfactants. This motion can be due to buoyancy or thermocapillary migration in microgravity applications. It is easier to consider the equivalent problem which has a coordinate system fixed at the center of the bubble, with a uniform stream, of velocity U_∞ say, far away. The fluid has density ρ and kinematic viscosity ν and we assume that the surfactant concentration is uniform far from the bubble and has value C_∞.

The motion is assumed to be axisymmetric and a spherical polar coordinate system r, θ, ϕ is used. All variables are independent of ϕ therefore, and $\theta = 0$ is taken to represent the front stagnation point which is the first point the oncoming stream meets as it approaches the bubble. The assumption of axisymmetry (i.e. a spherical bubble) is a reasonable one as long as inertial and viscous forces are small relative to surface tension forces. This requires the Weber number, $W_e = \frac{\rho U_\infty^2 a}{\sigma}$, and the capillary number, $C_a = \frac{\mu U_\infty}{sigma}$, to be small. Here σ is a representative value for the surface tension coefficient. Both these conditions are usually met in the applications we are considering here because bubble sizes and rise velocities are small and surface tension is relatively large.

Let the velocity field be $\mathbf{u} = u\mathbf{e}_r + v\mathbf{e}_\theta + 0\mathbf{e}_\phi = (u, v, 0)$ and denote the pressure by P. If we nondimesionalize lengths with the bubble radius a, velocities with U_∞, bulk concentration of surfactant with C_∞, and pressure with $\frac{\mu U_\infty}{a}$, then the dimensionless equations in the bulk, written in vector notation (for the component form of the equations see the [2], [18]), are the Navier-Stokes equations and a convection diffusion equation for the concentration:

$$R_e\left(\frac{\partial \mathbf{u}}{\partial t} + \mathbf{u}\cdot\nabla\mathbf{u}\right) = -\nabla P + \nabla^2\mathbf{u}, \tag{34}$$

$$\nabla\cdot\mathbf{u} = 0, \tag{35}$$

$$\frac{\partial C}{\partial t} + \mathbf{u}\cdot\nabla C = \frac{1}{P_e}\nabla^2 C, \tag{36}$$

where the Reynolds and Peclet number are given by

$$R_e = \frac{U_\infty a}{\nu}, \qquad P_e = \frac{U_\infty a}{a}. \tag{37}$$

Next we consider the boundary conditions paying particular attention to the surfactant balance. It is useful to motivate the nondimensionalizations by considering the kinetics. Surfactant on the interface is able to adsorb and desorb. The kinetic flux describing these effects is considered by assuming Langmuir kinetics,

$$\beta C_s(\Gamma_\infty - \Gamma) - \alpha\Gamma = j, \tag{38}$$

where α and β are the desorption and adsorption rate constants respectively, j is the kinetic flux, Γ is the surface concentration and Γ_∞ is the maximum packing concentration. The construction of boundary conditions is as follows: First a surfactant mass balance equation on the surface is written which allows for a flux of solute from the sublayer onto the interface (the latter represents the local surfactant adsorption rate). Second, the kinetic flux given by (38) is equated to the local surfactant adsorption rate since j represents the net flux of material from/onto the interface. The first condition (surfactant mass balance) gives (in dimensionless form)

$$\frac{\partial \Gamma}{\partial t} + \frac{1}{\sin\theta}\frac{\partial}{\partial\theta}(v_s \Gamma \sin\theta) = \frac{\chi_0 k}{P_e}\left(\frac{\partial C}{\partial r}\right)_{r=1}, \tag{39}$$

where Γ has been nondimensionalized by the maximum packing concentration Γ_∞, $v_s(\theta)$ represents the azimuthal component of $\mathbf{u}$ on the bubble surface; the other parameters are

$$\chi_0 = \frac{a\alpha}{\beta\Gamma_\infty}, \qquad k = \frac{\beta C_\infty}{\alpha}. \tag{40}$$

Equation (39) is written in the axisymmetric spherical coordinate system of interest here and the obvious nondimensionalizations are made, unless otherwise stated. For more general conditions the reader is referred to Stone [68]. The parameter k is a measure of bulk concentration and it plays a central role in our theoretical findings described later. The second condition, i.e. the balance of the kinetic flux with the local surfactant adsorption rate takes the following dimensionless form

$$\frac{\chi_0 k}{P_e}\frac{\partial C}{\partial r}\bigg|_{r=1} = B_i\left(k\, C|_{r=1}(1-\Gamma) - \Gamma\right), \tag{41}$$

where $B_i = \frac{a\alpha}{U_\infty}$ is the Biot number which is the ratio of the rate of kinetic exchange of surfactant at the gas/liquid interface to the rate at which it is convected from the front to the rear end by the flow.

The presence of surfactants creates a gradient in surface tension causing a force on tthe bubble surface that must be compensated by a viscous tangential stress on the interface. Denoting the surface tension coefficient by σ, this takes the dimensional form

$$\frac{1}{a}\frac{\partial\sigma}{\partial\theta} = -\,\tau_{r\theta}|_{r=a},$$

where $\tau_{r\theta}$ is the shear stress. Combining this with the equation of state derived using Frumkin adsorption (see [41]),

$$\sigma = \sigma_0 + RT\Gamma_\infty\left(\ln(1-\Gamma) - \frac{K}{2}\Gamma^2\right),$$

we obtain the following boundary dimensionless boundarry condition at $r = 1$:

$$\tau_{r\theta}|_{r=1} = r\frac{\partial}{\partial r}\left(\frac{v}{r}\right)\bigg|_{r=1} = Ma\left(\frac{1}{1-\Gamma} + K\Gamma\right)\frac{\partial\Gamma}{\partial\theta}, \tag{42}$$

where $Ma = \frac{RT\Gamma_\infty}{\mu\Gamma_\infty}$ is the Marangoni number, R is the gas constant, T is the temperature, and K is a nondimensional constant. Physically, the Marangoni number represents a measure of forces due to surface tension gradients relative to viscous forces. In addition, the radial velocity is zero on the bubble surface since the particle remains perfectly spherical,

$$u = 0 \quad \text{at} \quad r = 1. \tag{43}$$

Far from the bubble we match with the uniform stream to obtain

$$\mathbf{u} \to (-\cos\theta, \sin\theta, 0) \quad \text{as} \quad r \to \infty, \tag{44}$$

while (due to the nondimensioinalization adopted) the concentration satisfies

$$C \to 1 \quad \text{as} \quad r \to \infty. \tag{45}$$

In addition we have symmetry conditions at $\theta = 0, \pi$ which are

$$v = 0, \quad \frac{\partial C}{\partial \theta} = 0 \quad \text{at} \quad \theta = 0, \pi, \tag{46}$$

Specification of initial conditions completes the statement of the problem. We emphasize that little can be done analytically (we will refer to what can be done later), and in general numerical solutions are required. The problems between the fluid dynamics and the surfactant concentration are coupled nonlinearly and we describe numerical methods and results later. In addition to the nonlinear nature of the problem, certain dimensionless groups are large and make the equations stiff, in the sense that either small time-steps are needed or boundary layer structures need to be resolved. For example, the Peclet numbers tend to be large in applications (see earlier) and solute boundary layers develop near the surface. Also, the bulk concentration constant k which appears in the mass balance equation (39) can be of order $10^2 - 10^3$, making that equation stiff. These issues are taken up in Section ??.

3 Jets: Linear Theory

3.1 Introduction and Background

In this Section we consider the stability of some exact solutions of the governing equations for axisymmetric jets. The viscous system is (2)-(5) along with the inerfacial boundary conditions given in Section 2.1. It is clear that a perfectly cylindrical interface ($r = R$ in dimensional terms) along with a velocity field $\mathbf{u} = (U, 0, 0)$ where U is a constant, is an exact solution of the equations of motion and boundary conditions. This is clearly true also for inviscid flows having $\mu = 0$. (In fact, inviscid flows allow exact solutions for the velocity field of the form $\mathbf{u} = (U(r), 0, 0)$ with $U(r)$ any suitably differentiable function.)

The classical approach to linear stability theory for liquid jets is over a century old and was carried out by Rayleigh [59], [60] (see also the texts by Batchelor [2], Chandrasekhar [8], Drazin and Reid [17] and Middleman [48]). The analysis proceeds by decomposing the disturbance into spatially periodic normal modes and solving an eigenvalue problem to determine the *temporal* stability. That is, given a disturbance wavenumber k (equivalently wavelength), a solution proportional to $\exp(i(kz - \omega(k)t))$ is sought with k real; stability, marginal stability or instability is defined when the imaginary part of $\omega(k)$ is less than, equal to or greater than zero, respectively. An alternative, and in some cases physically more relevant, approach, is to develop a *spatial* stability theory: Here, the frequency is fixed (i.e. ω is real and given) and disturbances can grow, remain neutral or decay in space depending on whether the imaginary part of k is less than, equal to or greater than zero. The connection between the two approaches is through solution of an initial boundary value problem of a linear system of PDEs. This is usually achieved by taking a Fourier-Laplace transform and performing inversions. It is seldom possible to perform these inversions analytically, but progress can be made by studying solutions asymptotically for large times or distances away from the source of the disturbance. The problem, then, is to determine the evolution in space and time of a disturbance which is imposed for all time, for example, at a fixed point in he flow; variations of this are also possible, but the description given is seen to be a faithful reproduction of experiments. In the case of liquid jets, a disturbance is usually imposed at the nozzle where the jet issues by creating n oscillatory pressure wave inside the nozzle through a piezo-crystal assembly. It is also hoped that in the absence of forcing, the dominant instability will be that given by the most unstable eigenmodes of linear theory, thus providing a description of natural modes of evolution. Linear theory perform surprisingly well in predictions of jet breakup lengths and drop sizes, even though such predictions rely on empirical applications of the theory beyond its range of validity (see [17], [32], [48]).

The objective of this Setion is to describe solutions of the initial/boundary value problem of the linear system of equations. The main objective is to give the techniques involved and to this end we will consider inviscid flows with $\mu = 0$ in (2)-(4) as well as the interfacial conditions. Extension to viscous flows is straightforward - the price to pay is an implicit dispersion relation and hence lengthened and perhaps unnecessary exposition at the introductory level. A further assumption of constant surface tension σ (i.e. zero Γ) is made in the equations of Section 2.1. Spatial capillary instability of single jets was considered by Keller, Rubinow and Tu [35] who treat a doubly infinite inviscid jet and show that for large jet velocities (i.e. large values of U introduced above) one of the spatial modes approaches the temporal solution found by Rayleigh. Berger [4] carried out the stability analysis of a single doubly infinite jet as an initial value problem and showed some discrepancies between this and the normal mode approach, in particular with regard to empirically determined drop sizes. In experiments, however, jets are produced at a nozzle and so are semi-infinite. Leib and Goldstein [42] considered the linear stability of a jet flow which is in a pipe from $-\infty$ to 0 and then forms a free flowing jet in $z > 0$. They showed that below a critical Weber number (defined by $W = \rho R U^2/\sigma$) of about 3, the jet becomes *absolutely unstable*, that is the disturbance will grow exponentially at any point in space for large enough times. This is in contrast to a spatial or *convctive* instability where the disturbance is convected downstream as it grows. It is important in applications to determine conditions for a switch over from convective to absolute instability. One of the first complete expositions dealing with convective versus absolute instabilities is that of Briggs [6], in the context of plasma physics back in the 60s. The reader is referred to that monograph for many of the details which will be omitted in the sequel.

In this Section we present the theory for inviscid compound jets. The motivation comes from printing technologies as described in [32]; other applications include production of compound particles and coating flows. Sanz and Meseguer [65] have carried out a linear stability analysis of a reduced one-dimensional system of equations which retains the full curvature term. We will describe how to derive and analyze such long wave models in a formal asymptotic manner in the following Section. Most of the linear analysis described next has been carried out by Chauhan, Maldarelli, Rumschitzki and Papageorgiou [9] and many of the details can be found there. In addition see Chauhan et al. [10], [11] for a discussion of viscous jets.

3.2 Convective and Absolute Instability of Inviscid Compound Jets

3.2.1 Formulation of linear stability theory

Consider a doubly infinite jet of an inviscid fluid 1 of density ρ_1 which is surrounded by an inviscid fluid 2 of density ρ_2. The cylindrical coordinate system of Section 2.1 is used, and the undsturbed inner and outer interfaces are $r = R_1, R_2$ respectively, while corresponding interfacial tensions are σ_1 and σ_2. The equations of motion and boundary conditions are made dimensionless by choosing inertial rather than the viscous scales of Section 2.1. Hence, we nondimensionalize lengths by R_1, time by $(\rho R_1^3/\sigma_1)^{1/2}$, pressure by σ_1/R_1 and velocities by $(\sigma_1/\rho_1 R_1)^{1/2}$. The following dimensionless parameters emerge: a gap ratio $a = R_2/R_1$; surface tension ratio $\gamma = \sigma_2/\sigma_1$; density ratios $\beta_2 = \beta = \rho_2/\rho_1$, $\beta_1 = 1$. The Euler equations are (subscripts 1,2 denote the inner and outer fluids respectively):

$$\beta_i\left(\frac{\partial \mathbf{u}_i}{\partial t} + \mathbf{u}_i\cdot\nabla\mathbf{u}_i\right) = -\nabla P_i + \mathbf{F}_i(r,t)\delta(z-z_0), \tag{47}$$

$$\nabla\cdot\mathbf{u}_i = 0 \quad (i=1,2), \tag{48}$$

where $\mathbf{u}_i$ is the velocity field, P_i is the pressure and $\mathbf{F}_i$ is an externally applied force concentrated at a point $z = z_0$. The boundary conditions are those of normal stress balance (note that the tangential stress balance is trivial for inviscid flows), continuity of velocities and the kinematic condition, given in Section 2.1 but applied at each of the dimensionless interfacial positions $r = S_i(z,t)$ and for inviscid conditions. In addition, the velocity and pressure must be bounded at the jet axis $r = 0$. These are

$$[P]_i = \gamma_i\nabla\cdot\mathbf{n}_i, \quad \text{on} \quad r = S_i(z,t) \quad i=1,2, \tag{49}$$

$$[\mathbf{n}\cdot\mathbf{u}]_i = 0, \quad \text{on} \quad r = S_1, \tag{50}$$

$$\left(\frac{\partial}{\partial t} + \mathbf{u}_i\cdot\nabla\right)(r - S_i) = 0, \quad \text{on} \quad r = S_i(z,t) \quad i=1,2, \tag{51}$$

where $[]_i$ denotes difference in function values between the inside and outside of interface i; $\mathbf{n}_i$ is an outward pointing normal on interface i, and $\gamma_1 = 1$, $\gamma_2 = \gamma$.

In the absence of forcing the base state ia taken to be a uniform flow of value V where (due to the nondimensionalization) $V^2 = \rho R U^2/\sigma_1 = W$, i.e. the dimensionless base velocity is equal to the square root of the Weber number. The choice of no slip between the two fluid velocities in the baseflow, avoids the presence of Kelvin-Helmholtz instability and enables us to study the capillary instability. In a real flow viscous dissipation tends to give this type of baseflow a few diameters downstream of the nozzle, as seen in the experiments of Hertz and

Hermanrud [32]. The exact dimensionless solution is, then,

$$\mathbf{u}_i^0 = (V,0,0), \quad P_2^0 = \frac{\gamma}{a}, \quad P_1^0 = 1 + \frac{\gamma}{a}. \tag{52}$$

Next, add a disturbance of infinitessimal size ϵ to (52),

$$\begin{aligned} \mathbf{u}_i(z,r,t) &= \mathbf{u}_i^0 + \epsilon\tilde{\mathbf{u}}_i + O(\epsilon^2), \\ S_1(z,t) &= 1 + \epsilon\zeta_1(z,t) + O(\epsilon^2), \\ S_2(z,t) &= a + \epsilon\zeta_2(z,t) + O(\epsilon^2), \\ P_i(z,r,t) &= P_i^0 + \epsilon\tilde{P}_i(z,r,t) + O(\epsilon^2). \end{aligned}$$

Substitution into the governing equations (47) and (48) and linearization with respect to ϵ, we get the following equations for the perturbation (the tildes are dropped)

$$\beta_i\left(\frac{\partial \mathbf{u}_i}{\partial t} + \mathbf{u}^0\cdot\nabla\mathbf{u}_i\right) = -\nabla P_i + \mathbf{F}_i(r,t)\delta(z), \tag{53}$$

$$\nabla\cdot\mathbf{u}_i = 0 \quad (i=1,2), \tag{54}$$

where the forcing has been scaled with ϵ and is at $z=0$.

3.2.2 Temporal instability

In this case the forcing is absent from (53) and the initial conditions drive the instability. We will proceed using transform methods rather than the standard normal mode analysis. It is useful to analyze the stability in a frame which moves with the jet velocity V, thus making the base flow zero. The initial conditions for equations (53), (54) are taken to be of the form

$$\mathbf{u}_i(z,r,0) = 0, \qquad \zeta_i(z,0) = \zeta_{i,0}(z). \tag{55}$$

The divergence of (53) and use of (54) gives a Laplace equation for the pressures

$$\nabla^2 P_i = 0. \tag{56}$$

Defining Fourier and Laplace transforms through

$$\hat{f}(k) = \int_{-\infty}^{\infty} f(z)e^{-ikz}dz, \quad \tilde{f}(s) = \int_0^{\infty} f(t)e^{-st}dt,$$

equation (56) in the Fourier-Laplace domain becomes

$$\left(\frac{d^2}{dr^2} + \frac{1}{r}\frac{d}{dr} - k^2\right)\tilde{\hat{P}}(r,k,s) = 0, \tag{57}$$

whose general solution is

$$\tilde{\hat{P}} = A_i(k,s)I_0(kr) + B_i(k,s)K_0(kr), \tag{58}$$

where I_0 and K_0 are the modified Bessel functions of zero order and A_i, B_i are to be found.

Next we write down the linearized versions of the boundary and kinematic conditions (49)-(51) in transform space:

$$\begin{aligned}
\tilde{\hat{P}}_1 - \tilde{\hat{P}}_2 &= -(1-k^2)\tilde{\hat{\zeta}}_1 \quad \text{on} \quad r=1, && (59)\\
\tilde{\hat{P}}_2 &= -\gamma\left(\frac{1}{a^2} - k^2\right)\tilde{\hat{\zeta}}_2 \quad \text{at} \quad r=a, && (60)\\
\tilde{\hat{u}}_1 &= \tilde{\hat{u}}_2 \quad \text{at} \quad r=1, && (61)\\
\tilde{\hat{u}}_2 &= s\tilde{\hat{\zeta}}_2 - \hat{\zeta}_{2,0} \quad \text{at} \quad r=a, && (62)\\
\tilde{\hat{u}}_1 &= s\tilde{\hat{\zeta}}_1 - \hat{\zeta}_{1,0} \quad \text{at} \quad r=1, && (63)\\
& && (64)
\end{aligned}$$

and in addition flow variables are finite at $r = 0$.

The vector of unknown functions of k and s that we need in order to completely specify the solution is $\mathbf{X}^t = (A_1, A_2, B_2, \tilde{\hat{\zeta}}_1, \tilde{\hat{\zeta}}_2)$. The five boundary conditions (59)-(63) lead to the matrix problem

$$A(k,s)\mathbf{X} = \mathbf{b}(k,s), \tag{65}$$

where A is a 5×5 matrix with known entries (see [9]) and $\mathbf{b}^t = (0, \hat{\zeta}_{1,0}, \hat{\zeta}_{2,0}, 0, 0)$. Solutions of (65) can be written down using elementary linear algebra. For example the solutions for the transformed interfacial perturbations have the form

$$\tilde{\hat{\zeta}}_1(k,s) = \frac{1}{det(A)}(C_{22}\hat{\zeta}_{1,0} - C_{12}\hat{\zeta}_{2,0}), \qquad \tilde{\hat{\zeta}}_2(k,s) = \frac{1}{det(A)}(-C_{21}\hat{\zeta}_{1,0} + C_{11}\hat{\zeta}_{2,0}), \tag{66}$$

where $det(A)$ is the determinant of matrix A and C_{11} etc. are rather long but explicitly known expressions involving Bessel functions (see [9]). The solution is achieved by inversion,

$$\zeta_i(z,t) = \frac{1}{4i\pi^2}\int_{-\infty}^{\infty} e^{ikz}\left[\int_{c(k)-i\infty}^{c(k)+i\infty} \tilde{\hat{\zeta}}_i(k,s)e^{st}ds\right]dk, \tag{67}$$

where $c(k)$ is real and lies to the right of any singularities of the integrand in the s-plane for every k. Th Laplace inversion can be done by using a Bromwich contour and the residue theorem. The residues have to be calculated for every k, therefore, before performing the Fourier inversion. Inspection of (66) shows that poles possibly arise when $det(A) = 0$ and/or any of the C_{ij} terms are singular. It is easy to establish that the only poles come from solution of the *dispersion equation* $det(A) \equiv f(s,k) = 0$. The form of the dispersion equation is

$$Q_1(k)s^4 + Q_2(k)s^2 + Q_3(k) = 0, \tag{68}$$

where the $Q_i(k)$ are known expressions; this relation is derivable directly using normal modes, also. It can be proven (see [9]) that s^2 is real when k is real. Hence, stability is not damping but oscillatory in time as expected by the inviscid nature of the problem (see Drazin an Reid

[17] also). Equation (68) has exactly four roots, $\{s_n(k)\}, n = 1,4$ say, and when the poles are simple we have, for example,

$$\hat{\zeta}_1(k,t) = \sum_{n=1}^{4} \frac{1}{\frac{\partial \det(A)}{\partial s}}\Bigg|_{k,s_n(k)} \left[C_{22}(k,s_n)\hat{\zeta}_{1,0}(k) - C_{12}(k,s_n)\hat{\zeta}_{2,0}(k)\right] e^{s_n t}. \tag{69}$$

A similar expression follows for $\hat{\zeta}_1(k,t)$. Fourier inversion gives, next,

$$\zeta_i(z,t) = \frac{1}{2\pi}\int_{-\infty}^{\infty} \hat{\zeta}_i(k,t) e^{ikz} dk. \tag{70}$$

The poles in the integrands of (70) in the k-plane, arise from the poles k_0, say, of the Fourier transforms of the initial conditions, or when some of the C_{ij} become infinite. It can be shown (details in [9]) that as long as $\beta > 0$, the only poles are the poles k_0 of the initial conditions. If these are not on the real axis, then, the method of stationary phase furnishes the large time solution

$$\zeta_i \sim \sum_n \Lambda(k_m, s_n(k_m)) \frac{e^{ik_m z + s_n(k_m)t}}{\sqrt{t}}, \qquad \text{as} \quad t \to \infty, \tag{71}$$

where Λ is a coefficient coming from the expressions (66) and k_m is defined by $\frac{ds}{dk}(k_m) = 0$.

As an example, which enables direct comparison with the normal mode solution, consider initial conditions in the form

$$\zeta_{i,0} = \sum_{k_0} \Delta_{i,k_0} \sin(k_0 z),$$

whose transform is $\hat{\zeta}_{i,0} = \sum_{k_0} \Delta_{i,k_0}\delta(k - k_0)$. The only poles are now on he real axis at $k = k_0$ and the Fourier inversion can be done exactly to give

$$\zeta_i(z,t) = \sum_{k_0}\sum_n \Lambda(k_0, s_n(k_0)) e^{ik_0 z + s_n(k_0)t}. \tag{72}$$

If all values of k_0 are present (e.g. a white noise initial condition) then the dominant effect in (72) comes from a wave with wavenumber which maximizes the growth rate $|\mathcal{R}](s_n(k_0))|$. This is the usual conclusion of the normal mode analysis also.

3.2.3 Spatial and absolute instability

Spatial instabilities arise from disturbances localized in space and in the present problem these can be in the bulk (as formulated) or at the interfaces. The latter produces a slightly different formulation which is analyzable in exactly the same way as for bulk disturbances. The appropriate reference frame is a fixed laboratory frame since we are interested in instabilities that grow as they convect downstream.

We seek a solution by introducing a streamfunction ψ given by

$$u = \frac{1}{r}\frac{\partial \psi}{\partial z}, \quad w = -\frac{1}{r}\frac{\partial \psi}{\partial r},$$

so that (54) is satisfied identically and substitution into (53), elimination of the pressure and taking a Fourier-Laplace transform gives

$$\left[r\frac{d}{dr}\left(\frac{1}{r}\frac{d}{dr}\right)-k^2\right]\tilde{\hat{\psi}}_i=\frac{r}{\beta_i(s+ikV)}\left[\frac{d\tilde{F}_{i,z}(r,s)}{dr}-ik\tilde{F}_{i,r}(r,s)\right], \tag{73}$$

where $F_{i,z}, F_{i,r}$ are the z and r components of the force $\mathbf{F}_i$. In solving (73), therefore, we add a particular solution corresponding to the forcing. In what follows, we write

$$\mathbf{F}_i(r,t)=\mathbf{H}_i(r)T(t),$$

The solution is, then,

$$\tilde{\hat{\psi}}_i=R_i(k,s)rI_1(kr)+Q_i(k,s)rK_1(kr)+\tilde{\hat{\psi}}_{i,p}, \tag{74}$$

where the particular solutions are more suitably represented as

$$\tilde{\hat{\psi}}_{i,p}=\frac{\tilde{T}(s)}{s+ikV}Z_{i,p}(r,k), .$$

with $Z_{i,p}$ known in terms of integrals of Bessel functions (see [9]). It turns out that the only pole of $Z_{i,p}$ in the k-plane s $k=0$. the boundary conditions are the same as before, and the solution of the problem reduces to a similar matrix problem as in the previous subsection. The problem is

$$A_s(s,k)\tilde{\hat{\mathbf{x}}}=\frac{\tilde{T}}{s+ikV}\mathbf{c}, \tag{75}$$

where the vector of unknowns is $\tilde{\hat{\mathbf{x}}}^t=(R_1\bar{s},\beta R_2\bar{s},-\beta Q_2\bar{s},-i\tilde{\hat{\zeta}}_1,-i\tilde{\hat{\zeta}}_2)$, $\bar{s}=s+ikV$ and the vector $\mathbf{c}$ involves components of the forcing and its derivatives evaluated at $r=1,a$. The matrix A_s is the same as A in (65) but with s replaced by $s+ikV$, i.e. $A_s=A(k,s+ikV)$. If A_s is non-singular the solution of (75) can be found by inversion and we have, therefore,

$$\mathbf{x}(z,t)=\frac{1}{4i\pi^2}\int_{c-i\infty}^{c+i\infty}e^{st}\tilde{T}(s)\left[\int_{-\infty}^{\infty}\frac{A_s^{-1}(k,s)\mathbf{c}(k,s)}{s+ikV}e^{ikz}dk\right]ds. \tag{76}$$

Note that the Fourier integral can be evaluated by contour integration in the complex k-plane; the axis $k_i=0$ is part of the contour and to use the residue theorem one closes the contour above or below the real axis for $z>0$ or $z<0$ respectively, to ensure decay at infinity. We proceed in general terms by introducing a cofactor matrix D and so can write the Fourier inversion integral as

$$\tilde{\mathbf{x}}(z,s)=\tilde{T}(s)\int_{-\infty}^{\infty}\frac{D(k,s)\mathbf{c}(k,s)}{(s+ikV)det(A_s)}e^{ikz}dk. \tag{77}$$

It can be shown that the relevant singularities in the integrand of (77) come from zeros of $det(A_s)$, i.e. they are solutions of the dispersion relation. Assuming simple poles in the integrand of (77), the residue theorem gives he solution

$$\tilde{\mathbf{x}}(z,s)=i\tilde{T}(s)\sum_n\lim_{k\to k_n}\left[(k-k_n)\frac{D(k,s)\mathbf{c}(k,s)}{(s+ikV)det(A_S)}\right]e^{ik_nz}\equiv\tilde{T}(s)\phi(s,z). \tag{78}$$

Given the expression (78), it usually not possible to perform the Laplace inversion exactly and obtain closed form solutions. Since we are usually interested in the stability of the system the large time asymptotic behavior is sufficient and in what follows we describe the systematic way of determining spatial and absolute intability by mapping contours of the dispersion relation $det(A_s) = 0$.

The Laplace integral in (76) and (77) is inverted by using a Bromwich contour with vertical line $s_r = c$ which lies to the right of any singularities of $\tilde{\mathrm{x}}(z,s)$; such singularities arise from $\tilde{T}(s)$ or $\phi(z,s)$. As an example of practical interest, if $T(t) = sin(\omega t)$, then $\tilde{T}(s) = 1/(s^2+\omega^2)$ and the poles are at $s = \pm i\omega$. Next, we need to search for the poles of $\phi(z,s)$ as $s = s_r + is_i$ varies. This can be done for all values of s_r with $0 < s_r < c$, and varying s_i. Considering $z > 0$, the Fourier integral in (77) is evaluated by closing the contour above the real k-axis. It can be seen from (77) that as s varies so do the positions of the poles given by the dispersion relation. By definition of c, as s varies, the poles of the Fourier integral remain in the region $k_i > 0$ of th k-plane. As c is decreased and the vertical line of the Bromwich contour is traversed, however, it is possible for poles in (78), that were above or below the real k-axis at a slightly larger value of c, to cross the real k-axis produce a discontinuity in $\phi(s,z)$ and hence a singularity in $\tilde{\mathrm{x}}(z,s)$.

To keep the solution continuous, we need to define an analytic continuation of $\phi(s,z)$ by deforming the Fourier integral so as to exclude or include poles which cross the real k-axis as c is decreased (see Briggs [6]). A problem arises when two poles, coming from different sides of the deformed contour, merge and so prevent us from finding a contour that keeps one pole but not the other. A *pinch* singularity forms and gives rise to a second order pole inside the Bromwich contour. If at the same time $s_r > 0$, then an *absolute* instability ensues. If there are no pinch singularities with $s_r > 0$, then the system responds to the sinusoidal frequency of the forcing.

A necessary condition for a pinch-type singularity follows from the fact that it is a double root (or higher) in k of the dispersion relation. Hence, using the definition $f_s(k,s) = det(A_s(k,s))$ we have

$$f_s(k_0,s_0) = 0, \qquad \left.\frac{\partial f_s}{\partial k}\right|_{k_0,s_0} = 0.$$

Differentiating the dispersion relation with respect to k and using the chain rule, we have $\frac{\partial f_s}{\partial k} + \frac{\partial f_s}{\partial s}\frac{ds}{dk} = 0$ and as long as $\left.\frac{\partial f_s}{\partial s}\right|_{k_0,s_0} \neq 0$, we have the necessary but not sufficient condition for absolute instability

$$\frac{ds}{dk} = -\frac{\frac{\partial f_s}{\partial k}}{\frac{\partial f_s}{\partial s}} = 0 \quad \text{at} \quad (k,s) = (k_0,s_0). \tag{79}$$

The condition (79) is not sufficient because the two merging poles have to be inside and outside the Fourier contour before they coalesce.

Finally we discuss the onset of *spatial* instability. If the system is not absolutely unstable, then it can be seen from (78) that the sign of the imaginary part of the poles inside the deformed Fourier contour, determine spatial stability or instability. For example, if the imaginary part of k is negative we have exponential growth for $z > 0$. That is, spatial instability can only arise if poles are in the upper half k-plane for sufficiently large s_r, and they cross into the lower half plane as s_r is reduced in the range $0 < s_r < c$.

3.2.4 Results

Temporal stability

Here we need to study the variations of $s = s_r + is_i$ with real k as connected by the dispersion relation (68). The relation is quadratic in s^2 and explicit solutions are available. There are four roots, with two of them the negative of the other two. It has also been mentioned, that s^2 is real and so s is either purely real (growth or decay) or purely imaginary (oscillations in time). It is found that two of the modes are stable for all wavenumbers while the other two have bands of unstable wavenumbers. These bands extend from $k = 0$ to $k = 1, 1/a$ respectively, and beyond the cutoff the modes become purely imaginary (we term these modes I and II in the sequel). A representative set of results is given in Figure 1 which shows the two unstable modes and their asymptotic behavior for small k (this is linear and readily found from the dispersion relation). The modes are termed *stretching* and *squeezing*. This is due to the presence of two interfaces and the fact that the interfaces can grow in or out of phase. To see this mathematically we calculate for each mode (see (66)) the ratio

$$\frac{\bar{\hat{\zeta}}_2}{\bar{\hat{\zeta}}_1}(k, s_n) = -\frac{C_{11}}{C_{12}} = -\frac{C_{21}}{C_{22}}. \tag{80}$$

This amplitude ratio is real for both unstable modes, and so the interfaces are either exactly in or out of phase. Mode I grows in phase where unstable but oscillates out of phase where stable, while mode II grows out of phase where unstable and in phase where stable. In phase interfacial growth tends to stretch the annular film and so is termed a stretching mode, while out of phase growth tends to squeeze the film and so is called a squeezing mode. Such definitions were introduced by the studies of Taylor [74] and Felderhof [25].

Figure 1 also shows that there is a maximum growth rate, s_{max} say, for each mode with corresponding wavenumber k_{max}. The s_{max} for the stretching mode is greater than that for the squeezing mode, and assuming that linear theory provides a relatively accurate description of breakup features such as drop sizes (this is the case for single jets), then s_{max} can be used to estimate breakup times (and breakup lengths) while k_{max} gives an estimate for drop sizes. In the compound jet instability dominated by the stretching mode, the core region breaks first followed by the annular region, forming compound drops. If the squeezing mode is dominant, however, either core or film can break first, depending on relative core to annulus thickness. It can be suggested that if the annular film is thin enough then it can rupture first producing dewetting of the core.

We now turn to applications of the theoretical results. In most applications one aims to control breakup times and/or drop size. The physical variables that we can vary are γ, β, a (outer to inner surface tension ratio, density ratio and undisturbed radius ratio). The following results have been established (see [9]): (i) Given β and a, the growth rate f the squeezing mode increases monotonically with γ for all k in the unstable range $0 < k < 1/a$; (ii) As γ is increased (this implies that the outer surface tension σ_2 increases) the growth rate of the stretching mode increases for $0 < k < 1/a$ and decreases for $k > 1/a$ - this is because the outer interface perturbations have a net destabilizing effect for $k < 1/a$ and stabilizing for $k > 1/a$. This leads to the maximum growth rate of the stretching mode to shift to values of $k < 1/a$ and hence to larger drops on breakup. (iii) Increasing the outer fluid's density increases its inertia hence decreasing the growth rate for both modes.

Some limit cases are discussed next. First consider the case of a thin annular film, i.e. $a = 1 + \epsilon$ where $0 < \epsilon << 1$. The stability theory can be completed in two equivalent ways: (i) start with the momentum equations and boundary conditions, rescale film variable through a stretched variable $y = (r - 1)/\epsilon$ $(y = O(1))$, and match solutions in the film and core to solve the dispersion relation asymptotically in a series in ϵ; (ii) start with the exact dispersion relation (68) and analyze this for $a = 1 + \epsilon$ as $\epsilon \to 0$. We note that the former approach is more illuminating physically, it gives a basis for nonlinear theories and is a feasible approach in the absence of an exact dispersion relation. We refer the reader to [9] for details using both approaches, but we present the salient results. The main conclusion of this asymptotic limit is that the squeezing mode has a growth rate which scales as $\epsilon^{1/2}$ whereas the corresponding stretching mode growth rate is of order ϵ^0. Thus, the stretching mode will dominate for thin films. We also note that computations indicate that this asymptotic result also holds for annuli which are not asymptotically thin.

The opposite limit is that of a very thick annular region, i.e. $a >> 1$. Physically this is the problem of the capillary instability of a fluid thread immersed in an infinite fluid as first studied by Tomotika [77] Numerical solutions of the dispersion relation indicate that the squeezing mode disappears as a increases (this is expected since the squeezing mode has a band of unstable waves $0 < k < 1/a$), and the stretching mode reaches a limit which is independent of γ - this is because the outer interface is at infinity and has no effect on the stretching mode, to leading order.

Spatial stability

The objective here is to determine the system's response to a time harmonic disturbance of given frequency ω (this generates two poles $s = \pm\omega$ in the s-plane), for given jet velocity V and other flow parameters. In the absence of absolute instability the response will be phase-locked with the disturbance and spatial instability arises if the poles in the k-plane corresponding to $s = \pm\omega$ cross below the real k-axis (note that we are in the region $z > 0$). The non-zero poles in the k-plane come from either $(s + ikV) = 0$ or the dispersion relation $det(A_s) = 0$ (see equations (77) and (78)). By symmetry of eigenvalues in this problem, it is sufficient to consider the pole $s = -i\omega$ alone. As discussed in Section 3.2.3, we need to map out the poles (i.e. roots of the dispersion equation) in the complex k-plane. Ultimately, for a given frequency ω, we need to obtain the $k_i - k_r$ plot for $s_r = 0$. Following the discussion of Section 3.2.3 we achieve this by mapping out the poles for $s = s_r - i\omega$ and systematically reduce s_r to zero. (This is the procedure to determine absolute instability also.) It is convenient to take pairs of roots in order to determine, in the case of absolute instability, if a pinching of roots from different sides of the Fourier contour takes place. If no pinching double roots appear as s_r becomes zero *and* some $k_i(k_r)$ are less than zero, then we have a spatial instability.

This procedure is illustrated in Figure 2 which depicts variations for the first two roots at parameters $\beta = 1, \gamma = 2, a = 2, V = 2$. For each fixed value of s_r, a curve is generated in the complex k-plane by increasing the frequency ω - the arrows on the curves indicate the direction of increasing ω. There is spatial instability for $V = 2$ (see root 1, $s_r = 0$ curve) with a well defined maximum growth rate at a maximally growing wavelength. The two roots seem to intersect in the k-plane but these intersections do not represent double roots since the value of ω at the intersections is different. Root 1 shown in Figure 2 provides the most unstable spatial mode at the given parameter values. Note that there is a cutoff frequency above which the mode is stable. This is defined as the cutoff Strouhal number $St = |s_i|/V$,and for mode 1 this is about 1, even though the exact value depends on V (see [42] also for the single jet case). Higher modes (not shown here) are stable except one which exhibits similar characteristics as mode 1, the main difference being a lower maximum growth rate and a lower cutoff Strouhal number of about $1/a$ (again, this is V dependent).

It is easy to see that the poles corresponding to $(s + ikV) = 0$ are neutral; for large s_r, the pole of $(s + ikV)$ is in the upper half k-plane and hits the real axis when $s_r = 0$; at this point, the dispersion relation is $f_s(-ikV, k) = f(0, k) = 0$, which has no unstable modes (see earlier for definitions of f_s and f).

It can also be shown that as the jet velocity V decreases the spatial growth rates for the two unstable modes described above (termed primary and secondary from now on) increase and eventually absolute instability obtains. The situation for large V is quite different, however, with the spatial instability being equivalent to the temporal instability with the latter being viewed from a Galilean frame of reference travelling with the jet velocity. Ths can be shown asymptotically and also confirmed numerically (see [9] for details). This last result is critical in the evaluation of temporal over spatial instability. Even though spatial instability is more relevant for jet problems, the ability of temporal instability to capture well (albeit empirically) features such as breakup lengths and drop sizes along with the relative ease of its applicability, make it a useful engineering tool. The theory is expected to work well as long as the jet velocity is sufficiently large. Small jet velocities lead to absolute instabilities, as we describe next.

Absolute instability

As the jet velocity decreases, merging of two roots (a root which is spatially growing in $z > 0$ and an *evanescent* mode, a root which is spatially growing in $z < 0$) leads to absolute instability. This happens for all V below a critical value. Figure 3 shows results of absolute instability for parameter values $V = 1.7$, $\beta = 1$, $\gamma = 2$, $a = 2$. What is shown is the merging of the secondary growing root with the evanescent root. The two roots are apart at a value of $s_r = 0.005$ as shown on the Figure, but by $s_r = 0$ a merging and a change in the topology of the curves has occurred. (This topology is different from what was seen for $V = 2$, and is a feature to look out for in studying absolute instabilities.) It can be surmised, therefore, that a merging occurs between 0.005 and 0 signalling absolute instability. We note that at this value of $V = 1.7$, the primary mode remains convectively unstable, but on further reduction of V that also undergoes a merge with an evanescent mode.

In this particular problem a more direct method for finding criteria for absolute instability is available due to the explicit form of the dispersion relation, namely $s(k) = g(k) - ikV$. A necessary condition for the onset of absolute instability is

$$\frac{ds}{dk} = \frac{\partial s_r}{\partial k_r} + i\frac{\partial s_i}{\partial k_r} = 0. \tag{81}$$

As a consequence of this, then, another way of graphically determining absolute instability is to generate plots from the dispersion relation of constant k_i and variable k_r, compute curves in the s plane and look for a cusp formation. At the cusp we have $\frac{\partial s_r}{\partial k_r} = \frac{\partial s_i}{\partial k_r} = 0$ and hence (81) is satisfied. Results generated using this method, and corresponding exactly to the parameters of Figure 3, are shown in Figure 4. The frequency (i.e. s_i) at the cusp location agrees with that found using the method of Figure 3.

A general criterion for the maximum jet velocity below which the compound jet becomes absolutely unstable, can be found by use of the dispersion relation along with condition (81). From

$$\frac{ds}{dk} = g'(k;V) - iV = 0,$$

we calculate the value $V = V_c$ where s_r at absolute instability becomes zero. Above V_c the value of s_r at root mergers is negative and so the instability is convective rather than absolute. As an example, for the set of parameters $\beta = 1, \gamma = 2, a = 2$ we find $V_c = 1.77$.

Comparison of theory with experiment

In what follows we make a comparison of the temporal and spatial theories presented above, with the experimental findings of Hertz and Hermanrud [32]. In particular we compare with the set of results in their Figure 5.

The primary (i.e. core) fluid in the experiment is a water soluble ink while the secondary annular fluid is a silicone oil. The corresponding physical parameters are: $\sigma_2 = 20^{-3}\ N/m$, $\sigma_1 = 50 \times 10^{-3}\ N/m$, $\rho_2 = \rho_1 = 1000\ kg/m^3$, $R_1 = 75 \times 10^{-6}\ m$, $R_2 = 150 \times 10^{-6}\ m$, $U = 1.98\ m/s$. The corresponding dimensionless values used in the present theory are: $\gamma = 0.4$, $\beta = 1$ and $V = 2.426$. Using drop sizes from the experiments we can estimate an average wavenumber which has value 0.6. The experiment was performed without a frequency forcing and so the maximally growing wave is expected to be the observable one. Theoretical results of temporal and spatial stability theory, show that the two are close together but the velocity is small enough to allow for small discrepancies. Temporal theory gives a maximally growing wavenumber of 0.66 while spatial theory gives a value 0.68. The theory predicts that the jet is convectively unstable. It is interesting to note that the 1-D temporal linear theory of Sanz and Meseguer [65], predicts a maximally growing wavenumber of 0.7 and lends support to the utility of 1-D long wave theories. Such theories are taken up in the following Section and analyzed in the nonlinear regime.

4 Jets: Nonlinear Theory

4.1 Introductory remarks

Undoubtedly, direct numerical simulations are ultimately desirable in the study of these flows. Inviscid and highly viscous jets governed by the Stokes equations, can be formulated in terms of the boundary integral method because Green's functions for both these flows are known and the field equations are linear. The reader is referred to the studies of Chen and Steen [13], Day, Hinch and Lister [15], Mansour and Lundgren [43] for inviscid jets; for Stokes jets see the work of Stone and Leal [69], Tjahjadi, Stone and Ottino [76] (as well as later work by Stone, Lister and coworkers not referenced here), Pozrikidis [56] (for an excellent introduction to the boundary integral method for Stokes flow see the text by Pozrikidis [57]). For some recent calculations based on the Navier-Stokes equations see Richards, Lenhoff and Beris [62], and Kroger, Berg, Delgado and Rath [38] among others.

Our concern here is not with the direct numerical simulation approach but rather with a complementary one, where we seek to describe severe conditions through nonlinear asymptotic theories. One of the phenomena of relevance in liquid jets and bridges is that of jet breakup under capillary instability. The origin of the instability can be understood in overall physical terms as the quest for the jet to minimize surface area (and thus surface energy) so that it breaks into droplets which eventually become spherical due to energy dissipation. The event of breakup is significant mathematically also because the assumed equations of motion become singular at some point and usually at a finite time. Velocities diverge and direct numerical simulations become hard to implement and follow. It is precisely this separation of scales and times which is causing the difficulties and our objective in this Section is to present some asymptotic methods which can be used to extract almost all the relevant information in many cases. Details of a lot that follows can be found in Papageorgiou [50], [51], [52] and Papageorgiou and Orellana [53], Eggers [19], [20].

4.2 Breakup of single jets: Viscous theory

The linear stability of inviscid jets has been presented in the previous Section, and it is straightforward to extend the results to the viscous equations of motion and boundary conditions. Of relevance in this Section are the temporal results since we want to describe pinching in the absence of a convective flow - this can arise either from a Galilean transformation to remove the convective uniform component of the flow, or we can think of examples such as the breakup of liquid bridges. The viscous dispersion relation has been given by Rayleigh and further discussion can be found in the texts of Chandrasekhar [8], Lamb [39] and Middleman [48], for example.

An important subset of the equations given in Section 2.1 and which features in the discussion that follows, is the limit of highly viscous fluid threads governed by the Stokes equations. The nondimensonalization of Section 2.1 is such as to enable the Stokes limit to be taken by formally setting $R_e = 0$ in the momentum equations (25)-(26). In dimensional form, we drop the unsteady and inertia terms (on the left hand side) of the momentum equations (2)-(4). The boundary conditions are unaltered in both formulations. We begin, however, by examining breakup of jets with constant surface tension, that is σ in Section 2.1 is taken to be constant

Beginning with the *Stokes* equations, then, and introducing an interfacial disturbance so that the jet has shape of the form $r = R + \delta \exp(ikz + \omega t)$, where k is real and ω is to be found, one is led to an implicit dispersion relation whose long wave limit (in dimensional form) gives

$$\omega = \frac{1}{6}\frac{\sigma}{\mu R}\left[1 - (kR)^2\right] + \dots \qquad (82)$$

In deriving (82) velocities are scaled with σ/μ and time with $\mu R/\sigma$. The main feature of (82) is that the most unstable wave has $k = 0$. This has led Entov and Hinch [24] to make an estimate of the rate of jet breakup as follows: Terms $(kR)^2$ and higher are neglected giving a growth rate $\omega \approx \frac{\sigma}{6\mu R}$. It is then assumed that this growth rate is valid quasi-statically in time so that if $\eta(t)$ is the minimum jet radius we have

$$-\frac{1}{\eta}\frac{d\eta}{dt} = \frac{\sigma}{6\mu\eta} \quad \Rightarrow \quad \eta(t) = \left(\frac{1}{6}\right)\frac{\sigma}{\mu}(t_s - t), \tag{83}$$

where t_s is a constant which denotes time to breakup and depends on initial conditions. Physically for this result to hold, the jet is required to break as a thinning fluid cylinder of uniform but quasi-statically varying radius; the longitudinal curvature is neglected and the price to pay for this is the prefactor (1/6) in (83).

There are several other theories, also, which allow for longitudinal curvature effects. The jet in this case breaks at a point with a well defined minimum radius, $R_{min}(t)$ say. All theories predict that the minimum radius tends to zero linearly in time, that is,

$$R_{min}(t) = C\frac{\sigma}{\mu}(t_s - t), \tag{84}$$

and each theory has its own value of C. The value of C is an important theoretical quantity which, if known, can be used to predict the ratio σ/μ for a given fluid. The results described constitute an example where different contenting theories predict varying results, but a resolution of the problem is cleanly decided by a set of careful experiments. The approach will be:

1. Analysis/numerics to predict C.
2. Experimental results which select the relevant theory.
3. Practical applications.

4.2.1 Long wave models, similarity solutions and scaling predictions

The breakup phenomenon in liquid threads can be divided into three stages starting with a linear stage where imposed or naturally selected disturbances grow as described in Section 3, then a nonlinear stage which takes the system into the final stage of pinching or topological singularity formation. We describe the last two stages by the following approach:

1. Consider the pinching phenomenon as mathematical finite-time singularity of the governing equations.
2. Look for similarity solutions as the jet breaks and extract scaling functions and scaling laws.
3. Confirm the similarity solutions (which are local in time and space) by numerical solution of the initial boundary value problem.

The central assumption that allows considerable analytical progress is that of long wave modelling. The assumption is that as the jet radius tends to zero, the axial length scale is larger than the radial one. That is, if a typical axial wavelength is of order D, then the parameter $\epsilon^2 = R/D$ is small and can be used to develop an asymptotic theory to obtain 1-D models and so reducing the mathematical complexity significantly. We must check *a posteriori*, of course, that the long wave assumption is not violated during the evolution.

We postulate an asymptotic expansion of the form

$$\begin{aligned} u(z,r,t) &= u_0 + \epsilon^2 u_1 + \ldots, \\ w(z,r,t) &= \frac{1}{\epsilon} w_0 + \epsilon w_1 + \ldots \\ p(z,r,) &= p_0 + \epsilon^2 p_1 + \ldots \\ S(z,t) &= S_0 + \epsilon^2 S_1 + \ldots \end{aligned}$$

These expressions are substituted into the momentum and continuity equations (25)-(27) as well as the tangential stress (29), normal stress (30) and kinematic (31) boundary conditions. In order to keep inertial terms, for the time being, the scaling $R_e = \epsilon^2 \kappa$ is required where $\kappa = O(1)$ and $\kappa = 0$ for Stokes flows. Note that the size of w is picked by the continuity equation. The advantage of the asymptotic limit is that the Laplace operators loose their ellipticity and closed form solutions in terms of unknown functions of z and t are possible. More precisely, the first two orders of the axial momentum equation (25) are

$$w_0 \equiv w_0(z,t), \tag{85}$$

$$\kappa(w_{0t} + w_0 w_{0z}) = -p_{0z} + w_{1rr} + \frac{1}{r} w_{1r} + w_{0zz}, \tag{86}$$

while the leading order of the radial momentum equation (26) and the continuity equation (27) give

$$-p_{0r} + u_{0rr} + \frac{1}{r} u_{0r} - \frac{u_0}{r^2} = 0, \tag{87}$$

$$u_0 = -\frac{1}{2} r w_{0z}. \tag{88}$$

Using (88) into (87) verifies that $p_{0r} \equiv 0$ which implies that $p_0 \equiv p_0(z,t)$ as would be expected from a lubrication theory. The leading order pressure throughout the jet can therefore be determined by evaluating it at the interface through the leading order terms of the normal stress balance equation (30). We find

$$p_0(z,t) = \frac{1}{S_0} - w_{0z}. \tag{89}$$

Next, the tangential stress condition (29) is used to obtain the first evolution equation. The axial velocity correction w_1 follows from (86) since the only function of r is w_1, and is found in terms of w_0 and S_0. This is then substituted along with the expression (88) for u_0 into (29) to yield. The leading order contribution of the tangential stress balance is the consistency condition $w_{0r} = 0$, while the next order yields the first desired evolution equation. The second equation follows from the leading order contributions of the kinematic condition (31). The system to be addressed is:

$$\kappa(w_{0t} + w_0 w_{0z}) = \frac{3(S_0^2 w_{0z})_z}{S_0^2} - \left(\frac{1}{S_0}\right)_z, \tag{90}$$

$$S_{0t} + \frac{1}{2} S_0 w_{0z} + w_0 S_{0z} = 0. \tag{91}$$

Note that the order one parameter κ can be scaled out of the problem if inertia is present.

For Stokes flow we have $\kappa = 0$ in (90). Manipulation of (90) in this case, gives the equation

$$(3S_0^2 w_{0z} + S_0)_z = 0. \tag{92}$$

The quantity being differentiated in (92) can be identified as the leading order long wave limit of the force at any position in the jet. As pointed out by Renardy [61], if there is no inertia then the force is constant in the axial direction and (92) can be written down (see also Papageorgiou [51]).

Continuing with Stokes flows, we look for similarity solutions of equations (92) and (91). Suppose that S_0 goes to zero at time t_s at the point z_s. Defining $\tau = t_s - t > 0$, we seek asymptotic solutions for small τ and around the position z_s so that $|z - z_s| << 1$. Balancing terms in (92) and (91) suggests the following exact transformation

$$S_0(z,t) = \tau f(\xi), \quad w_0(z,t) = \tau^{\beta-1} g(\xi), \quad \xi = \frac{z - z_s}{\tau^\beta}, \tag{93}$$

where $0 << \beta < 1$ (the upper bound is required for consistency with the long wave approximation) and the scaling functions $f(\xi)$, $g(\xi)$ are to be found. Substitution into (92) and (91) and use of the chain rule gives the following set of ODEs (all terms of the PDE are present and so this is an exact solution)

$$(g + \beta\xi)f' + (\frac{1}{2}g' - 1)f = 0, \tag{94}$$

$$\frac{d}{d\xi}(3f^2 g' + f) = 0, \tag{95}$$

where primes denote $d/d\xi$. Integration of (95) gives

$$g' = -\frac{1}{3f} + \frac{k}{f^2}, \tag{96}$$

where the constant k can be found by a further integration and using the fact that $|g(\xi)| \to 0$ as $|\xi| \to \infty$. (An alternative derivation of the expression for k begins with (92), integrates once to obtain $w_{0z} = (1/3)[(\lambda(t)/S_0^2) - (1/S_0)]$; integration over a periodic domain yields $\lambda(t)$ and substitution of the ansatz (93) provides the result below.) The expression is

$$k = \frac{1}{3}\frac{\int_{-\infty}^{\infty}(1/f)d\xi}{\int_{-\infty}^{\infty}(1/f^2)d\xi}. \tag{97}$$

The following useful asymptotic forms far from the pinch region, emerge for large $|\xi$.

$$f(\xi) \sim |\xi|^{1/\beta}, \quad g(\xi) \sim |\xi|^{-(1-\beta)/\beta}. \tag{98}$$

The objective is to determine the constant k, the scaling exponent β and the scaling functions $f(\xi)$, $g(\xi)$. From (98) we see that g vanishes at infinity and so there must exist at least one point, ξ_0 say, such that $g(\xi_0) + \beta\xi_0 = 0$. This is seen to be a singular point of (94) and to keep the solutions smooth we must impose conditions that make this point a regular singular point. This is readily achieved by making use of the transformations

$$f \to f, \quad G(\eta)G + \beta\xi_0, \quad \eta = \xi - \xi_0,$$

which shift ξ_0 to the origin. The conditions for a regular singular point are seen to be

$$G(0) = G'0) - 2 = 0.$$

A series solution for small η is

$$f(\eta) = f_0 + \eta^2 f_2 + \eta^4 f_4 + \ldots, \qquad G(\eta) = 2\eta + \eta^3 g_3 + \ldots,$$

where

$$f_0 = \frac{1}{12(1+\beta)}, \qquad k = \frac{3+2\beta}{72(1+\beta)^2}. \tag{99}$$

Combining (99) with (97) gives the following eigenvalue problem for *beta*

$$\frac{3+2\beta}{72(1+\beta)^2} = \frac{1}{3}\frac{\int_{-\infty}^{\infty}(1/f)d\xi}{\int_{-\infty}^{\infty}(1/f^2)d\xi}. \tag{100}$$

The value of β must obtained by numerical iteration; a value of β is guessed and the functions $f(\eta)$, $g(\eta)$ are obtained by integrating the ODEs using the appropriate initial conditions at $\eta = 0$, or by using a closed form implicit solution given in [50]. Equation (100) is then used to update β until convergence. The choice $f_2 \neq 0$ gives a unique value of $\beta \approx 0.175$. Other solutions with larger values of β are possible by choosing f_2, f_4 etc. to be zero. This way a countably infinite set of solutions is found but they are all unstable. (Brenner, Lister and Stone [5] for details of the higher solutions and their stability.) The value found here is the one that emerges from solutions of the initial value problem - the higher solutions do not appear because they are unstable. Such numerical solutions are described next.

Numerical solutions of the IVP

Pinching is a local phenomenon and is expected to be insensitive to the exact far field boundary conditions. In particular we choose to solve the problem on an axially periodic domain. We are concerned with Stokes flows ($\kappa = 0$) where the relevant 1-D equations are (92) and (91). Integrating (92) once and using periodicity gives the results

$$\frac{\partial w_0}{\partial z} = \frac{1}{3}\left(\frac{\lambda(t)}{S_0^2} - \frac{1}{S_0}\right) \quad \Rightarrow \quad \lambda(t) = \frac{\int_0^{2\pi}(1/S_0)dz}{\int_0^{2\pi}(1/S_0^2)dz}. \tag{101}$$

Equations (101) show that it is sufficient to prescribe initial conditions on S_0 alone. Numerical solutions are carried out using pseudospectral methods where all integrations are done by taking transforms of appropriate functions. The strategy is to first calculate λ at a given time from knowledge of S_0 at that time. Equation (101) can then be used to calculate w_0 at the current time and equation (91) is used to advance S_0 to the next time level. The process is then continued. We used predictor corrector or Runge-Kutta methods for the time integrations - see [50] for details.

Initial conditions can be chosen that are even about $z = \pi$ (general initial conditions lead to the same structures locally near the pinch point - see [50]). In the results that follow we have $S_0(z,0) = 0.5 + 0.1\cos(z)$. The numerical solutions monitor the evolution of the minimum jet radius $S_{min}(t)$, the evolution of $\lambda(t)$ and the maximum value of the axial velocity $w_{max}(t)$. Using $\beta = 0.175$, the similarity solutions presented earlier predict the following forms for these quantities near the singular time t_s,

$$S_{min}(t) \approx 0.07092(t_s - t), \qquad \lambda(t) \approx 0.01011(t_s - t), \qquad w_{max}(t) \sim (t_s - t)^{-0.825}. \qquad (102)$$

The value f t_s is not known and depends, in fact, on initial conditions. The way this is predicted is to use the analytical finding of the linear evolution of $S_{min}(t)$ and $\lambda(t)$ and perform least squares fits to the data. This turns out to be very efficient and in addition, for the initial conditions used the evolution gets attracted to the linear laws (102) after about a third of the computational times. (See Figure 8 of reference [50].) With a method of estimating t_s, it is possible to test the exponent of w_max which is blowing up. This is one way of extracting a value for β from solutions of the initial value problem. The data set $[\log(t_s - t), \log(w_max(t))]$ is generated, plotted and a least squares fit is carried out to find the exponent (see Figure 7 of [50]). We find a value $\beta - 1 = -0.823$ as opposed to the theoretical value of -0.825 (see (102)), thus confirming the theory to within the accuracy of the computations.

Other non-symmetric solutions give singularity formation at other points besides $z = \pi$. It can be established numerically, however, that locally to the off-center pinch point, the solution is symmetric and the same scaling laws apply. Considering symmetric but different initial conditions, then, the numerics predicts breakup according to the theory presented and the scaling laws found using similarity solutions. Initial conditions with larger interfacial perturbations pinch at smaller t_s as expected. A collection of such results is given in Figure 16 of [50]. Using these results it is also possible to construct the scaling functions $f(\xi)$ using data from runs with different initial conditions. The objective is to use (93) to numerically construct $f(\xi)$. This is done as follows: (i) Estimate t_s by least squares fits. (ii) For a time near t_s and $-\pi < z < \pi$ compute and store $\xi = z/(t_s - t)^\beta$ for all grid points. (iii) From the numerical solution $S(z,t)$ at this time, construct $f(\xi) = S(z,t)/(t_s - t)$ or each grid point. Plots of different scaling functions corresponding to different initial conditions are given in Figure 17 of [50]. These are universal to within a constant axial rescaling (the mathematical origin of this is the constant f_2 in the series solution near the pinch point).

Similarity solutions using natural length and time scales

The approach of Eggers [19], [20], [21], proceeds on more physical grounds by using the three physical parameters in the problem, namely μ, ρ and σ, to construct both length and time scales for the similarity solutions. The results described above assume an externally imposed axial length scale as in an experiment of a liquid bridge, for instance. We present the theory of Eggers next, and in what follows we compare the two theories with experiments.

Natural length and time scales for the problem are

$$l_E = \frac{\mu^2}{\rho\sigma}, \qquad t_E = \frac{\mu^3}{\rho\sigma^2}. \qquad (103)$$

The length scale is the ratio of viscosity to velocity, the latter being σ/μ as before, while the time scale is that of capillary instability based on l_e, i.e. $sqrtl_e^3\rho/\sigma$ (see [17] for example).

We note in passing that the Ohnesorge number is $Oh = l_E/R$ where R is the undisturbed jet radius. In dimensional form, then, Eggers seeks a solution of the 1-D system of equations in the form

$$S(z,t) = l_E \tau f(\xi), \quad w(z,t) = \frac{\sigma}{\mu}\tau^{-1/2} g(\xi), \tag{104}$$

where

$$\xi = \frac{z}{l_E \tau^{1/2}}, \qquad \tau = \frac{t_s - t}{t_E}.$$

This ansatz preserves all terms in the 1-D equations and represents a balance between viscous and inertial terms. It is equivalent to the ansatz (93) with $\beta = 1/2$ when substituted into (90) with κ scaled out. The scaling functions satisfy a nonlinear set of ODEs which we need not give here. Eggers analyzed these equations and in particular constructed numerical solutions which are *universal.* The scaling function $f(\xi)$ is not symmetric; its behavior at infinity is

$$f(\xi) \sim 4.635\xi^2 \quad \text{as} \quad \xi \to +\infty,$$

$$f(\xi) \sim 6.074 \times 10^{-4}\xi^2 \quad \text{(as)} \quad \xi \to -\infty.$$

Of particular interest in making comparisons with some recent experiments, is the rate at which $S(t)$ tends to zero (this is the constant C in the expressions (84)) which is found to be

$$R_{min}(t) = 0.0304\frac{\sigma}{\mu}(t_s - t). \tag{105}$$

4.2.2 Comparison with experiments

We conclude this Section by evaluating the two theories outlined above by comparing with some recent experiments by McKinley and Tripathi [45]. The experiments are motivated by the construction of an extensional rheometer device, the Capillary Breakup Extensional Rheometer (CABER). The experiment we consider here consists of two vertically aligned discs of diameter 6 mm, which are initially a distance 2.2 mm apart and containing a blob of viscous fluid between them. At $t = 0$, the discs are extended to a distance of 6.6 mm apart so that the liquid bridge is beyond its stability length and begins to thin under capillary forces. The evolution is monitored by digitizing the bridge shape in real time and in particular recording the evolution of the minimum thread radius. The fluids used are Glycerin (GLY; $\mu = 1.03$ Pa s), Silicone oil (PDMS; $\mu = 10.5$ Pa s) and Polybutene H100 (PB; $\mu = 24.0$ Pa s). The experiment is a stringent and careful evaluation of the behavior (84). The reader is referred to [45] for several figures of results. For example, in the case of the Glycerol experiment a breakup time of approximately 0.33 seconds is found and the authors superimpose onto the experimental results the results of Entov and Hinch with $C = 1/6$, Papagergiou with $C = 0.0709$ and Eggers with $C = 0.0304$. The viscous solution described in detail above (see Papageorgiou [50], [51]) is almost indistinguishable from the experiment. In addition, it captures about the last 25% of the evolution before breakup almost perfectly. On the other hand, the Entov and Hinch prediction is quite poor whereas the Eggers universal solution is not seen to be approached at all - it is possible that the Eggers solution is seen when the thread gets very thin, but these experiments did not reveal any such trends. We discuss the reasons for the shortcomings of the universal solution later.

First, we discuss the practical applications of CABER and the need for a good value of C. As described above, the experiment monitors and plots the minimum thread radius as a function of time and in particular the stages where the minimum radius tends to zero linearly with $t_s - t$. This provides a measurement for the slope, therefore, which can be used along with (84) and the *correct* value of C, to obtain a measurement for the ratio σ/μ. This is the objective of CABER. It is found that for the fluids used the Papageorgiou solution gives the correct value to use n applications. Independent measurement of σ/μ was also compared with the CABER results described above and agreement was found to be to within 5% or better.

We close this Section o viscous jets with a discussion as to the reasons why the natural length and time scales of Eggers are not observed in the experiments described here. The reason for this seems to lie with the fact that for the fluids used in [45], the Reynolds number $R_e = 1/Oh$ as defined in (28) is small, or equivalently Oh is large. If we consider (25), for example, and use the Stokes singularity scalings (93), then we see that the left hand side of the equation containing unsteady and inertial terms will remain asymptotically small compared to the viscous terms on the right, as long as

$$\frac{1}{Oh}\tau^{\beta-2} << \tau^{-1-\beta}, \tag{106}$$

where $\tau = (t_s - t)/t_p$ is the dimensionless time to the singularity and $t_p = \frac{\mu R}{\sigma}$ is the time scale used by Papageorgiou. Using $\beta = 0.175$, we can rearrange (106) to give a time above which the Stokes theory is expected to be valid. That is, we expect to see $C = 0.0709$ as long as

$$\tau >> Oh^{-1/(1-2\beta)} \approx Oh^{-1/0.65}. \tag{107}$$

We can use (107) to define a *crossover* time τ^+ which can be used to guide experimental observations:

$$\tau^+ \sim Oh^{-1/0.65}. \tag{108}$$

In the McKinley and Tripathi experiments, the undisturbed thread radius can be taken to be the plate radius which is $R = 0.296$ cm. Using the Stokes timescale, (108) becomes

$$\tau^+ \sim \frac{\mu R}{\sigma} Oh^{-1/0.65},$$

and these values are tabulated in Table 1 for the three different fluids along with the estimates of the natural length and time scales l_e and t_E.

Table 1	Glycerol	Silicone oil	Polybutene oil
t_p [s]	0.05s	1.5s	2.4s
Oh	4.38	1806	6549
$\tau^+ t_p = \frac{\mu R}{\sigma} Oh^{-1./0.65}$	4.8×10^{-3}s	1.5×10^{-5} s	3.2×10^{-6} s
$l_E = \frac{\mu^2}{\rho\sigma}$	1.3 cm	535 cm	1938 cm
$t_E = \frac{\mu^3}{\rho\sigma^2}$	0.21 s	45 min	4.3 hours

The results of Table 1 suggest some clear reasons why the scaling exponents based on natural length scales are not seen in the present experiments. The crossover times τ^+ are of the order of 10^{-5} seconds, and it is not surprising, therefore, that the Stokes theory predicts the measurable features of the experiment. In addition, the natural length scales get very large, as large as almost 3000 times the liquid bridge length for the most viscous fluid (PB) for any hope of the exponents to be seen, while at the same time the natural time scales are huge compared to the observed breakup times (as much as over three hours for the most viscous fluid (PB) ut of the same order of magnitude for GLY). The larger the Ohnesorge numbers, the more pronounced is the inability of the natural length scales theory to be observed in practice. It is reasonable to suppose that as the jet thins to radii which are of the order of milliseconds from the rupture time, the natural length scales would become important. Besdes the fact that other neglected physical effects such Van der Waals forces would have to be retained then, it is both unlikely and impractical for rheometer devices such as the CABER to be operational at such molecular levels. In any case, it can extract the rheological information required from "on-the-fly" measurements along with the scalings predicted using the theory for Stokes flows.

4.3 Breakup of single jets: Inviscid theory

We describe briefly some results regarding efforts to describe inviscid jet breakup using 1-D models derived from the Euler equations. There have been numerous works on this problems; one of the earlier works is that of Lee [40], who uses what is called a "slice" model to describe breakup even though no detailed study of singularity formation is carried out. Meseguer [47] considers the parallel problem of liquid bridges using similar ideas and Schulkes [66], [67] develops weakly nonlinear evolution equations and makes a comparative study of different models. None of these works has studied singularity formation carefully. The question to be addressed is whether a long wave inviscid mode is even consistent during its evolution or if the long wave approximation is violated at a finite time by an infinite slope singularity, for example. Such violations are found in leading order 1-D models. Such models were first suggested by Ting and Keller [75] and suggestions of ill-posedness were put forward by Forest and Wang [26]. Analytical and numerical evidence of this is given by Papageorgiou and Orellana [53], who use the complexification method to cast the elliptic equations into hyperbolic ones and follow intersection of envelopes of characteristics with the real axis; such events herald formation of physical infinite slope singularities and thus the violation of the slender jet approximation.

Papageorgiou and Orellana [53], attempt a regularization of the ill-posed leading order system of Ting and Keller by retention of higher order terms. Mathematically, this introduce a dispersive regularization to the equations and these are studies asymptotically and numerically. The details of the derivation as well as inclusion of a second annular phase in a tube, are given in [53]. We just state the systems of PDEs to be addressed and in particular two models to be compared: (i) The asymptotic model, (ii) The full curvature model. These are

$$(S^2)_t + (WS^2)_z = 0, \tag{109}$$

and either of

$$W_t + WW_z = \frac{S_z}{S^2} + \epsilon^2 S_{zzz} + \frac{1}{2}\epsilon^2 \left(\frac{S_z^2}{S}\right)_z, \quad (110)$$

$$W_t + WW_z = -\left[\frac{1}{(1+\epsilon^2 S_z^2)^{1/2}}\left(\frac{1}{S} - \frac{\epsilon^2 S_{zz}}{1+\epsilon^2 S_z^2}\right)\right]_z. \quad (111)$$

Equation (109) is a statement of mass conservation in the long wave limit. The model (110) is seen to be the two terms Taylor expansion in ϵ^2 of model (111) which retains the full curvature term. In [53] a careful comparative study of these two models is made and we describe the results while referring the reader to the article for details.

The asymptotic slice model (110) is found to pinch, i.e. $S \to 0$, after a finite time. Starting from symmetric initial conditions for S and odd ones for W, it is found that pinching occurs simultaneously at two points symmetrically placed about the center (periodic boundary conditions are used) thus giving rise to a satellite drop. This behavior is fairly for the models an was also observed by Lee in his original study. On resolution of the structures near pinching, it is found that touchdown is accompanied by a cusp formation with S_z blowing up as $S \to 0$ at the pinch points. In fact, analysis and numerics indicate the following asymptotic behavior as the jet pinches:

$$\eta = S^2 = \tau F(\xi) + \ldots, \quad u = S^2 W = \tau^{5/8} G(\xi) + \ldots, \quad \xi = \frac{z - z_s}{\tau^{5/8}}, \quad (112)$$

where $\tau = t_s - t$ is the time from the singularity and z_s is the position of the singularity. This suggests, then, that even though $S \to 0$ as $\tau \to 0$, we also have $S_z \sim \tau^{-1/8}$ indicating cusp formation and violation o the sender jet ansatz. These results suggest an introduction of an inner Euler region with relaxation of the long wave model.

Curiously enough, and contrary to the findings of work on viscous jet, retention of the full curvature terms can make the situation worse in the sense that the long wave ansatz can terminate in an infinite slope singularity before pinching can occur. This behavior is found to be strongly dependent on initial conditions (see [53] for details). If the initial conditions are of relatively large amplitude, the dynamics terminates in an infinite slope singularity after a finite time, whereas for small amplitudes the growth takes place through the linear regime and the nonlinear capillary instability stage rather than the inertia dominated dynamics of large initial disturbances.

5 Motion of Bubbles In Liquids Containing Surfactants

The governing equations and boundary conditions have been developed in Section 2.2. The system to be solved consists of the Navier-Stokes and continuity equations (34), (35), the concentration equation in the bulk (36) as well as the boundary conditions (39), (41) and (42) which are the surface mass balance, kinetic flux balance ad tangential stress balance respectively. As reviewed in Section 2.2, most studies have examined the effect of dilute concentrations, i.e. trace amounts of surfactants. At low concentrations the drag is found to increase linearly with bulk concentration and the amount of surfactant on the interface is far below the maximum packing concentration Γ_∞. The aim of this Section is to consider what happens as the bulk concentration, and hence the amount of surfactant adsorbing onto the interface, increases. To achieve this theoretically, we need to incorporate a nonlinear adsorption isotherm. In order to set the stage for the different regimes addressed numerically, we present some scaling arguments in order to identify the relevant physics.

Begin with a given concentration far away from the bubble in the bulk. The average surfactant concentration on the bubble surface at steady state scales with Γ_0, and the value given by the nonlinear isotherm being used. Using the notation of Section 2.2 we estimate the rate of convective flux on the surface to be $\Gamma_0 aU$, where U is the terminal velocity, while the rate of diffusive flux off the surface per unit area is $a^2 \times (D\Delta C/a)$ (ΔC is the characteristic difference between the sublayer and bulk concentration - see Section 2.2). The following estimate is made, therefore,

$$\Lambda_D = \frac{rate\ of\ diffusion}{rate\ of\ convection} = O\left[\frac{\Delta C \chi_0 (1+k)}{C_0 P_e}\right]. \tag{113}$$

In writing the form above we use the expression (38 for the kinetic flux to obtain the equilibrium concentration Γ_0. Setting $j = 0$ in (38) and noting that the sublayer concentration is equal to the bulk concentration, we obtain

$$\Gamma_0 = \Gamma_\infty \frac{k}{1+k} \quad \text{where} \quad k = \frac{\beta C_0}{\alpha}.$$

With this observation and the definitions of P_e, χ_0 from Section 2.2, (113) follows. Note that for trace amounts of surfactants $k << 1$ and $\Lambda_D \sim \chi_0/P_e$.

Next we estimate the ratio between the characteristic rate of kinetic flux to the rate of convection. To obtain the former, we linearize the kinetic flux j around the equilibrium surface concentration Γ_0 and sublayer concentration (i.e. write $\Gamma = \Gamma_0 + \Delta\Gamma$ and $C_s = C_0 + \Delta C$ and linearize; the leading order is zero by definition of equilibrium). We find

$$\Lambda_K = \frac{rate\ of\ kinetic\ exchange}{rate\ of\ convection} = O\left[Bi\left(\frac{\Delta C}{C_0} - (1+k)\frac{\Delta\Gamma}{\Gamma_0}\right)\right]. \tag{114}$$

The Biot number is $Bi = \alpha a/U$. There are two regimes identified with the above scales.

1. **Negligible surface/bulk exchange, $\Lambda_D, \Lambda_K << 1$.**

   This regime is achieved when $\chi_0(1+k)/P_e \ll 1$ and $Bi \ll 1$. The surfactant behaves as if it is insoluble in this case and if in addition the surface Peclet number $Pe_s = Ua/D_S$ is infinite, no diffusion to the front is possible. A stagnant cap forms in the vicinity of the back end on which the no-slip condition is satisfied, while the front end is free of surfactant and stress free. The size of the cap is characterized by an angle which depends on k, $\phi(k) say$, and its size is determined by an overall mass balance on the surface which requires that the net flux of surfactant to the surface be zero at steady state (see Harper [30], He,Dagan and Maldarelli [31]. An exact solution (in the form of a series) has been given by Sadhal and Johnson [64] in the creeping flow regime.

They demonstrate that the drag varies from that for a clean bubble to that of a solid sphere as the cap size increases to cover the whole surface. In the presence of inertia, Navier-Stokes computations have been carried out by Bel Fdhila and Duineveld [3] who computed the drag on a buoyantly rising bubble by fixing the cap angle; wakes are found to form at sufficiently high Reynolds numbers. McLaughlin [46] allowed for deformation and again found downstream wakes at large enough Reynolds numbers. In this case, then, drag increases as a function of k and eventually reaches that for a solid sphere or particle.

2. **Finite surface/bulk exchange, $\Lambda_D, \Lambda_K = O(1)$.**

In this case the exchange of surfactant between the bulk and the surface opens the possibility for reduction of the Marangoni force. When the rates of bulk diffusion and surfactant desorption are of the same order as surface convection (i.e. the present regime), the surfactant exchange mechanisms can be used to identify factors affecting the Marangoni force. The non-dimensional Marangoni force is

$$\tau_m = \frac{interfacial\ tension\ gradient}{viscous\ stress} = O\left[\frac{1}{\mu U}\left(\frac{\partial\sigma}{\partial\Gamma}\right)_{\Gamma_0}\Delta\Gamma\right]. \tag{115}$$

From the expressions (113) and (114) and using the fact that the ratios are order one, we can obtain scales for $\Delta\Gamma$ and ΔC as follows:

$$\frac{\Delta C}{C_0} = O\left[\frac{P_e}{\chi_0(1+k)}\right], \quad \frac{\Delta\Gamma}{\Gamma} = O\left[\frac{1}{Bi(1+k)} + \frac{P_e}{\chi_0(1+k)^2}\right].$$

From the Langmuir adsorption isotherm, for example, we have

$$\left[\frac{\partial\sigma}{\partial\Gamma}\right]_{\Gamma_0} = -\frac{RT}{1-\frac{\Gamma_0}{\Gamma_\infty}},$$

the retarding Marangoni force can be estimated to be

$$\tau_m = O\left[Ma\left(\frac{k}{Bi(1+k)} + \frac{kP_e}{\chi_0(1+k)^2}\right)\right], \tag{116}$$

where Ma is the Marangoni number defined in Section 2.2. The expression (116) shows how the bulk concentration and the rate of kinetic exchange to surface convection reflected by Bi, affect the retarding Marangoni force. In the dilute limit regime $k << 1$, the Marangoni force is seen to increase linearly with k and the terminal velocity decreases. In the high concentration limit, however, the retarding force becomes of the order of Ma. This has been found by Chen and Stebe [12] who find limiting terminal velocities at large k, which decrease to the clean value in the further limit $Bi \to \infty$. This is the regime of *kinetically controlled* uniform retardation.

The opposite regime of *bulk diffusion control* of the Marangoni force has $\chi_0(1+k)/P_e = O(1)$ and $Bi >> 1$, has received little attention. We can estimate from (116), therefore, that as $Bi \to \infty$ $\tau_m = O(\frac{MakP_e}{\chi_0(1+k)^2})$. Using this result we have the following paradigm for *remobilization*: For large k, τ_m scales as $P_e/(\chi_0 k)$, so as the concentration increases, P_e and χ_0 held fixed, the Marangoni force should disappear. The numerical solutions described in what follows, lend support for this possibility for both zero and order one Reynolds numbers.

5.1 Numerical Solutions

Two different numerical methods will be described. The first uses a streamfunction-vorticity formulation and the second method uses primitive variables. Each method has its relative advantages and disadvantages. For example, the former method does not require the pressure but requires vorticity boundary conditions instead, while the primitive variables method requires a Poisson solver per time step in order to compute the pressure.

5.1.1 Streamfunction-vorticity method

Begin with equations (34)-(35) and using the usual spherical coordinate system $\mathbf{x} = (r, \theta, \phi)$ fixed to the bubble, with corresponding velocity field $\mathbf{u} = (u_r, u_\theta, u_\phi)$. Due to axisymmetry (no ϕ dependence), the continuity equation (35) can be satisfied by introducing a streamfunction $\psi(r, \theta, t)$ such that (see [2]):

$$u_r = -\frac{1}{r^2 \sin\theta}\frac{\partial \psi}{\partial \theta}, \quad u_\theta = \frac{1}{r\sin\theta}\frac{\partial\psi}{\partial r}. \tag{117}$$

Corresponding to this velocity field we have a vorticity $(0, 0, \omega(r, \theta, t))$ where

$$\omega = \frac{1}{r}\left[\frac{\partial(r u_\theta)}{\partial r} - \frac{\partial u_r}{\partial \theta}\right] = \frac{1}{r}E^2\psi, \tag{118}$$

where

$$E^2 \equiv \frac{1}{\sin\theta}\frac{\partial^2}{\partial r^2} + \frac{1}{r^2}\frac{\partial}{\partial\theta}\left(\frac{1}{\sin\theta}\frac{\partial}{\partial\theta}\right). \tag{119}$$

The system to be solved numerically becomes, then,

$$\frac{\partial\omega}{\partial t} + \frac{\partial}{\partial r}(r u_r \omega) + \frac{\partial}{\partial\theta}(U_\theta\omega) = \frac{1}{R_e}E^2(r\omega\sin\theta), \tag{120}$$

$$\omega = \frac{1}{r}E^2\psi, \tag{121}$$

$$\frac{\partial C}{\partial t} + bfu\cdot\nabla C = \frac{1}{P_e}\nabla^2 C, \tag{122}$$

where ∇^2 is the spherical Laplacian. The boundary conditions for the velocity field given in Section 2.2 are easily converted into corresponding ones involving the streamfunction:

$$\psi = 0 \quad \text{at} \quad \theta = 0, \pi, \tag{123}$$

$$\psi = 0 \quad \text{at} \quad r = 1, \tag{124}$$

$$\psi = \frac{1}{2}r^2\sin^2\theta \quad \text{as} \quad r\to\infty, \tag{125}$$

$$\omega = 0 \quad \text{at} \quad \theta = 0, \pi, \tag{126}$$

$$\omega = 0 \quad \text{as} \quad r \to \infty, \tag{127}$$

$$\omega = \frac{2}{\sin\theta}\frac{\partial\psi}{\partial r}\bigg|_{r=1} + \frac{Ma}{1-\Gamma}\frac{\partial\Gamma}{\partial\theta}, \tag{128}$$

$$\frac{\partial C}{\partial \theta} = 0 \quad \text{at} \quad \theta = 0, \pi, \tag{129}$$

$$C = 1 \quad \text{as} \quad r \to \infty, \tag{130}$$

$$\frac{\partial \Gamma}{\partial t} + \frac{1}{\sin\theta}\frac{\partial}{\partial \theta}(u_\theta \Gamma \sin\theta) = \left.\frac{\chi_0 k}{P_e}\frac{\partial C}{\partial r}\right|_{r=1} \tag{131}$$

Condition (128) is the tangential stress balance written as a boundary condition for the vorticity. A final condition is the kinetic flux balance (41). In the first set of results to be discussed we compute the diffusion control case (see above) and so take the limit $Bi \to \infty$. This implies the following equation of state from (41):

$$\Gamma = \left.\frac{kC}{1+kC}\right|_{r=1}. \tag{132}$$

Integration of the stresses over the bubble surface gives the following expression for the drag coefficient

$$C_D = \frac{F_z}{\pi a^2 \rho U^2} = \int_0^\pi \sin^2\theta \frac{\partial p}{\partial \theta} d\theta - \frac{1}{R_e}\int_0^\pi \left(2\frac{\partial \psi}{\partial r} + \frac{Ma\sin\theta}{1-\Gamma}\frac{\partial \Gamma}{\partial \theta}\right)\sin\theta d\theta. \tag{133}$$

where F_z denotes the force along the axis of symmetry.

The numerical method uses centered finite differences for spatial derivatives and time integrations are performed using the ADI method. The ADI scheme splits each time step into two and a semi-implicit Crank-Nicolson scheme is used treating implicitly the r-direction over half a time step and then the θ-direction over the second half. In addition, a pseudo-unsteady system is solved which includes a term $\partial\psi/\partial t$ on the left hand side of (121) and integrating forward to steady state (see Peyret and Taylor [55]). The physical domain is mapped onto a rectangular computational domain by the transformation $r = r_\infty^x$, $\theta = \pi y$, where x, y are the new independent variables and r_∞ is a large number of the order of 50-100 times the bubble radius. The boundary conditions at infinity are applied either directly for non-zero Reynolds numbers or through an asymptotic approximation which increases accuracy for zero Reynolds number as we explain next. One way to improve the boundary condition at infinity is to analyze the equations for large r and include a correction term. For Stokes flow we find (see Happel and Brenner [29]) that the asymptotic result is

$$\psi = \frac{1}{2}r^2\sin^2\theta + \frac{1}{8}F_D r \sin^2\theta \quad \text{as} \quad r \to \infty, \tag{134}$$

where F_D is the total drag on the bubble non-dimensionalized by the creeping flow scale $\pi\mu U a$. The condition (134) is implemented numerically at a finite value of r_∞ with the advantage that higher resolution runs can be completed faster. As the computation proceeds, the drag is monitored (see equation (133) for its calculation) and used to update the boundary condition (134) in a time dependent manner. For full details of the numerical methods as well as code validation using certain asymptotic theories, see Wang [79].

Results

The representative results that follow exhibit conclusively the possibility of removing the Marangoni force and completely remobilizing the surface. (For more details see [80], [81].)

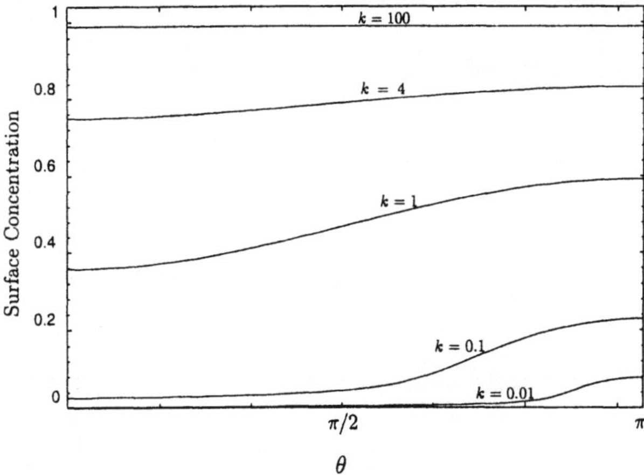

Figure 1: The surface concentration distribution, for Re = 50, Ma = 5 and $\chi = 1$, and $k = \frac{\beta C_\infty}{\alpha}$ is the measurement of bulk concentration.

We begin with zero Reynolds numbers. Figure 5.1 shows the surfactant concentration $\Gamma(\theta)$ on the bubble surface for different increasing bulk concentrations. The Peclet number is 10 and the other parameters are $Ma = 5$, $\chi_0 = 1$; $\theta = 0$ represents the leading end of the bubble. It can be seen from the results that or any k, the surface concentration at the back end is higher that at the leading end. For small values of k there is a sharp increase in surface concentration at the back end and the Marangoni force is significant. As k increases, however, the remobilization mechanisms discussed above come into play. The average surfactant concentration increases but the distribution becomes increasingly uniform, the surface tension gradient is reduced and the Marangoni force disappears. The bubble is remobilized.

These results are reiterated in Figure 5.2 which shows a collection of our numerical results regarding the drag experienced by the bubble as a function of bulk concentration k and Peclet number P_e. All cases show an increase in drag as k increases from a clean system; this initial increase is larger for larger Peclet numbers, and the drag coefficient lies between the values 4 and 6 corresponding to a clean bubble and a solid sphere respectively. Remobilization is clearly seen as k increases for each Peclet number. We see that there is a maximum drag at an order one value of k, but further increase starts bringing the drag down and eventually the Marangoni force disappears and the drag returns to its clean value providing remobilization. For additional results including bulk concentration variations see [80].

Next, we consider non-zero Reynolds numbers. The physics here is quite different because of the possibility of inertia inducing reversed flow and wakes behind surfactant adsorbing bubbles. The negative consequences of this for applications have been discussed in Section 2.2 and in what follows we provide conclusive evidence of remobilization at non-zero Reynolds numbers. Results similar to those of Figure 5.1 have been obtained for $R_e = 50$, $Ma = 5$ and Peclet numbers of 100 and 200. Increasing k removes the surface tension gradients on

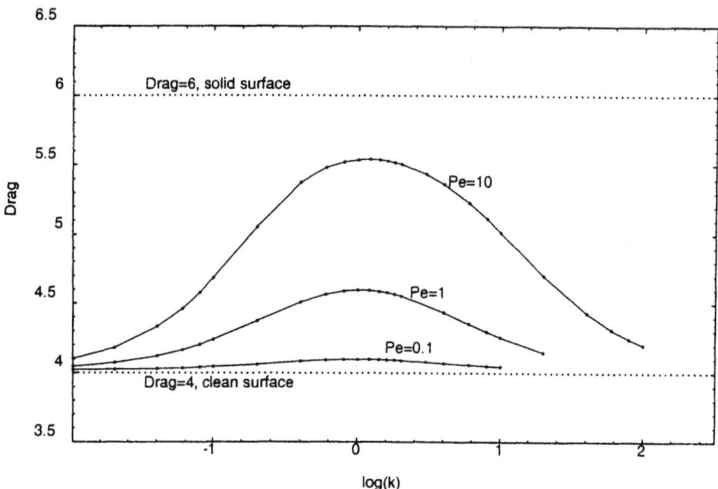

Figure 2: The effect of concentration on the drag, for Re = 50, Ma = 5 and $\chi = 1$, and $k = \frac{\beta C_\infty}{\alpha}$ is the measurement of bulk concentration.

the bubble surface and th bubble remobilizes. The reader is referred to [81] for extensive results, but here we conclude with the flow fields in order to evaluate the possibility of wake formation and control.

It is well known that the flow of a uniform stream past a solid sphere undergoes a separation and wake formation at a Reynolds number (based on the radius) of about 12. The wakes grow with Reynolds number and symmetry breaking and unsteady wake shedding ensue at higher R_e. For a recent computation of unsteady effects in the flow past a solid sphere see Johnson and Patel [34]. Undeformed clean bubbles, on the other hand, maintain attached flow and wakes can form only at non-zero Weber numbers; see Ryskin and Leal [63]. For bubble motion in surfactant solutions the accumulation of surfactant at the back end makes the surface there less mobile and this change in boundary condition decelerates the surface flow and can cause reversed flow if the Reynolds number is large enough. This is illustrated in Figure 5.3 which shows computed steady state streamlines for a fixed bulk concentration $k = 5$, Peclet number $P_e = 100$, $Ma = 5$ and $\chi_0 = 1$. At low Reynolds numbers the flow is attached and remains so until a Reynolds number between 15 and 20, which, as expected, is larger than the separation Reynolds number for a solid sphere. As the Reynolds number increases further, the wake gets longer in line with knowledge of solid sphere results. Hence, at a fixed bulk concentration we expect to see flow separation and wake formation if the Reynolds number is large enough. We note that the highest Reynolds number in Figure 4.3 is 55 which is safely inside the axisymmetric steady flow regime we are assuming.

Figure 5.4 is a theoretical confirmation of the mechanism of remobilization discussed earlier. Here, a Reynolds number of 50 is chosen to ensure that a wake is present at order one values of k. The other parameters are $P_e = 200$, $Ma = 5$, $\chi = 1$. The first set of results has $k = 5$ and a wake is seen to be present with length about the bubble diameter. As the

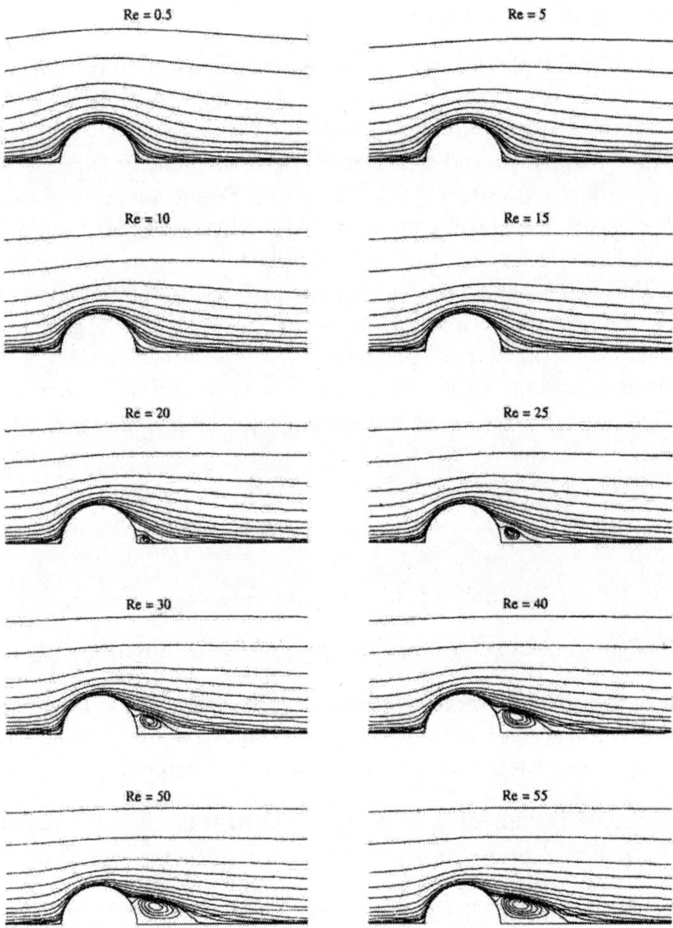

Figure 3: Flow around the bubble for Pe = 100, Ma = 5, $\chi = 1$ and $k = 5$.

bulk concentration is increased, the wake is seen to decrease in size and to disappear at a Reynolds number between 45 and 50. As k increases further, the interface becomes more mobile and the flow approaches that for flow past a clean bubble at a Reynolds number of 50. The effect of increasing Peclet number (not shown here) is to delay the wake disappearance to larger values of k, all other parameters being fixed.

5.1.2 Primitive variables method

A finite volume method is used to discretize the equations of motion which are first written in conservative form by use of the continuity equation. A staggered grid is used between velocities and pressure in order to maintain stability; for details of these methods see Patankar [54] and Tannehill, Anderson and Pletcher [73]). Our main interest here is to extend the computations described above to high values of the Peclet number, of the order of 10^5. As can be seen from the concentration equation (36), solute boundary layer will form near the bubble surface at high Peclet numbers and a finer grid near the surface is needed to resolve these features. The grid is chosen by numerically estimating the solute boundary layer thickness on a fluid-fluid interface (this is of order $P_e^{-1/2}$), and inserting at least 10 grid points within this layer. Note that for a solid particle, the solute boundary layer thickness is of order $P_e^{-1/3}$ which is asymptotically larger than that on a fluid interface. In stagnant cap regimes, therefore, a grid which over refines the fluid-fluid boundary layer is used to keep the coding relatively simple.

The following set of steps is a possible algorithm:

1. Initialize u, v, C, Γ.

2. Advance in time using the projection method of Chorin [14].

3. First half of the time step advances the velocity in the absence of the pressure gradient.

4. In the second step the convective terms are dropped and a Poisson equation for the pressure emerging from the zero divergence of the velocity field, must be solved. In this step, the tangential stress balance evaluates Γ explicitly.

5. The bulk concentration equation is solved using the kinetic flux condition (41) and treating Γ explicitly.

6. Advance Γ from the surface equation (39).

7. Repeat until convergence.

Some preliminary results as well as code validation tests will be presented during the Lecture Series.

References

[1] Aris, R. 1962 *Vectors, Tensors, and the Basic Equations of Fluid Mechanics.* Prentice Hall, New Jersey.

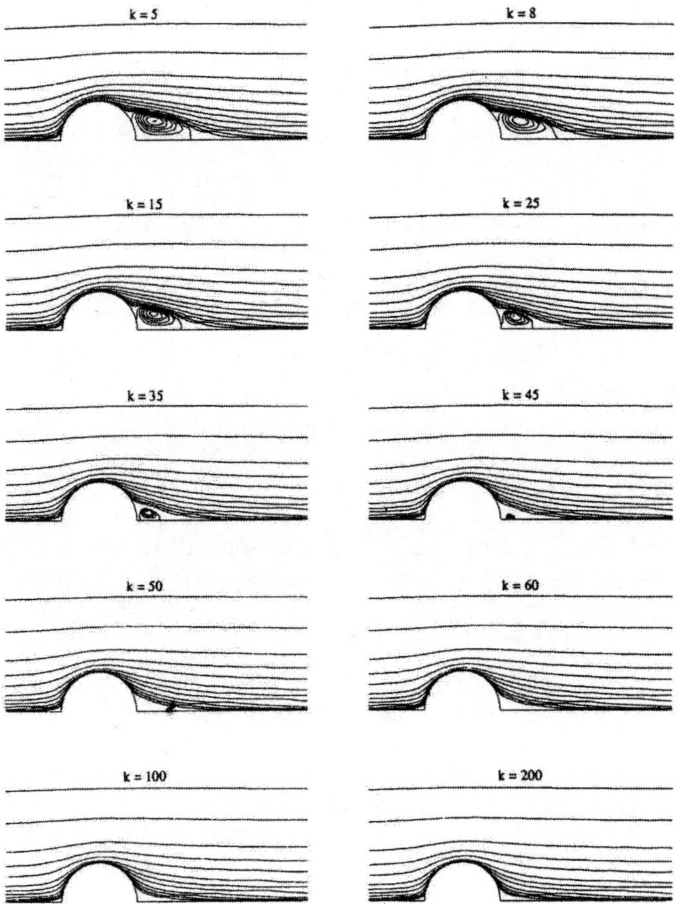

Figure 4: Flow around the bubble for Pe = 200, Ma = 5, $\chi = 1$ and Re = 50.

[2] Batchelor, G.K. 1967 *An Introduction to Fluid Dynamics.* Cambridge University Press, Cambridge, England.

[3] Bel Fdhila, R. and Duineveld, P.C. 1996 The effect of surfactants on the rise of a spherical bubble at high Reynolds and Peclet numbers. *Phys. Fluids*, **8**, pp. 310-321.

[4] Berger, S.A. 1988 Initial-value stability analysis of a liquid jet. *SIAM J. Appl. Math.*, **48**, pp. 973-991.

[5] Brenner, M.P., Lister, J.R. and Stone, H.A.

[6] Briggs, R.J. 1964 *Electron Stream Interaction With Plasmas.* MIT Press, Cambridge, MA.

[7] Clift, R., Grace, J.R. and Weber, M.E. 1978 *Bubbles, Drops and Particles.* Academic Press, New York.

[8] Chandrasekhar, S. 1961 *Hydrodynamic ad Hydromagnetic Stability.* Clarendon Press, Oxford.

[9] Chauhan, A., Maldarelli, C., Rumschitzki, D.S. and Papageorgiou, D.T. 1996 Temporal and spatial instability of an inviscid compound jet. *Rheol Acta*, **35**, pp. 567-583.

[10] Chauhan, A., Maldarelli, C., Papageorgiou, D.T. and Rumschitzki, D.S. 1999 Linear instability of a two-phase compound jet. IUTAM Symposium on Nonlinear Singularities in Deformation and Flow, D. Durbn and J.R.A. Pearson eds, Kluwer Academic.

[11] Chauhan, A., Maldarelli, C., Rumschitzki, D.S. and Papageorgiou, D.T. 2000 The capillary instability of a viscous compound jet. *J. Fluid Mech.*, in the press.

[12] Chen, J. and Stebe, K. 1997 Surfactant-induced retardation of the thermocapillary migration of a droplet. *J. Fluid Mech.*, **340**, pp. 35-60.

[13] Chen, Y.-J. and Steen, P. 1997 Dynamics of inviscid capillary breakup: Collapse and pinchoff of a film bridge. *J. Fluid Mech.*, **341**, pp. 245-267.

[14] Chorin,A.J. 1968 *Math. Comput.*, **22**, pp. 745-762.

[15] Day, R.F., Hinch, E.J. and Lister, J.R. 1998 Self-similar capillary pinchoff of an inviscid fluid. *Phys. Rev. Lett.*, **80**, p. 704.

[16] Denn, M.M. 1980 Drawing of liquids to form fibers. *Annu. Rev. Fluid Mech.*, **12**, p. 365.

[17] Drazin, P.G. and Reid, W.H. 1981 *Hydrodynamic Stability.* Cambridge University Press, London.

[18] Edwards, D.A., Brenner, H. and Wasan, D.T. 1991 *Interfacial Transport Processes and Rheology.* Butterworth-Heinemann Series in Chemical Engineering, Massachusetts.

[19] Eggers, J. 1993 Universal pinching of 3D axisymmetric free-surface flow. *Phys. Rev. Lett.*, **71**, pp. 3458-3460.

[20] Eggers, J. 1995 Theory of drop formation. *Phys. Fluids*, **7**, pp. 941-953.

[21] Eggers, J. 1997 *Rev. Modern Phys.*.

[22] Elzinga, E.R. and Banchero, J.T. 1961 Some observations on the mechanics of drops in liquid-liquid systems, *AIChE J.*, **7**, pp. 394-399.

[23] Ennis, B.J., Li, J., Tardos, G.I. and Pfeffer, R. 1990 The influence of viscosity on the strength of an axially strained pendular liquid bridge, *Chem. Eng. Sci.*, **45**, p. 3071.

[24] Entov and Hinch, E.J. 1997

[25] Felderhof, B.U. 1968 Dynamics of free liquid films. *J. Chem. Phys.*, **49**, pp. 44-51.

[26] Forest, M.G. and Wang, Q. 1990 Change-of-type behavior in viscoelastic slender jet models. *Theoret. Comput. Fluiud Dynam.*, **2**, pp. 1-25.

[27] Frumkin, A.N. and Levich, V.G. 1947 On surfactants and interfacial motion, *Zhur. Fiz. Khim.*, **21**, p. 1183. (In Russian.)

[28] 1955 Garner, F.H. and Skelland, H.P. 1955 Some factors afffecting droplet behavior in liquid-liquid systems, *Chem. Engng. Sci.*, **4**, pp. 149-158.

[29] Happel, J. and Brenner, H. 1962 *Low Reynolds Number Hydrodynamics*. Prentice Hall.

[30] Harper, J.F. 1973 On bubbles with small immobile adsorbed films rising in liquids at low Reynolds numbers. *J. Fluid Mech.*, **58**, pp. 539-545.

[31] He, Z., Dagan, Z. and Madarelli, C. 1991 The size of stagnant caps of bulk soluble surfactant on interfaces of translating fluid droplets. *J. Colloid and Interface Sci.*, **146**, pp. 442-451.

[32] Hertz, C.A. and Hermanrud, B. 1983 A liquid compound jet. *J. Fluid Mech.*, **131**, pp. 271-287.

[33] Huang, W.S. and Kintner, R.C. 1969 Effects of surfactants on mass transfeer inside drops, *AIChEJ.*, **15**, pp. 735-744.

[34] Johnson, T.A. ad Patel, V.C. 1999 Flow past a sphere up to a Reynolds number of 300. *J. Fluid Mech.*, **378**, pp. 19-70.

[35] Keller, J.B., Rubinow, S.L. and Tu, Y.O. 1973 Spatial instability of a jet. *Phys. Fluids*, **16**, pp. 2052-2055.

[36] Kim, H. and Subramanian, R. 1989a Thermocapillary migration of a droplet with insoluble surfactant, I. Surfactant cap, *J. Colloid and Int. Sci.*, **127**, pp. 417-430.

[37] Kim, H. and Subramanian, R. 1989b Thermocapillary migration of a droplet with insoluble surfactant, II. General case, *J. Colloid and Int. Sci.*, **130**, pp. 112-125.

[38] Kroger, R., Berg, S., Delgado, A. and Rath, H.J. 1992 Stretching behavior of large polymeric and Newtonian liquid bridges in plateau simulation. *J. Non-Newtonian Fluid Mech.*, **45**, p. 385.

[39] Lamb, H. 1932 *Hydrodynamics.* Cambridge University Press.

[40] Lee, H.C. 1974 Drop formation in a liquid jet. *IBM J. Res. Dev.*, **18**, pp.364-369.

[41] Levich, V.G. 1962 *Physicochemical Hydrodynamics.* Englewood Cliffs, New Jersey: Prentice Hall.

[42] Leib, S.J. and Goldstein, M.E. 1986 The generation of capillary instabilities on a liquiud jet. *J. Fluid Mech.*, **168**, pp. 479-500.

[43] Mansour, N.N. and Lundgren, T.S. 1990 Satellite formation in capillary jet breakup. *Phys. Fluids A*, **2**, pp. 1141-1144.

[44] McConnell, A.J. 1957 *Applications of Tensor Analysis.* Dover, New York.

[45] McKinley, G.H. and Tripathi, A. 2000 *J. Rheol.*, in the press.

[46] McLaughlin, J.B. 1996 Numerical simulation of bubble motion in water. *J. Colloid Interface Sci.*, **184**, pp. 614-625.

[47] Meseguer, J. 1983 The breaking of axisymmetric liquid bridges. *J. Fluid Mech.*, **130**, pp. 12-151.

[48] Middleman, S. 1995 *Modeling Axisymmetric Flows. Dynamics of Films, Jets ad Drops.* Academic Press, San Diego, CA.

[49] Nadim, A. and Bohran, A. 1989 Effects of surfactant on the motion and deformation of a droplet in thermocapillary migration, *PhysicoChemical Hydrodynamics*, **11**, pp. 753-764.

[50] Papageorgiou, D.T. 1995a On the breakup of viscous liquid threads. *Phys. Fluids*, **7**(7), pp. 1529-1544.

[51] Papageorgiou, D.T. 1995b Analytical description of the breakup of liquid jets. *J. Fluid Mech.*, **301**, pp. 109-132.

[52] Papageorgiou, D.T. 1996 Description of jet breakup. *Advances in Multi-Fluid Flows*, Coward, Papageorgiou, Renardy and Sun eds, SIAM Proceedings Series, pp. 171-198.

[53] Papageorgiou, D.T. and Orellana, O. 1998 Study of cylindrical jet breakup using one-dimensional approximations of the Euler equations. *SIAM J. Appl. Math.*, **59** No 1, pp. 286-317.

[54] Patankar, S.V. 1980 *Numerical Heat Transfer and Fluid Flow.* Hemisphere Publishers.

[55] Peyret, R. an Taylor, T.D. 1983 *Computational Methods for Fluid Flow.* Springer-Verlag.

[56] Pozrikidis, C. 1999

[57] Pozrikidis, C. 1996 *Introduction to the Boundary Integral Method.* Cambridge Texts in Applied Mathematics Serries, Cambridge Universiity Press.

[58] Raymond, D.R. and Zieminski, S.A. 1971 *AIChEJ*, **17**, pp. 57-65.

[59] Rayleigh, Lord 1879 On the instability of jets. *Proc. Lond. Math. Soc.*, **10**, p. 4.

[60] Rayleigh, Lord 1879 On the instability of a cylinder of viscous liquid under capillary forces. *Phil. Mag.*, **34**, p. 145.

[61] Renardy, M. 1994 Some comments on the surface-tension driven breakup (or lack of it) of viscoelastic jets. *J. Non-Newtonian Fluid Mech.*, **51**, p. 97.

[62] Richards, J.R., Lenhoff, A.M. and Beris, A.N. 1994 Dynamic breakup of liquid-liquid jets. *Phys. Fluids*, **6**, p. 2640.

[63] Ryskin, G. and Leal, L.G. 1982 Numerical solution of free boundary problems in fluid mechanics. *J. Fluid Mech.*, **148**, pp. 1-17.

[64] Sadhal, s. and Johnson, R. 1983 Stokes flow past bubbles and drops partially coated with thin films. Part 1. Stagnant cap of surfactant film - exact solution. *J. Fluid Mech.*, **126**, pp. 237-250.

[65] Sanz, A. and Meseguer, J. 1985 One-dimensional linear analysis of a compound jet. *J. Fluid Mech.*, **159**, pp. 55-68.

[66] Schulkes, R.M.S.M.S. 1993 Dynamics of liquid jets revisited. *J. Fluid Mech.*, **250**, pp. 635-650.

[67] Schulkes, R.M.S.M.S. 1993 Nonlinear dynamics of liquid columns. *Phys. Fluids*, **A 5**, pp. 2121-2130.

[68] Stone, H.A. 1990 A simple derivation of time dependent convective diffusion equation for surfactant transport along a deforming interface, *Phys. Fluids A*, **2**, p. 111.

[69] Stone, H.A. and Leal, L.G. 1989 Relaxation and breakup of an initially extended drop in an otherwise quiescent fluid. *J. Fluid Mech.*, **198**, p. 399.

[70] Subramanian, R.S. 1992 The motion of bubbles and drops in reduced gravity, in *Transport Processes in Bubbles, Drops and Particles*, Subramanian, Chabra and DeKee, Eds., Hemisphere Publishing, New York, NY.

[71] Sweet, R.G. 1964 Stanford University Technical Report No. **1722**.

[72] Takemura, F. and Yabe, A. 1999 Rising speed and dissolution rate of a carbon dioxide bubble in slightly contaminated water, *J. Fluid Mech.*, **378**, pp. 319-334.

[73] Tannerhill, C., Anderson, D.A. and Pletcher, R.H. 1997 *Computational Fluid Mechanics and Heat Transfer.* Taylor & Francis. Washington, DC.

[74] Taylor, G.I. 1959 The dynamics of thin sheets of fluid. II. Waves on fluid sheets. *Proc. Roy. Soc.*, **A 253**, pp. 296-312.

[75] Ting, L. and Keller, J.B. 1990 Slender jets and thin sheets with surface tension. *SIAM J. Appl. Math.*, **50**, pp. 1533-1546.

[76] Tjahjadi, M., Stone, H.A. and Ottino, J.M. 1992 Satellite and subsatellite formation in capillary breakup. *J. Fluid Mech.*, **243**, p. 297.

[77] Tomotika, S. 1935 On the instability of a cylindrical thread of a viscous liquid surrounded by another viscous liquid. *Proc. Roy. Soc.*, **A 150**, pp. 322-337.

[78] Tsamopoulos, J., Chen, T.-Y. and Borkar, A. 1992 Viscous oscillations of capillary bridges. *J. Fluid Mech.*, **235**, p. 579.

[79] Wang, Y. 1999 Theoretical study of bubble motion in surfactant solutions. *Ph.D. Thesis*, New Jersey Institute of Technology.

[80] Wang, Y., Papageorgiou, D.T. and Maldarelli, C. 1999 Increased mobility of aa surfactant-retarded bubble at high bulk concentrations, *J. Fluid Mech.*, **390**, pp. 251-270.

[81] Wang, Y., Papageorgiou, D.T. and Maldarelli, C. 2000 Using surfactants to control the formation and size of wakes behind moving bubbles at order one Reynolds numbers, *J. Fluid Mech.*, submitted.

[82] Wong, H., Rumschitzki, D.S. amd Maldarelli, C.

BENARD LAYERS WITH HEAT OR MASS TRANSFER

M.G. Velarde
Instituto Pluridisciplinar, UCM, Madrid, Spain

Abstract. I have taken the Benard liquid-layer set-up to succintly illustrate the variety of phenomena and mathematical problems that the Marangoni stress can originate already in the simplest case of (1+1)D geometry. This surface (tangential) stress may be due to heat or mass transfer creating a surface tension gradient that ultimately leads to flow motion and/or instability and subsequent convection (Marangoni effect). For the (1+1)D geometry I provide, first, under drastically simplifying approximations the conditions for the onset and nonlinear evolution of steady convection (as a caricature of Benard cells). The evolutionary problem is presented for two different effective gravity levels, one of them corresponding to microgravity (the kind of effective gravity existing in the free fall of an Earth's orbital space lab. or station). Then I present the theory when instability leads to oscillatory motions and waves. I give discussion of physical mechanisms leading to either capillary-gravity (transverse) waves or dilational (sound-like, longitudinal) waves, and even internal waves, and their possible resonance. For the transverse mode I discuss its nonlinear evolution by deriving a ("long wave", shallow layer) generalized (dissipation-modified) Boussinesq-Korteweg-de Vries equation. I discuss how solitary waves and solitons are expected to appear in the Benard layer. Finally, I comment on a few experimental results about dissipative solitons.

1 Evolution equations and boundary conditions

We consider the case of a horizontal liquid, layer confined between planes located at heights $z = 0$ and $z = d$, respectively. For simplicity we limit ourselves to the case $d \ll L$, where L is either of the two horizontal scales of the layer. Later on we shall even restrict consideration to a (1+1)D two-dimensional geometry thus disregarding one of the two horizontal scales. The layer is assumed to be heated or cooled from below (Bénard layer). In the simplest Newtonian and Boussinesquian approximations (to be defined more precisely) the evolution of the liquid layer is governed by the following balance equations:

$$\frac{\partial v_i}{\partial x_i} = 0, \tag{1}$$

$$\varrho^* \left(\frac{\partial v_i}{\partial t} + v_j \frac{\partial v_i}{\partial x_j}\right) = -\frac{\partial p}{\partial x_i} + \eta \nabla^2 v_i + \varrho g e_i, \tag{2}$$

$$\varrho^* c_p \left(\frac{\partial T}{\partial t} + v_j \frac{\partial T}{\partial x_j}\right) = -\frac{\partial (J_q)_i}{\partial x_i} \tag{3}$$

together with the equation of state

$$\varrho = \varrho^*[1 + \alpha(T^* - T)]. \tag{4}$$

The summation convention over repeated indices is assumed. Here an asterisk denotes some constant reference value taken, say, at the undisturbed interface. Besides, ϱ denotes density and p, pressure. v_i and T denote the i-th component of the velocity field and temperature, respectively, η is the shear or dynamic viscosity, $\eta = \varrho\nu$ with ν the kinematic viscosity, c_p the specific heat at constant pressure. g is the acceleration due to gravity and e the unit normal vector along z. α is the coefficient of thermal expansion. In eq. (3) we have also used the simplest caloric equation, of state between energy and temperature. To a first approximation the heat (J_q) flux is

$$(J_q)_i = \lambda \frac{\partial T}{\partial x_i}, \tag{5}$$

λ is the thermal conductivity of the liquid. Note that for simplicity we have used $\partial T/\partial x_i$ rather than $\partial(1/T)/\partial x_i$ and corresponding sign and T^2 factor, although the latter is the proper thermodynamic force.

As we consider the liquid layer open to the ambient air, we use the following equation of state for the surface tension, σ, at the upper plane (initially at $z = d$) taken as an open deformable surface (liquid-air interface)

$$\sigma = \sigma^* + (\partial\sigma/\partial T)(T - T^*). \tag{6}$$

As a consequence of the thermal gradient across the layer, and for moderately low values of ΔT, the fluid establishes itself in a motionless steady state denoted by the superscript s

$$v_i^s = 0, \tag{7}$$

$$T^s = T^0 - (\Delta T/d)z, \tag{8}$$

$$p^s = p^* - \varrho^* g(z - d) - \varrho^* g\alpha\Delta T(z - d)^2/2d \tag{9}$$

with $\Delta T = T^0 T^*$, where the superscript ^{0} denotes values at $z = 0$.

We are interested in the instability of the motionless steady state when the thermal constraint takes higher and higher, albeit controlled values. A reasonable assumption is that under increased constraint the thermohydrodynamic fields would tend to depart from their steady values (7)-(9) and the upper surface would not remain at $z = d$. The latter would move such that $z = d + \zeta(x,t)$ which shows inhomogeneity in the deformation, $\zeta(x,t)$, along the horizontal. For simplicity we restrict consideration to a two-dimensional problem (x, z) and disregard any y-dependence along the horizontal. Thus we exclude explicit consideration of hexagonal tessellation at the upper surface in the Bénard layer. The remaining quantities are $v_i = 0 + \delta v_i$ $(i = 1, 2)$, $T + \delta T$ and $p + \delta p$, where all disturbances depend on x, z and t. These disturbed fields also obey the original thermohydrodynamic equations. We have

$$\partial v_i/\partial x_i = 0, \tag{10}$$

$$\partial v_i/\partial t + v_j(\partial v_i/\partial x_i) = -(\varrho^*)^{-1}\partial\delta p/\partial x_i + \nu\nabla^2 v_i + \alpha\delta T g e_i, \tag{11}$$

$$\partial\delta T/\partial t + v_i(\partial\delta T/\partial x_i) = \kappa\nabla^2\delta T + (\Delta T/d)v_i e_i. \tag{12}$$

On the other hand, eq. (6) becomes

$$\sigma = \sigma^* + (\partial\sigma/\partial T)[\delta T - \zeta(\Delta T/d)] \tag{13}$$

with $\kappa = \lambda/\varrho^* c_p$ being the thermal diffusivity or thermometric conductivity of the liquid and $e_i = (0,1)$. If the problem (10)-(13) together with the appropriate boundary conditions (b.c.) has nontrivial solutions, then the steady state (7)-(9) will be unstable.

In accordance with the considerations given above, we shall consider the following b.c. The lower surface located at $z = 0$ is taken mechanically rigid, that is

$$v_i = 0 \quad \text{at} \quad z = 0. \tag{14}$$

Here heat is assumed to flow across the boundary following Newton's law of cooling (Robin, Biot or mixed condition). We have

$$\lambda\partial\delta T/\partial z = q^0\partial T \quad \text{at} \quad z = 0. \tag{15}$$

where q^0 is a parameter that accounts for the transfer characteristics of the boundary. The limits q^0 going to zero and to infinity, respectively, correspond to the cases of a poor and a perfectly conducting surface.

At the upper surface we have an interface liquid-air, and we assume that it follows the velocity field (no cavitation exists). Thus we have the kinematic b. c.

$$\partial\zeta/\partial t = N v_i n_i \tag{16}$$

We also have continuity in the pressure (stress) field

$$\begin{aligned}(-p^s - \delta p + p^*)n_i + \eta(\partial v_i/\partial x_j + \partial v_j/\partial x_i)n_j &= \\ = k\sigma n_i + (t_j\partial\sigma/\partial x_j)t_i, \quad i,j = 1,2.\end{aligned} \tag{17}$$

Note that writing eq. (17) we have tacitly assumed that the ambient air at the upper level is some kind of a large reservoir mechanically passive (air has a dynamic shear viscosity two orders of magnitude lower than that of standard liquids) and that the interface liquid-air although deformable is mechanically ideal. Thus we have

$$\begin{aligned}-n_i\delta p + n_i[\varrho^* g\zeta + \varrho^* g\alpha\Delta T\zeta^2/2d] + \eta n_j(\partial v_i/\partial x_j + \partial v_j/\partial x_i) &= \\ = k\sigma n_i + (t_j\partial\sigma/\partial x_j)t_i\end{aligned} \tag{18}$$

with σ given by eq. (13). We have introduced the following notation, $\mathbf{n}$ is the outward unit normal vector to the liquid-air surface, given by

$$\mathbf{n} = (-\partial\zeta/\partial\mathbf{x}, \mathbf{1})/\mathbf{N} \tag{19}$$

and the unit tangent vector $\mathbf{t}$ is

$$\mathbf{t} = (\mathbf{1}, \partial\zeta/\partial\mathbf{x})/\mathbf{N}. \tag{20}$$

The curvature k

$$k = N^{-3}\partial^2\zeta/\partial x^2 \tag{21}$$

with

$$N = [1 + (\partial\zeta/\partial x)^2]^{\frac{1}{2}}. \tag{22}$$

We also assume that temperature disturbances follow, at $z = d + \zeta(x,t)$, similar b.c. to (15). That is, for the temperature the b.c. becomes

$$\lambda n_i \partial(T + \delta T)/\partial x_i = -\lambda \Delta T/d - q^*(\delta T - \zeta \Delta T/d). \tag{23}$$

For universality in the argument we now rescale the variables using the following units: d for length, d^2/κ for time, κ/d for velocity, $\eta\kappa/d^2$ for pressure, ΔT for temperature and denote $\theta = \delta T/\Delta T$. With these units the system (10)-(12) becomes in dimensionless form

$$\partial v_i/\partial x_i = 0 \tag{24}$$

$$\mathrm{Pr}^{-1}(\partial v_i/\partial t + v_j \partial v_i/\partial x_j) = \partial \tau_{ij}/\partial x_j + \mathrm{Ra}\theta e_i \tag{25}$$

$$\partial\theta/\partial t + v_j(\partial\theta/\partial x_j) = \nabla^2\theta + w \tag{26}$$

with $\mathbf{v} = (u, w)$ and the stress tensor of the liquid $\tau_{ij} = -p\delta_{ij} + (\partial v_i/\partial x_j + \partial v_j/\partial x_i)$. δ_{ij} is the Kronecker delta. The following dimensionless groups have been introduced:

$$\text{Rayleigh number}: \ \mathrm{Ra} = \frac{\alpha g d^3 \Delta T}{\kappa\nu}, \tag{27}$$

and

$$\text{Prandtl number}: \ \mathrm{Pr} = \nu/\kappa. \tag{28}$$

At $z = 0$, the b.c. (14) and (15) become

$$u = w = 0, \tag{29}$$

$$\partial\theta/\partial z = \mathrm{Bi}^0\theta, \tag{30}$$

whereas at $z = 1 + \xi(\xi = \zeta/d)$, we have, from eq. (16),

$$\partial\xi/\partial t = N v_i n_i. \tag{31}$$

From the continuity of stress at the interface, eq. (18), we have

$$\begin{aligned}\tau_{ij}n_j = -(\mathrm{Bo}/C)\left(\xi + \tfrac{d_T}{2}\xi^2\right)n_i + (K/C)[1 - \mathrm{Ma}C(\theta - \xi)]n_i - \\ -\{t_j\partial[\mathrm{Ma}(\theta - \xi)]/\partial x_j\}\,t_i.\end{aligned} \tag{32}$$

Note that the term $\frac{K}{C}[1 - \mathrm{Ma}C(\theta - \zeta)]$ tends to avoid the vertical displacement (deformation) of the interface. For the temperature field, the boundary condition at the free surface, eq. (23), becomes

$$n_i\partial\theta/\partial x_i = -\mathrm{Bi}^*(\theta - \xi) + (1 - N)/n. \tag{33}$$

The dimensionless curvature of the interface, which appears in eq. (32), is given by

$$K = N^{-3}\partial^2\xi/\partial x^2 \tag{34}$$

and

$$\mathbf{n} = (-\partial\xi/\partial x, 1)/N, \tag{35}$$

$$\mathbf{t} = (1, \partial\xi/\partial x)/N, \tag{36}$$

$$N^2 = 1 + (\partial\xi/\partial x)^2. \tag{37}$$

Also, we have introduced the dimensionless groups

$$\text{Bond number}: \quad \text{Bo} = \varrho^* g d^2/\sigma^*, \tag{38}$$

$$\text{Capillary or crispation number}: \quad C = \eta\kappa/\sigma^* d \tag{39}$$

and

$$\text{Marangoni number}: \quad \text{Ma} = -\left(\frac{\partial\sigma}{\partial T}\right)\frac{\Delta T d}{\eta\kappa}. \tag{40}$$

The heat transfer groups (Bi^0 and Bi^*, with Bi$= qd/\lambda$) are all nonnegative. With the Bo number the effect of surface tension in making a meniscus (spherical) is compared with the effect of gravity in keeping the surface *flat* (equipotential) over length scales of order *d*.

In view of b. c. (32) and possible different experimental circumstances, like variable effective-g, another useful group is the ratio of the Bond and the capillary numbers

$$\text{Galileo number}: \quad \text{Ga} = \text{Bo}/C = g d^3/\kappa\nu. \tag{41}$$

Lastly, we note that the Boussinesquian approximation demands that

$$\alpha_T \equiv \alpha\Delta T = (\varrho^* - \varrho^0)/\varrho^* = \text{Ra}C/\text{Bo} = \text{Ra}/\text{Ga} \ll 1.$$

In standard liquids $\alpha \approx 10^{-3} - 10^{-4}\ \text{K}^{-1}$. Also $\rho^* g \chi_T d \ll 1$, with χ_T the isothermal compressibility, and $\alpha g d/c_p \ll 1$. Accordingly, temperature and pressure effects on parameters are neglected. Thus we have in mind moderate thermal gradients and moderate layer thicknesses. Note that when the Rayleigh number is taken into consideration together with surface interface deformation the Boussinesquian approximation demands Ra$\ll$ Ga. For large values of Ga(Ga $\to \infty$) deformation is not relevant. To help the reader we give now the b.c. (18) or (32) in explicit form. For the normal component

$$\begin{aligned} &p - \text{Ga}(\xi + \alpha_T\xi^2/2) + [1/C - \text{Ma}(\theta - \xi)](\partial^2\xi/\partial x^2)/N^3 = \\ &= 2[(\partial u/\partial x)(\partial\xi/\partial x)^2 - (\partial u/\partial z + \partial w/\partial x)(\partial\xi/\partial x) + (\partial w/\partial z)]/N^2 \end{aligned} \tag{42}$$

and for the tangential component

$$\begin{aligned} &\text{Ma}[\partial\theta/\partial x - \partial\xi/\partial x + (\partial\xi/\partial x)(\partial\theta/\partial z)] = \\ &= -\{(\partial u/\partial z + \partial w/\partial x)[1 - (\partial\xi/\partial x)^2] + 2(\partial w/\partial z - \partial u/\partial x)(\partial\xi/\partial x)\}/N \end{aligned} \tag{43}$$

with $u \equiv v_1$ and $w \equiv v_2$. Eq. (43) accounts for the surface tension gradient-driven stress (Marangoni stress; the eventual subsequent flow is the Marangoni effect) and was ultimately responsible for cellular convection in the seminal work by Bénard.

2 Scaling and stability analysis. Ground-based conditions and steady convection in (1+1)D geometry

We are not going to consider the most general problem. Rather for tutorial purposes we shall consider the simplest albeit significant case corresponding to one possibility with long wave disturbances in a (1+1)D geometry. Indeed, when the heat transfer across the horizontal boundaries of the liquid layer is low enough, it is known according to experiment, that the pattern at the onset of convective instability tends to be long horizontally elongated cells which dissipate less. We shall take advantage of this fact and set

$$\mathrm{Bi}^0 = \varepsilon^2 \ \text{ and } \ \mathrm{Bi}^* = 0. \tag{44}$$

which, defines a smallness and ordering parameter, ε. This is not the only possible case of long wave length instability but we shall not dwell on this question here. For a more complete study of this problem the reader is adviced to proceed to the resent monographs by Colinet *et al* (2001) and by Nepomnyashch *et al* (2001). Thus we now redefine the units of time and length in the appropriate manner. We set

$$\tau = \varepsilon^2 t, \tag{45}$$

$$X = \varepsilon^{1/2} x \tag{46}$$

and

$$Z = z. \tag{47}$$

Then we assume that

$$\mathrm{Ga} = g/\varepsilon, \tag{48}$$

i.e. we assume $\mathrm{Ga}\alpha_T = O(\varepsilon^0)$ for liquid layers not really thin. Tins is not the only scaling that can be introduced for the gravitational acceleration. Another case of interest is

$$\mathrm{Ga} = \varepsilon^0 g. \tag{49}$$

We shall later on comment on the use of (49), when discussing the case of very thin liquid layers or variable, and eventually low effective gravity or slightly altered free fall conditions.

Together with (45)-(48) we now assume the following expansions:

$$\xi = \varepsilon\xi_1 + \varepsilon^2\xi_2 + \ldots, \tag{50}$$

$$u = \varepsilon^{1/2}[u_0 + \varepsilon u_1 + \ldots] \equiv \varepsilon^{1/2}\tilde{u}, \tag{51}$$

$$w = \varepsilon[w_0 + \varepsilon w_1 + \ldots] \equiv \tilde{w}, \tag{52}$$

$$p = p_0 + \varepsilon p_1 + \ldots, \tag{53}$$

$$\theta = \theta_0 + \varepsilon\theta_1 + \ldots, \tag{54}$$

$$\mathrm{Ra} = R_0 + \varepsilon R_1 + \ldots \tag{55}$$

and

$$\mathrm{Ma} = M_0 + \varepsilon M_1 + \ldots \tag{56}$$

Note that we do not consider here any non-Boussinesquian effect thus restricting consideration to low enough values of d_T. For the time being we shall not bother about the restriction on the values taken by Ra. In due time we shall recall that Ra$\ll$ Ga, as earlier stated.

Then the disturbance equations (24)-(26) become

$$\partial\tilde{u}/\partial X + \partial\tilde{w}/\partial Z = 0, \tag{57}$$

$$\begin{aligned} &\varepsilon^2\partial\tilde{u}/\partial\tau + \varepsilon\tilde{u}\partial\tilde{u}/\partial X + \varepsilon\tilde{w}\partial\tilde{u}/\partial Z = \\ &= -\mathrm{Pr}\partial p/\partial X + \mathrm{Pr}(\varepsilon\partial^2\tilde{u}/\partial X^2 + \partial^2\tilde{u}/\partial Z^2), \end{aligned} \tag{58}$$

$$\begin{aligned} &\varepsilon^2\partial\tilde{w}/\partial\tau + \varepsilon^2\tilde{u}\partial\tilde{w}/\partial X + \varepsilon^2\tilde{w}\partial\tilde{w}/\partial Z = \\ &= -\mathrm{Pr}\partial p/\partial Z + \mathrm{Pr}(\varepsilon^2\partial^2\tilde{w}/\partial X^2 + \varepsilon\partial^2\tilde{w}/\partial Z^2) + \mathrm{PrRa}\theta, \end{aligned} \tag{59}$$

and

$$\varepsilon^2\partial\theta/\partial\tau + \varepsilon\tilde{u}\partial\theta/\partial X + \varepsilon\tilde{w}(\partial\theta/\partial Z - 1) = \varepsilon\partial^2\theta/\partial X^2 + \partial^2\theta/\partial Z^2 \tag{60}$$

i) *Zeroth-order problem (linear stability analysis).* Inserting the expansion (50)-(56) in these equations and keeping in mind that ε is an *ordering* parameter, the evolution equations must be satisfied identically whatever the value of ε. To the zeroth-order approximation we have

$$\frac{\partial u_0}{\partial X} + \frac{\partial w_0}{\partial Z} = 0, \tag{61}$$

$$\frac{\partial p_0}{\partial X} - \frac{\partial^2 u_0}{\partial Z^2} = 0, \tag{62}$$

$$\frac{\partial p_0}{\partial Z} - R_0\theta_0 = 0, \tag{63}$$

$$\frac{\partial^2\theta_0}{\partial Z^2} = 0 \tag{64}$$

together with the b.c. at $Z = 0$

$$u_0 = w_0 = \frac{\partial\theta_0}{\partial Z} = 0 \tag{65}$$

and at $Z = 1$

$$w_0 = 0, \tag{66}$$

$$p_0 = g\xi_1, \tag{67}$$

$$M_0\frac{\partial\theta_0}{\partial X} = -\frac{\partial u_0}{\partial Z}, \tag{68}$$

$$\frac{\partial\theta_0}{\partial Z} = 0 \tag{69}$$

Note that the h.c. are at $Z = 1 + \xi$. We shall be consistent, however, with the expansion procedure introduced and take for any function

$$\begin{aligned} &f(Z = 1 + \xi) = f(1 + \varepsilon\xi_1 + \varepsilon^2\xi_2 + \ldots) = \\ &= f(1) + \varepsilon\left(\tfrac{\partial f}{\partial Z}\right)\xi_1 + \varepsilon^2\xi_2\left(\tfrac{\partial f}{\partial Z}\right) + \tfrac{\varepsilon^2}{2}\xi_1^2\left(\tfrac{\partial^2 f}{\partial Z^2}\right) + O(\varepsilon^3). \end{aligned}$$

Thus incorporating the ε-expansion we are allowed to take the b.c. at $Z = 1$, according to the ordering indicated by the power of ε. Besides the following relation also holds:

$$\int_0^{1+\xi} f\mathrm{d}Z = \int_0 1 f\mathrm{d}Z + \varepsilon\xi_1 f(1) + O(\varepsilon^2).$$

The solution of (64) is

$$\theta_0 = F(X, \tau). \tag{70}$$

Then from (63) and (67)

$$p_0 = R_0 F[Z-1] + g\xi_1. \tag{71}$$

From (62) and (68) we get

$$u_0 = R_0 F' \left[\frac{Z^3}{6} - \frac{Z^2}{2} + \frac{Z}{2}\right] + g\xi_1' \left[\frac{Z^2}{2} - Z\right] - M_0 F' Z, \tag{72}$$

where the *dash* denotes a derivative with respect to X. From (61) we now get

$$w_0 = R_0 F'' \left[-\frac{Z^4}{24} + \frac{Z^3}{6} - \frac{Z^2}{4}\right] + g\xi_1'' \left[-\frac{Z^3}{6} - + \frac{Z^2}{2}\right] + M_0 F'' \frac{Z^2}{2}. \tag{73}$$

Using b.c. (66) we get

$$\frac{g}{3}\xi_1'' + F'' \left[\frac{M_0}{2} - \frac{R_0}{8}\right] = 0 \tag{74}$$

which is in fact $\left[\int_0^1 u_0 \mathrm{d}z\right]'$ and, as there is no net velocity of the liquid layer as a whole, the bracketed term vanishes. We have

$$\frac{g}{3}\xi_1' + F' \left[\frac{M_0}{2} - \frac{R_0}{8}\right] = 0 \tag{75}$$

Now we note that another relation can be established between ξ_1 and F, i.e. between ξ_1 and θ_0. This can be obtained from the energy equation integrated over the liquid depth. It plays the role of solvability condition for the ϵ-hierarchy of equations. To the lowest-order approximation in ϵ we get

$$F'' \left[1 + \frac{M_0}{6} - \frac{R_0}{20}\right] + \frac{g}{8}\xi_1'' = 0, \tag{76}$$

that together with (74) yields

$$\frac{R_0}{320} + \frac{M_0}{48} = 1, \tag{77}$$

which defines the line of neutral stability. Thus the zeroth-order problem gives

$$\theta_0 = F = -\chi\xi_1 + \phi(\tau), \tag{78}$$

$$p_0 = \chi R_0 \xi_1[-Z+1] + g\xi_1 + R_0\phi[Z-1], \tag{79}$$

$$u_0 = \chi\xi_1' \left[R_0 \left(-\frac{Z^3}{6} + \frac{Z^2}{2} - \frac{13}{20}Z\right) + 48Z\right] + g\xi_1' \left(\frac{Z^2}{2} - Z\right) \tag{80}$$

and

$$w_0 = \chi\xi_1'' \left[R_0 \left(\frac{Z^4}{24} - \frac{Z^3}{6} + \frac{13}{40} Z^2 \right) - 24Z^2 \right] + g\xi_1'' \left(-\frac{Z^3}{6} + \frac{Z^2}{2} \right) \quad (81)$$

with

$$\chi \equiv \frac{g}{72} \, \frac{1}{1 - R_0/120}.$$

ii) *First-order problem (nonlinear evolution of disturbances).* The next approximation in theε-expansion is

$$\frac{\partial u_1}{\partial X} + \frac{\partial w_1}{\partial Z} = 0, \quad (82)$$

$$\frac{\partial^2 u_1}{\partial Z^2} - \frac{\partial p_1}{\partial X} = -\frac{\partial^2 u_0}{\partial X^2} + \frac{1}{\mathrm{Pr}} \left[u_0 \frac{\partial u_0}{\partial X} + w_0 \frac{\partial u_0}{\partial Z} \right], \quad (83)$$

$$\frac{\partial p_1}{\partial Z} - R_0\theta_1 = R_1\theta_0 + \frac{\partial^2 w_0}{\partial Z^2}, \quad (84)$$

$$\frac{\partial^2 \theta_1}{\partial Z^2} = -\frac{\partial^2 \theta_0}{\partial X^2} + u_0 \frac{\partial \theta_0}{\partial X} - w_0 \quad (85)$$

together with the b.c. At $Z = 0$

$$u_1 = w_1 = \partial\theta_1/\partial Z = 0 \quad (86)$$

and at $Z = 1$

$$w_1 = \xi_1' u_0 - \xi_1 \frac{\partial w_0}{\partial Z}, \quad (87)$$

$$\frac{\partial u_1}{\partial Z} = -M_0 \frac{\partial \theta_1}{\partial X} + M_0\xi_1' - M_1 \frac{\partial \theta_0}{\partial X} - \xi_1 \frac{\partial^2 u_0}{\partial Z^2} - \frac{\partial w_0}{\partial X}, \quad (88)$$

$$p_1 = g\xi_2 + 2\frac{\partial w_0}{\partial Z} - \xi_1 \frac{\partial p_0}{\partial Z} \quad (89)$$

and

$$\frac{\partial \theta_1}{\partial Z} = 0. \quad (90)$$

We see that the left-hand sides of (82)-(85) and (61)-(64) are identical and will be so at all orders in ε. Thus instead of the energy relation (76) to solve the problem we could have used the standard Fredholm's alternative. Note also that in (89) we have ξ_2 as in (67) we had ξ_1.

A similar procedure to the method sketched for the zeroth-order problem, using again the energy equation to the appropriate order in ε, after elementary but lengthy algebra, provides

$$\begin{aligned} \theta_1 = {} & \chi^2(\xi_1')^2 \left[\left(\tfrac{Z^5}{120} - \tfrac{Z^4}{60} + \tfrac{Z^3}{120} \right) R_0 - 3Z^4 + 4Z^3 \right] + \\ & + \chi\xi_1'' \left[R_0 \left(-\tfrac{Z^6}{720} + \tfrac{Z^5}{300} - \tfrac{Z^4}{480} \right) + \tfrac{3}{5} Z^5 - Z^4 + \tfrac{Z^2}{2} \right] + \Gamma(X, \tau), \end{aligned} \quad (91)$$

$$\begin{aligned} p_1 = {} & \chi(\xi_1')^2 \left[\left(\tfrac{Z^6}{720} - \tfrac{Z^5}{300} + \tfrac{Z^4}{480} \right) R_0^2 + R_0 \left(-\tfrac{3Z^5}{5} + Z^4 \right) \right] + \\ & + \chi\xi_1'' \left[\tfrac{R_0^2}{120} \left(-\tfrac{Z^7}{42} + \tfrac{Z^6}{15} - \tfrac{Z^5}{20} \right) \right] + \\ & + R_0 \left[\left(\tfrac{Z^6}{10} - \tfrac{Z^5}{5} + \tfrac{Z^3}{3} - \tfrac{Z^2}{5} + \tfrac{Z}{20} \right) - 36Z^2 + 24Z \right] + \\ & + R_0 \Gamma Z + \Pi(X, \tau) + R_1 \phi(\tau) Z - R_1 \chi \xi_1 Z, \end{aligned} \quad (92)$$

$$
\begin{aligned}
u_1 = {} & \chi^2[(\xi_1')^2]' \left[\left(\tfrac{Z^8}{8\cdot7\cdot720} - \tfrac{Z^7}{7\cdot25\cdot72} + \tfrac{Z^6}{14400} \right) R_0^2 + R_0 \left(-\tfrac{Z^7}{70} + \tfrac{Z^6}{30} \right) \right] + \\
& + \chi\xi_1'' \left[R_0^2 \left(\tfrac{Z^9}{9\cdot8\cdot7\cdot720} + \tfrac{Z^8}{7\cdot14400} - \tfrac{Z^7}{7\cdot14400} \right) + \right. \\
& \left. + R_0 \left(\tfrac{Z^8}{8\cdot70} - \tfrac{Z^7}{210} + \tfrac{Z^5}{40} - \tfrac{Z^4}{30} + \tfrac{Z^3}{60} \right) - 6Z^4 + 8Z^3 \right] + \\
& + \tfrac{\chi^2}{\mathrm{Pr}}[(\xi')^2]' \left[\left(\tfrac{Z^8}{7\cdot32\cdot72} - \tfrac{Z^7}{42\cdot120} + \tfrac{37Z^6}{144000} - \tfrac{Z^5}{6000} + \tfrac{Z^4}{19200} \right) R_0^2 + \right. \\
& \left. + R_0 \left(-\tfrac{Z^7}{28} + \tfrac{29}{300} Z^6 - \tfrac{11}{100} Z^5 + \tfrac{Z^4}{20} \right) + \tfrac{36Z^6}{5} - \tfrac{72Z^5}{5} + 12Z^4 \right] + \\
& + \mathrm{Ra}\Gamma' \tfrac{Z^3}{6} + \Pi' \tfrac{Z^2}{2} - R_1 \xi_1' \tfrac{Z^3}{6} + VZ
\end{aligned} \tag{93}
$$

and

$$
\begin{aligned}
w_1 = {} & \chi^2[(\xi_1')^2]'' \left[\left(-\tfrac{Z^9}{72\cdot7\cdot720} + \tfrac{Z^8}{7\cdot72\cdot200} - \tfrac{Z^7}{7\cdot72\cdot200} \right) R_0^2 + R_0 \left(\tfrac{Z^8}{560} - \tfrac{Z^7}{210} \right) \right] + \\
& + \chi\xi_1'' \left[R_0^2 \left(\tfrac{Z^{10}}{72\cdot7\cdot7200} - \tfrac{Z^9}{18\cdot7\cdot7200} + \tfrac{Z^8}{16\cdot7\cdot7200} \right) + \right. \\
& \left. + R_0 \left(-\tfrac{Z^9}{72\cdot70} + \tfrac{Z^8}{24\cdot70} - \tfrac{Z^6}{240} + \tfrac{Z^5}{150} - \tfrac{Z^4}{240} \right) + \tfrac{6}{5} Z^5 - 2Z^4 \right] + \\
& + \tfrac{\chi^2}{\mathrm{Pr}}[(\xi_1')^2]'' \left[\left(-\tfrac{Z^9}{7\cdot4\cdot72\cdot72} + \tfrac{Z^8}{7\cdot8\cdot720} - \tfrac{37Z^7}{7\cdot72\cdot2000} + \tfrac{Z^6}{72\cdot500} - \tfrac{Z^5}{96000} \right) R_0^2 + \right. \\
& \left. + R_0 \left(\tfrac{Z^8}{7\cdot32} - \tfrac{29}{7\cdot300} Z^7 + \tfrac{11}{600} Z^6 - \tfrac{Z^5}{100} \right) - \tfrac{36}{7\cdot5} Z^7 + \tfrac{12}{5} Z^6 - \tfrac{12}{5} Z^5 \right] - \\
& - R_0 \Gamma'' \tfrac{Z^4}{24} - \Pi'' \tfrac{Z^3}{6} + R_1 \chi \xi_1'' \tfrac{Z^4}{24} - V' \tfrac{Z^2}{2}
\end{aligned} \tag{94}
$$

together with an equation for ξ_1 which comes from applying the solvability condition using the integrated energy equation (60) to order ε^2. This equation is

$$
\begin{aligned}
& \tfrac{\partial \xi_1}{\partial \tau} + \xi_1^{iv} + \left[-\tfrac{241}{11\cdot27\cdot35} q^2 + \tfrac{19\cdot2}{9\cdot105} q + \tfrac{1}{15} \right] + \\
& + \xi_1'' \left[\tfrac{1-q}{\chi} + \left(\tfrac{M_1}{48} + \tfrac{R_1}{320} \right) \right] + \xi_1 + [\xi_1^2]''[1+q] + \\
& + \chi[(\xi_1')^2]'' \left[-\tfrac{4}{7\cdot9} q^2 + \tfrac{397}{7\cdot90} q - \tfrac{13}{20} \right] + \\
& + \tfrac{\chi}{\mathrm{Pr}}[(\xi_1')^2]'' \left[-\tfrac{8}{7\cdot5\cdot9} q^2 + \tfrac{11\cdot19}{9\cdot7\cdot20} q - \tfrac{1}{5} \right] + \\
& + \chi^2[(\xi_1')^3]' \left[-\tfrac{16\cdot8}{35\cdot81} q^2 + \tfrac{8}{35\cdot3} q - \tfrac{48}{35} \right] = 0
\end{aligned} \tag{95}
$$

with $q = R_0/320$. Thus, to obtain the temperature, velocity and pressure fields, we must solve eq. (95) which is in fact the *nonlinear* evolution equation for the deformable liquid-air interface. Recall that, strictly speaking, in the Boussinesquian approximation we are not allowed to use (95) for arbitrary values of Ra as Ra$\ll$ Ga.

iii) *Nature of the bifurcation or the nonlinear steady convective state* The nonlinear evolution equation of the liquid-air interface, eq. (95), can be written in a more compact and clear form by rescaling the space coordinate. Taking

$$
\tilde{x} = \left[-\frac{241}{11\cdot27\cdot35} q^2 + \frac{38}{945} q + \frac{1}{15} \right]^{-\frac{1}{4}} X, \tag{96}
$$

eq. (95) becomes

$$
\frac{\partial \xi_1}{\partial \tau} + \xi_1^{iv} + 2m\xi_1'' + D[(\xi_1)^2]'' + E[(\xi_1')^2]'' + H[(\xi_1')^3]' = 0 \tag{97}
$$

with

$$
m = \frac{1}{2} \left[\frac{1-q}{\chi} + \left(\frac{M_1}{48} + \frac{R_1}{320} \right) \right] \left[\frac{-241}{11\cdot27\cdot35} q^2 + \frac{38}{945} q + \frac{1}{15} \right]^{-\frac{1}{2}}, \tag{98}
$$

which plays the role of the control parameter. Note that $S = \mathrm{Ma}/48 + \mathrm{Ra}/320$, formally, defines the critical line. The critical state, for $\mathrm{Bi}^0 = 0$, corresponds to $S_0 = 1$, eq. (77), whereas $S_1 = M_1/48 + R_1/320$ accounts for the departure from the critical value. Then S_1 gives the first-order correction when Bi^0 is very small albeit nonzero. For the physical acceptable states the condition $\mathrm{Ra} \ll \mathrm{Ga}$ must also be satisfied. The other quantities in eq. (97) are

$$D \equiv [1+q]\left[\frac{-241}{11\cdot 27\cdot 35}q^2 + \frac{38}{945}q + \frac{1}{15}\right]^{-\frac{1}{2}}, \tag{99}$$

$$E \equiv \chi\left\{-\tfrac{4}{63}q^2 + \tfrac{397}{630}q + \tfrac{1}{\mathrm{Pr}}\left[-\tfrac{8}{315} + \tfrac{209}{1260}q - \tfrac{1}{5}\right]\right\} \cdot \left[-\tfrac{241}{11\cdot 27\cdot 35}q^2 + \tfrac{38}{945}q + \tfrac{1}{15}\right]^{-1} \tag{100}$$

and

$$H \equiv \chi^2\left[-\frac{128}{35\cdot 81}q^2 + \frac{8}{105}q - \frac{48}{35}\right]\left[-\frac{241}{11\cdot 27\cdot 35}q^2 + \frac{38}{945}q + \frac{1}{15}\right]^{-1} \tag{101}$$

Note that in view of the scaling used, $\mathrm{Ga} = O(\varepsilon^{-1})$, deformation here, apparently, plays a minor role in the thermoconvective problem.

To follow the evolution of the convective branch to a first approximation we now assume infinitesimal disturbances. We set

$$\xi_1 \sim \exp[ik\tilde{x} + \omega\tau] \tag{102}$$

Thus, neglecting the nonlinear terms, we obtain the dispersion relation

$$\omega - 2mk^2 + k^4 + 1 = 0. \tag{103}$$

At the neutral states $\omega = 0$. Then when $\mathrm{d}\omega/\mathrm{d}k = 0$ we have the critical state. It follows from eq. (103)

$$m_c = 1 \quad \text{and} \quad k_c = 1. \tag{104}$$

For a slight departure of Bi^0 from zero, these are the corrections to the neutral line, eq. (77), and to the wave number, respectively.

The nature of the bifurcation in the neighborhood of the critical state, $m_c = 1$, can be studied by considering small departures around m_c. We set

$$m = m_c(1+\delta^2) \tag{105}$$

with δ a smallness parameter yet to be determined.

To be consistent we rescale the time

$$\tilde{\tau} = \delta^2\tau. \tag{106}$$

We now seek solutions of eq. (97) in the form

$$\xi_1 = \delta\zeta_1(\tilde{\tau},\tilde{x}) + \delta^2\zeta_2(\tilde{\tau},\tilde{x}) + \delta^3\zeta_3(\tilde{\tau},\tilde{x}) + O(\delta^4). \tag{107}$$

As in the previous sections δ is an ordering parameter. To first-order approximation we have

$$\xi_1 = A_1(\tilde{\tau})\cos\tilde{x}, \tag{108}$$

whereas to the second order

$$\zeta_2 = A_2(\tilde{\tau}) \cos(\tilde{x} + \Phi) + A_1^2(D - E)\frac{2}{9} \cos 2\tilde{x}. \tag{109}$$

The solvability condition (Fredholm alternative) for the third-order equation in the δ-hierarchy originated from the nonlinear eq. (97) implies the following relationship:

$$\frac{\mathrm{d}A_1}{\mathrm{d}\tilde{\tau}} = 2A_1 + A_1^3 \left[\frac{3}{4}H + \frac{2}{9}(D - E)(D + 2E)\right], \tag{110}$$

which is a form of Landau equation in the slow time scale. Therefore, when the coefficient of A_1^3 is negative, the bifurcation is direct and, when it becomes positive, the bifurcation is inverted and to follow the bifurcated branch we ought to proceed to the high-order equation in the hierarchy in order to get the *fifth*-order power in (110). The sign of this coefficient depends on the actual values of the parameters in an experiment.

3 The case of very thin liquid layers and/or microgravity conditions and steady convection

Until now we have restricted ourselves to situations where $G = O(\varepsilon^{-1})$, that may correspond to conditions on Earth and, formally, arbitrarily thick liquid layers. However, for experiments conducted aboard a spacecraft like the Shuttle or the Space Station (with variable and, eventually, very low effective-g) or with very thin liquid layers, G could be of order unity (ε^0). Under these circumstances, the appropriate scaling is (we still keep $\mathrm{Bi}^0 \equiv \varepsilon^2$)

$$\xi = \sum_{i=1}^{\infty} \varepsilon^i \xi_i, \tag{111}$$

$$u = \varepsilon^{\frac{1}{2}} \sum_{i=1}^{\infty} \varepsilon^i u_i, \tag{112}$$

$$w = \varepsilon \sum_{i=1}^{\infty} \varepsilon^i w_i, \tag{113}$$

$$P = \sum_{i=1}^{\infty} \varepsilon^i P_i, \tag{114}$$

$$\theta = \sum_{i=1}^{\infty} \varepsilon^i \theta_i, \tag{115}$$

$$\mathrm{Ra} = R_0 + \varepsilon R_1 + \ldots, \tag{116}$$

$$\mathrm{Ma} = M_0 + \varepsilon M_1 + \ldots. \tag{117}$$

Again we note that only rather small values of Ra can be considered if we disregard all non-Boussinesquian effects (λ_T small). Accordingly, Ra is kept for purely formal manipulations and to cross-check with results with no surface deformation.

Note that with respect to the former scaling (50)-(54) the main difference is that all the variables have increased one order in ε, except the surface deformation. The leading order of the product $\mathrm{Ga}\xi = O(\varepsilon)$, whereas formerly it was of order unity.

Inserting this scaling into eqs. (24)-(26) and boundary conditions at $z = 0$ (29), (30) and at $z = 1$ (31)-(33), (42), (43), we obtain again a hierarchy of linear equations in powers of ε. Solving the first-order equation and accounting for the solvability condition to the following higher order, we get

$$(M_0/48 + R_0/320 - 1)2\mathrm{Ga}/3 + M_0(1 - R_0/120) = 0, \qquad (118)$$

which defines the neutral line for the transition to long wavelength steady cellular convection in the low Biot number region. Recall that there is another longwave instability threshold of different origin. As expected, in the formal limit of Ga going to infinity, it coincides with the result obtained in the previous section, eq. (77), thus indicating that deformation there was of minor importance. Besides, for a stress-free upper surface ($\mathrm{Ma} = 0$) (and no deformation), the onset of buoyancy-driven instability is at Ra = 320. Note that the latter critical value does not depend on Ga, whereas, if buoyancy is negligible, the critical Marangoni number is

$$M_0 = 48/(1 + 72/\mathrm{Ga}), \qquad (119)$$

which shows a drastic dependence on Ga. In particular, in the limit of vanishing Ga, M_0 goes to zero and a vanishing small heating (no threshold exists) suffices to have convection in the liquid layer. This is due to the neglect of the hydrostatic part in the transverse b.c. (18) or (42).

Finally, to the lowest-order approximation in the hierarchy we also obtain

$$(1 - R_0/120)\theta_1(\varepsilon^{\frac{1}{2}}x, \varepsilon^2 t) = -\mathrm{Ga}\xi_1(\varepsilon^{\frac{1}{2}}x, \varepsilon^2 t)/72 + \phi(\varepsilon^2 t), \qquad (120)$$

that relates the first-order temperature disturbance θ_1 to the interfacial deformation ξ_1. One is tempted to say that one quantity tightly *slaves* the other. We obtain from eq. (120) that θ_1 is maximum where ξ_1 is minimum provided Ra< 120, which agrees with Benard's finding. However, for higher values of the Rayleigh number (*i.e.* Ra > 120) we have the opposite structure for buoyancy-driven convection.

Solving the second-order approximation and accounting for the corresponding solvability condition, we get the nonlinear evolution equation for the interfacial deformation

$$\begin{gathered} Q_1\partial\xi_1/\partial\tau + 2S_1Q_2\xi_1'' + Q_3\xi_1^{iv} + Q_4\xi_1 + \\ +\tfrac{1}{2}Q_5(\xi_1')^2 + Q_6[(\xi_1)^2]'' - \left(\phi + \tfrac{d\phi}{d\tau}\right)Q_7 - \phi\xi_1''Q_8 = 0 \end{gathered} \qquad (121)$$

with

$$\begin{gathered} S_1 = (1 - R_0/120 + \mathrm{Ga}/72)3M_1/2\mathrm{Ga} + \\ +[(\mathrm{Ga}/120 - 1)/(1 - R_0/120 + \mathrm{Ga}/72)]R_1/192. \end{gathered} \qquad (122)$$

S_1 plays the role of the control parameter (compare to (98)). For convenience to have a compact form (121) we have used

$$Q_1 \equiv (1 - 8q)(1 - 16q) + \mathrm{Ga}(1/8 - (q/9) + \mathrm{Ga}^2/216, \qquad (123)$$

$$Q_2 \equiv (1 - 8q/3 + \mathrm{Ga}/72)\mathrm{Ga}/6, \qquad (124)$$

$$Q_3 \equiv [24/5 - 64\ldots 46/315q + 29\ldots 8q^2/315 + 320q^3/693+ \\ +\mathrm{Ga}(1/45 + 38q/2835 - 241\ldots 64q^2/6237)] + (1-q)(1-8q/3)/3C, \tag{125}$$

$$Q_4 \equiv (\mathrm{Ga}/3 - 40q)\mathrm{Ga}/72, \tag{126}$$

$$Q_5 \equiv (1 - 8q/3)(40q - \mathrm{Ga}/3), \tag{127}$$

$$Q_6 \equiv (q-1)(1-8q/3)320q/6+ \\ +(-12 - 268q/3 + 64q^2/3)\mathrm{Ga}/(72 + (q+1)\mathrm{Ga}^2/216, \tag{128}$$

$$Q_7 \equiv \mathrm{Ga}/3 - 40q, \tag{129}$$

$$Q_8 \equiv (q-1)320q/3 \tag{130}$$

and

$$q \equiv R_0/320. \tag{131}$$

Equation (121) is now the analog to eq. (95) when $G = O(\varepsilon^0)$. The condition that volume must be conserved

$$\int_0^L \xi_1 \mathrm{d}X = 0 \tag{132}$$

with periodic boundary conditions at $X = 0$ and $X = L$ determines $\phi(\tau)$.

Now we rewrite eq. (121) in a still more compact form. We define

$$\tilde{x} = (Q_4/Q_3)^{1/4} X, \tag{133}$$

$$\tilde{t} = (Q_4/Q_1)\tau, \tag{134}$$

$$\zeta = \xi_1 Q_5 (Q_4 Q_3)^{-1/2}, \tag{135}$$

and then eq. (121) becomes

$$\frac{\partial \zeta}{\partial \tilde{t}} + 2m\zeta'' + \zeta^{iv} + \zeta + \frac{1}{2}(\zeta')^2 + D(\zeta^2)'' - E\left(\Phi + \frac{Q_4}{Q_1}\frac{\mathrm{d}\Phi}{\mathrm{d}\tilde{t}}\right) - H\Phi\zeta'' = 0 \tag{136}$$

with

$$m \equiv S_1 Q_2 (Q_4 Q_3)^{-\frac{1}{2}}, \tag{137}$$

$$D \equiv Q_6/Q_5, \tag{138}$$

$$E \equiv Q_5 Q_7 (Q_4^3 Q_3)^{-\frac{1}{2}}, \tag{139}$$

$$H \equiv Q_8 (Q_4 Q_3)^{-\frac{1}{2}}. \tag{140}$$

Using (132) we obtain

$$\Phi = \exp[-Q_1\tilde{t}/Q_4]\left[\int_0^{\tilde{t}} \frac{\mathrm{d}\tilde{t}}{2EL}\int_0^L \mathrm{d}\tilde{x} Q_1 (\partial\zeta/\partial\tilde{x})^2 \exp[Q_1\tilde{t}/Q_4]/Q_4 + \Phi_0\right]. \tag{141}$$

Equation (136) looks very much like eq. (97) and again, as in sect. 3, we study the onset of convection by considering infinitesimal disturbances. We set

$$\zeta \sim \exp[ik\tilde{x} + \omega\tilde{t}] \tag{142}$$

Thus, neglecting the nonlinear terms, we get the same dispersion relation eq. (103) with critical values $m_c = 1$ and $k_c = 1$, respectively.

The nature of the bifurcation can also be studied by introducing (105) and (106),

$$m = m_c(1 + \delta^2), \tag{143}$$

$$\tilde{\tau} = \delta^2 \tilde{t} \tag{144}$$

and seeking solutions of eq. (136) in the form (107). We set

$$\zeta = \delta\zeta_1(\tilde{\tau}, \tilde{x}) + \delta^2\zeta_2(\tilde{\tau}, \tilde{x}) + \delta^3\zeta_3(\tilde{\tau}, \tilde{x}) + O(\delta^4), \tag{145}$$

where again δ is an ordering parameter. To first-order approximation in δ

$$\zeta_1 = A_1(\tilde{\tau}) \cos\tilde{x} \tag{146}$$

and to the second order

$$\zeta_2 = A_2(\tilde{\tau}) \cos(\tilde{x} + \Phi) + A_1^2(\tilde{\tau}) \left[\frac{1}{8} + D\right] \frac{2}{9} \cos 2\tilde{x} \tag{147}$$

To obtain ζ_2 we need the solvability condition (Fredholm alternative) for the third-order equation, which as in (110) yields a Landau equation that here is

$$\frac{dA_1}{d\tau} = 2A_1 + A_1^3 \left[\left(\frac{1}{8} + D\right)(D - 1)\frac{2}{9} - \frac{H}{4E}\right]. \tag{148}$$

Then, as in the previous case, the sign of the coefficient of A_1^3 determines the nature of the bifurcation, direct or inverted (also called soft or hard, respectively). For instance if $q = 0$ ($R_0 = 0$ and buoyancy is set to zero) then Eq. (136) reduces to

$$\frac{\partial\zeta}{\partial t} + 2m\zeta'' + \zeta^{iv} + \zeta + \frac{1}{2}(\zeta')^2 + D(\zeta^2)'' - \frac{1}{2L}\int_0^L (\zeta')^2 dx = 0 \tag{149}$$

with periodic b. c., $\zeta(x+L,t) = \zeta(x,t)$; $D = \frac{1}{2}(1 - \frac{\mathrm{Ga}}{36})$, and thus from the sign of the coefficient of A_1^3 in the reduced form of eq. (148) follows that bifurcation is direct for Ga < 45 and inverted for Ga > 45. If, however, $q = 1$, buoyancy is retained but the Marangoni number is set to zero (no surface stresses exist) then bifurcation is direct for Ga < 60 while Ga > 60 it is inverted.

When $q = 0$, eq.(149) provides an extension of the Kuramoto-Sivashinsky equation to the case presented here. There is a new term, $D(\zeta^2)''$, due to the surface tension gradient-driven Marangoni stress.

4 Oscillatory instability and waves. Heuristic arguments

Let us now discuss what oscillatory motions and waves could be expected in a Bénard layer. To do this we focus on the various time scales involved in the problem. When the liquid layer is heated from the air side or open to suitable mass adsorption of a "light" component from a vapor phase above, with subsequent absorption in the bulk, a (stabilizing) thermal gradient inside the liquid layer and, consequently, a stably stratified layer is created. Then, the problem with Marangoni stresses, gravity and buoyancy, involves several time scales. On the one hand we have the viscous and thermal scales, $t_{vis} = d^2/\nu$, $t_{th} = d^2/\kappa$, respectively. There also exist two other time scales associated with gravity and surface tension (Laplace overpressure) that tend to suppress surface deformation: $t_{gr} = (d/g)^{1/2}$ and $t_{cap} = (\rho d^3/\sigma)^{1/2}$. The time scale related to the Marangoni effect is $t_{mar} = (\rho d^2/ \mid \sigma_T \beta \mid)^{1/2}$. There is also another time scale related to buoyancy due to the stratification imposed by the temperature gradient, $t_{st} = (1/ \mid \alpha\beta \mid g)^{1/2}$. Note that we shall use here $\beta > 0$ when heating the liquid layer from below (relative to our earlier definition, this is going to alter the sign of the Marangoni number to be redefined soon) and that α is positive for standard liquids. The various ratios between time scales, and, accordingly, the ratios between forces involved in the dynamics, provide the earlier defined dimensionless groups (Prandtl, Marangoni, Rayleigh, Galileo and (static) Bond numbers, respectively). We have Pr $=t_{th}/t_{vis} = \nu/\kappa$, Ma$=(d\sigma/dT)t_{th}t_{vis}/t_{mar}^2 = (d\sigma/dT)\beta d^2/\eta\kappa$, Ra $= t_{th}t_{vis}/t_{st}^2 = \alpha\beta g d^4/\nu\kappa$, Ga$=t_{th}t_{vis}/t_{gr}^2 = gd^3/\nu\kappa$, and Bo$=t_{cap}^2/t_{gr}^2 = gd^2/\sigma$.

These time scales are not always of the same quantitative order. For example, for the simplest problem treated by Pearson (1958) when dealing with a liquid layer with undeformable surface, we have Ma$\approx$1, but Ga$\gg$1 and Ra$\ll$Ma. Indeed, although Pearson neglected gravity his assumption of undeformability was practically equivalent to gravity been able to keep the surface level, whatever flows and thermal inhomogeneities exist. The characteristic time scale of the problem is $t_{th} \approx t_{vis} \approx t_{mar}$(at Pr$\approx$ 1). For monotonic instability and hence the case leading to Benard cells when heating the liquid layer from the liquid side there exists a finite limit of the critical Marangoni number as Ga$\rightarrow \infty$. No oscillatory instability appears in the single-layer problem with undeformable surface. If such instability is possible the critical Marangoni number tends to infinity with Ga$\rightarrow \infty$. Thus the critical Marangoni number should better be scaled with Ga, as Ga becomes very large.

For high enough values of Ga the oscillatory mode is the capillary-gravity wave. The time scales t_{gr}, and t_{cap}, associated with this twofold wave are much smaller than the viscous and thermal timescales (at least for Pr$\approx$1, Bo$\approx$1). Then dissipative effects are relatively weak and the dispersion relation is

$$\omega^2 = \mathrm{GaPr}[1 + k^2/\mathrm{Bo}] \ \tanh(k) \tag{150}$$

(to nondimensionalize ω the thermal time scale is used hereafter, k is the dimensionless wave number in units of d^{-1}). Clearly, the higher is Ga (and the wave frequency), the stronger should be the work of the Marangoni stresses (i. e. the higher must be the critical Marangoni number) to excite and sustain capillary-gravity waves. Accordingly, for a standard liquid, $d\sigma/dT < 0$, this instability indeed appears when heating the liquid layer from the air side (Ma$<$ 0).

Let us now focus on the viscous, longitudinal or dilational mode. When a liquid element rises to the surface, it creates a cold spot there. Then, the surface tension gradient acts towards this spot, pushing the element back to the bulk, hence overstability. High values of Ma (in absolute value) ensure that the oscillations exist as earlier stated. Let their characteristic time scale be also t_{mar}. Calculation yields the following expression for the frequency of the longitudinal wave (in the limit Ma$\rightarrow -\infty$):

$$\omega^2 = -\mathrm{Ma}[\mathrm{Pr}/(\mathrm{Pr}^{1/2} + 1)]k^2. \tag{151}$$

Although this dilational wave is intrinsically dissipative, the damping rate is asymptotically smaller, $O(|\mathrm{Ma}|^{1/4})$, than its frequency. Up to some extent the flow field accompanying the dilational wave is qualitatively similar to that of the capillary-gravity wave. Potential flow can be assumed in the bulk of the layer, while vorticity is present only in boundary layers at the bottom rigid plate and at the upper free surface. The boundary layer thickness is of order of $O(|\mathrm{Ma}|^{-1/4})O(\mathrm{Ga}^{-1/4})$. For the dilational wave the horizontal velocity field in the surface boundary layer is stronger than the potential flow in the bulk [by $O(|\mathrm{Ma}|^{1/4})$] at variance with the capillary-gravity wave. Thus, for the viscous wave the flow motion is really concentrated near the surface. Noteworthy is that with an undeformable surface ($1 \ll |\mathrm{Ma}| \ll \mathrm{Ga}$), the dilational mode is always damped as earlier stated thus justifying that no oscillatory instability was expected in the one layer Marangoni problem without surface deformability. However, if the dilational wave is accompained by non-negligible surface deformation ($|\mathrm{Ma}| \geq \mathrm{Ga}$), it can be amplified, a striking result. Accordingly, at $\mathrm{Ga} \gg 1$, two tightly coupled thresholds for oscillatory Marangoni instability are expected with corresponding two (high-frequency) wave modes, capillary-gravity and dilational wave motions. As already said, to sustain the longitudinal wave one needs surface deformability. Alternatively, to sustain a capillary-gravity wave one needs the Marangoni stress. The most dramatic manifestation of the tight coupling occurs at resonance, when the frequencies are equal to each other. Near resonance there is mode-mixing. Namely, the capillary-gravity mode in the parameter half-space from one side of the resonance manifold is swiftly converted into the dilational one when crossing the manifold, and vice versa. Another feature of resonance is that the damping/amplification rates are drastically enhanced here, namely, $O(\mathrm{Ga}^{3/5})$ versus $O(\mathrm{Ga}^{1/4})$ far from resonance.

In view of the above given arguments, a thoroughly detailed analytical study of all these oscillatory instabilities demand due consideration of the role of the boundary layers near the solid bottom and the open surface or interface. Indeed, taking the Galileo number rather large, the layer can be divided into three portions: two boundary layers, one at the bottom where there is no slip and the other at the open surface, while the bulk of the layer can be well approximated by inviscid flow. Because the threshold for overstability is expected at relatively high values of the Marangoni number, one can use an inverse of a modified Marangoni number, $1/m = \mathrm{Ga}/\mathrm{Ma}$. Approximate analytical results and exact computer calculations show that the marginal curves found delineate a bag or bubble-like region whose lower boundary corresponds to the onset of capillary-gravity waves while the upper one gives the dilational waves. At a critical value of m, the stable dilational mode becomes the stable capillary-gravity mode.

Now let us turn to yet another oscillatory mode in the liquid layer. Indeed, if the liquid layer is deep enough, and hence stratified there is also the possibility of internal waves of frequency given by the Brunt- Väisälä frequency

$$\omega^2 = -\mathrm{RaPr}\frac{k^2}{k^2 + \pi^2 n^2} \quad (n = 1, 2, ...). \tag{152}$$

Disregarding capillary-gravity waves, hence surface deformation, the possibility exists of coupling dilational to internal waves with $|\mathrm{Ra}| \ll \mathrm{Ga}$. This may be called the Rayleigh-Marangoni problem and it is the natural extension to overstable motions of Nield's (1964) study of monotonic instability. The marginal curves are again in the form of closed bags or bubbles with the upper boundary providing the threshold for internal waves and the lower one for dilational (surface) waves. The bubbles collapse when lowering in (absolute value) the Rayleigh number. In the absence of the Marangoni effect, no unstable oscillatory motion persists undamped, which again stresses the crucial role played by the coupling of the two wave disturbances. Note that in this

Rayleigh-Marangoni case we have a countable number ($n = 1, 2, ...$) of internal wave modes, the dilational wave can be coupled to each of them and, hence, to a countable number of marginal stability conditions. Relative to the earlier case of coupling capillary-gravity to dilational waves the form of the marginal curves is qualitatively different. Furthermore, there exists the minimally possible Rayleigh number (in absolute value), below which there is no oscillatory instability. No such bound has been found for the Galileo number in the other case (at least in the region where Ga remains high). For further details, in more appropriate context, about the three wave motions discussed above together with a comparison with available experimental data the reader is adviced to proceed to Linde *et al* (2002).

5 Nonlinear surface waves in Benard layers in (1+1)D geometry

5.1 Asymptotic approach

The nonlinear evolution of either capillary-gravity/transverse or dilational/longitudinal surface waves poses formidable tasks. Let us then concentrate for one of the two possible waves on a simplified analysis although amenable to experimental test. Consider the horizontal liquid layer open to air and heated from above where "long" (a term to be made precise later on) capillary-gravity waves can be excited. The liquid layer is placed on a flat rigid support but the air or gas layer is bounded from above by a flat rigid top which fits well with experimental set-ups. For simplicity, we assume that the layers are of infinite horizontal extent and treat the problem again in the (1+1)D two-dimensional geometry. At rest, there exists a linear vertical temperature distribution.

Let h_j denote vertical depth and ρ_j density, ν_j kinematic viscosity, χ_j thermal diffusivity and κ_j thermal conductivity, where the subscripts $j = 1, 2$ refer to the liquid and gas layers, respectively. The corresponding symbols without subscript denote ratios: $\rho \equiv \rho_2/\rho_1$, $\nu \equiv \nu_2/\nu_1$, $\chi \equiv \chi_2/\chi_1$, and $\kappa \equiv \kappa_2/\kappa_1$. We assume that h is of order unity, while ν and χ are large enough, and ρ and κ are smaller than unity, in accordance with standard gas and liquid properties. The ratio of the dynamic viscosities, $\rho\nu$, is also small enough. The Prandtl number for both the liquid, $\mathrm{Pr} \equiv \nu_1/\chi_1$, and the gas, $P = \nu_2/\chi_2$, are also taken of order unity.

As we shall only consider "long" enough waves let us define a smallness parameter ϵ as the ratio of the depth to a characteristic wavelength. Then, there are two time scales in the problem. One of them is defined by heat diffusion, $t_{\mathrm{th}} = h_1^2/\chi_1$ (as Pr is assumed of order unity, the viscous time scale, $t_{\mathrm{vis}} = h_1^2/\nu_1$ is of order of the thermal one). The other time scale, $t_{\mathrm{gr}} = \epsilon(h_1/g^{1/2})$ is associated with "long" gravity waves, as g is the gravity acceleration. When $t_{\mathrm{th}} \ll t_{\mathrm{gr}}$ the heat and viscous effects are predominant which, in practice, corresponds to very shallow liquid layers or microgravity conditions. In the opposite situation, $t_{\mathrm{th}} \gg t_{\mathrm{gr}}$, the dissipation is limited to the boundary layers at the bottom and, if the Marangoni effect is significant, at the upper surface. In terms of the Galileo number, $\mathrm{Ga} = t_{\mathrm{th}}t_{\mathrm{vis}}/t_{\mathrm{gr}}^2$, the first case corresponds to $\mathrm{Ga}\epsilon^2 \ll 1$, while in the second $\mathrm{Ga}\epsilon^2 \gg 1$. We shall consider $\mathrm{Ga} \gg 1$.

The thickness d of the boundary layers can be estimated as follows. We have $t_{\mathrm{gr}} \approx t_{\mathrm{th}}$ where t_{gr} is as earlier defined while for t_{th} we here have $t_{\mathrm{th}} = d^2/\chi_1$. Then we get

$$\frac{d}{h_1} \approx \frac{1}{\epsilon^{1/2}\mathrm{Ga}^{1/4}}. \tag{153}$$

We define dimensionless quantities using suitable scales: h_1 for length, $(gh_1)^{1/2}$ for velocity, $(h_1/g)^{1/2}$ for time, $\rho_1 gh_1$ for pressure in the liquid, $\rho_2 gh_1$ for pressure in the air, βh_1 for temperature in the liquid, $\kappa^{-1}\beta h_1$ for temperature in the air, where β is again the vertical temperature gradient in the liquid layer but with the convention $\beta > 0$ corresponds to heating from above and $\beta < 0$ if heating is from below. By pressure and temperature we denote deviations of the corresponding quantities from their stationary distribution, linear with the vertical coordinate.

Let x and z be the horizontal and vertical coordinates, respectively. The bottom of the layer is taken at $z = -1$, the free surface at $z = \eta(x,t)$, and the top of the air layer at $z = h$, where t is time and $\eta(x,t)$ describes the surface deformation. Thus as already mentioned we restrict consideration to (1+1)D flow motions. To search for only "long" traveling wave motions in a shallow layer we redefine the horizontal variable, $\xi = \epsilon(x - Ct)$, where C is a phase velocity to be determined. In addition we scale horizontal velocity, pressure and deformation of the surface η with ϵ^2, vertical velocity with ϵ^3 and introduce the slow time scale $\tau = \epsilon^3 t$. The scale for temperature is determined by the leading convective contribution to the temperature field which is of order ϵ^2. Accordingly, the equations governing "long" wave disturbances are

$$u\xi + w_z = 0 \tag{154}$$

$$\epsilon^2 u_\tau - Cu_\xi + \epsilon^2 wu_z = -p_\xi + \epsilon^{-1}\left(\frac{\mathrm{Pr}}{\mathrm{Ga}}\right)^{1/2}\left(\epsilon^2 u_{\xi\xi} + u_{zz}\right), \tag{155}$$

$$\epsilon^4 w_\tau - \epsilon^2 Cw_\xi + \epsilon^4 uw_\xi + \epsilon^4 ww_z = -p_z + \epsilon\left(\frac{\mathrm{Pr}}{\mathrm{Ga}}\right)^{1/2}\left(\epsilon^2 w_{\xi\xi} + w_{zz}\right), \tag{156}$$

$$\epsilon^2 T_\tau - CT_\xi + \epsilon^2 uT_\xi + \epsilon^2 wT_z + w = \epsilon^{-1}\frac{1}{(\mathrm{Pr\,Ga})^{1/2}}\left\{\epsilon^2 T_{\xi\xi} + T_{zz}\right), \tag{157}$$

$$U_\xi + W_z = 0, \tag{158}$$

$$\epsilon^2 U_\tau - CU_\xi + \epsilon^2 UU_\xi + \epsilon^2 WU_z = -\Pi_\xi + \epsilon^{-1}\left(\frac{\mathrm{Pr}}{\mathrm{Ga}}\right)^{1/2}\nu\left\{\epsilon^2 U_{\xi\xi} + U_{zz}\right), \tag{159}$$

$$\epsilon^4 W_\tau - \epsilon^2 CW_\xi + \epsilon^4 UW_\xi + \epsilon^4 WW_z = -\Pi_z + \epsilon\left(\frac{\mathrm{Pr}}{\mathrm{Ga}}\right)^{1/2}\nu\left(\epsilon^2 W_{\xi\xi} + W_{zz}\right), \tag{160}$$

$$\epsilon^2 \theta_\tau - C\theta_\xi + \epsilon^2 U\theta_\xi + \epsilon^2 W\theta_z + W = \epsilon^{-1}\left(\frac{\mathrm{Pr}}{\mathrm{Ga}}\right)^{1/2}\frac{\nu}{P}\left(\epsilon^2\theta_{\xi\xi} + \theta_{zz}\right), \tag{161}$$

with the boundary conditions:
at $z = -1$:

$$u = w = T = 0, \tag{162}$$

at $z = h$:

$$U = W = \theta = 0, \tag{163}$$

and at $z = \eta(\xi, \tau)$:

$$\begin{aligned} &w = \epsilon^2\eta_\tau - C\eta_\xi + \epsilon^2 u\eta_\xi = W, \\ &u = U, \\ &p = \eta - \epsilon^2\left[\tfrac{1}{B} - \epsilon^2\tfrac{\mathrm{Ma}}{\mathrm{Ga}}(T + \eta)\right]\tfrac{\eta_{\xi\xi}}{N^3} + \\ &+ \tfrac{2}{N^2}\epsilon\left(\tfrac{\mathrm{Pr}}{\mathrm{Ga}}\right)^{1/2}\left[w_z - \epsilon^2\eta_\xi\left(u_z + \epsilon^2 w_\xi\right) + \epsilon^6 u_\xi\eta_\xi^2\right], \end{aligned} \tag{164}$$

$$\left(u_z + \epsilon^2 w_\xi\right)\left(1 - \epsilon^6 \eta_\xi^2\right) + 2\epsilon^4 \eta_\xi \left(w_x - u_\xi\right) + \frac{\mathrm{Ma} N \epsilon}{(\mathrm{Pr\,Ga})^{1/2}} \left(\eta_\xi + T_\xi + \epsilon^2 \eta_\xi T_z\right) = 0\,, \tag{165}$$

$$\eta + \theta = 0\,, \tag{166}$$

$$T_z - \epsilon^4 \eta_\xi T_\xi = \theta_z - \epsilon^4 \eta_\xi \theta_\xi\,, \tag{167}$$

with

$$\mathrm{Ma} \equiv -\frac{d\sigma}{dT}\frac{\beta h_1^2}{\rho_1 \nu_1 \chi_1}\,; \quad \mathrm{Bo} \equiv \frac{\rho_1 g h_1^2}{\sigma}\,; \quad N \equiv \left(1 + \epsilon^6 \eta_\xi^2\right)^{1/2}.$$

Here u, w, p and T denote the horizontal and vertical velocity components, pressure and temperature fields in the liquid layer. U, W, Π and θ denote the corresponding fields in the air layer. Note that we define Ma and Bo with the liquid layer properties. They are the Marangoni and (static) Bond numbers, respectively. Bo is assumed of order unity, while Ma is taken large enough as, indeed, the Marangoni effect is leading to instability past a (high enough) threshold [5,14]. The scaling of Ma with ϵ will be provided when solving the problem. The dynamic properties of air are neglected in the normal (164) and tangential (165) stress balances. The smallness of κ permitted to write the boundary condition representing continuity of temperature across the surface, $T+\eta = \kappa^{-1}(\eta+\theta)$ in the form (166). Thus, κ as well as ρ disappear from the equations.

5.2 Evolution equation for transverse (capillary-gravity) waves

Now let us discuss what should be the relation between the smallness parameters ϵ and Ga^{-1}. As long as we limit ourselves to the case $\epsilon^{-1/2}\mathrm{Ga}^{-1/4} \ll 1$, i.e. to the case when the liquid layer can be subdivided into the bulk where the flow is potential and the boundary layers, the most interesting asymptotics corresponds to the case when the boundary layer thickness and the deformation of the surface are of the same order. From (153) follows that $\epsilon^{-1/2}\mathrm{Ga}^{-1/4} \approx \epsilon^2$, i.e.

$$\epsilon = \mathrm{Ga}^{-1/10}\,. \tag{168}$$

(We write "=" to define ϵ in terms of Ga). Then the effects of energy output (due to heat and viscous dissipation) and input (due to the Marangoni effect) will be of the same order as nonlinearity and dispersion. The latter two are in appropriate (local) balance for the Bounssinesq-Korteweg-de Vries (BKdV) equation for long waves in shallow inviscid liquid layers.

Turning to the equations in the air layer, we assume that due to its relatively large kinematic viscosity ($\nu \gg 1$) and thermal diffusivity ($\chi \gg 1$), inertia in the air is no more dominating over dissipative effects. Then in the most general case

$$a^2 \equiv \epsilon^{-1}\,\mathrm{Pr}^{1/2}\,\mathrm{Ga}^{-1/2}\nu \approx 1\,.$$

The coefficients of the Laplacians in Eqs. (159,161) are a^2 and a^2/P, respectively. From (168), in the liquid layer, Eqs. (155,157), it follows that they are ϵ^4 and $\epsilon^4/\,\mathrm{Pr}$, respectively.

To solve the problem (154–167), all components of $f = (u, w, p, T, U, W, \Pi, \theta)$ are suitably expanded with ϵ (η is not yet expanded here). In the bulk of the liquid layer and in the air layer we set

$$f(\xi, z; \tau) = f_0 + \epsilon^2 f_1 + \ldots\,, \tag{169}$$

while in the boundary layer near the open surface

$$f(\xi, \bar{z}, \tau) = \bar{f}_0 + \epsilon^2 \bar{f}_1 + \dots , \tag{170}$$

and in the bottom boundary layer

$$f(\xi; \tilde{z}, \tau) = \tilde{f}_0 + \epsilon^2 \tilde{f}_1 + \dots , \tag{171}$$

where new vertical coordinates have been introduced:

$$\bar{z} \equiv -\frac{z}{\epsilon^2} \quad \text{and} \quad \tilde{z} \equiv \frac{z+1}{\epsilon^2}$$

in both the surface and bottom boundary layers.

After substituting (169–171) into (154–167) we get a hierarchy of linear problems corresponding to ϵ^n ($n = 0, 2, \dots$). At each step a solvability condition must be satisfied, that eventually yields an evolution equation for η. Without loss of generality, we restrict consideration to solutions with zero mean value of η.

In the boundary layers, Eqs. (154,156) yield $\bar{p}_{0\bar{z}} = \bar{w}_{0\bar{z}} = \tilde{p}_{0\tilde{z}} = \tilde{w}_{0\tilde{z}} = 0$, hence these functions are constants. Thus BC (162,164,164) can be directly applied to p_0 and w_0. Finally, the leading order solution in the bulk coincides with that in the inviscid liquid case:

$$p_0 = \eta , \quad u_0 = \eta , \quad w_0 = -\eta_\xi (1 + z) , \tag{172}$$

with $C = 1$. For simplicity here we consider only the right propagating wave ($C > 0$). The results for the left propagating wave ($C = -1$) can then be deduced by symmetry.

At the same time we can write

$$\bar{p}_0 = \tilde{p}_0 = \eta , \quad \bar{w}_0 = -\eta_\xi , \quad \tilde{w}_0 = 0 . \tag{173}$$

For the horizontal velocity in the bottom boundary layer, Eqs. (155,173) yield

$$-\tilde{u}_{0\xi} = -\eta_\xi + \tilde{u}_{0\tilde{z}\tilde{z}} .$$

Taking BC (162) into account, the solution is

$$\tilde{u}_0 = \eta - \frac{\tilde{z}}{2\pi^{1/2}} \int_\xi^\infty \frac{\eta(\xi')}{(\xi' - \xi)^{3/2}} \exp\left[-\frac{\tilde{z}^2}{4(\xi' - \xi)} \right] d\xi' .$$

Then using Eq. (154) and BC (162) $\tilde{w}_1$ is:

$$\tilde{w}_1 = -\int_0^{\tilde{z}} \tilde{u}_{0\xi}(\tilde{z}') d\tilde{z}' = -\eta_\xi \tilde{z} + \frac{1}{\pi^{1/2}} \frac{d}{d\xi} \int_\xi^\infty \frac{\eta(\xi')}{(\xi' - \xi)^{1/2}} \left\{ 1 - \exp\left[-\frac{\tilde{z}^2}{4(\xi' - \xi)} \right] \right\} d\xi' .$$

The matching condition between $\tilde{w}_0 + \epsilon^2 \tilde{w}_1$ and $w_0 + \epsilon^2 w_1$ yields the boundary condition for the function w_1 in the bulk at $z = -1$,

$$w_1 = \frac{1}{\pi^{1/2}} \frac{d}{d\xi} \int_\xi^\infty \frac{\eta(\xi')}{(\xi' - \xi)^{1/2}} d\xi' . \tag{174}$$

In the surface boundary layer, Eqs. (155,173) yield

$$-\bar{u}_{0\xi} + \eta_\xi \bar{u}_{0\bar{z}} = -\eta_\xi + \bar{u}_{0\bar{z}\bar{z}} \,.$$

This equation poses a difficult task due to the appearance of the variable coefficient, which moreover is the unknown function η_ξ. However, redefining the vertical coordinate (Prandtl transformation) as

$$\phi = \bar{z} + \eta(\xi)$$

yields

$$-\bar{u}_{0\xi} = -\eta_\xi + \bar{u}_{0\phi\phi} \,,$$

thus greatly simplifying the problem. Then

$$\bar{u}_0 = \eta_\xi + V(\xi, \phi) \,, \tag{175}$$

with

$$V = \frac{\phi}{2\pi^{1/2}} \int_\xi^\infty \frac{c(\xi')}{(\xi' - \xi)^{3/2}} \exp\left[-\frac{\phi^2}{4(\xi' - \xi)}\right] d\xi' \,,$$

where $c(\xi)$ is yet to be found.

The continuity equation (154) becomes

$$\bar{u}_{0\xi} + \bar{u}_{0\phi}\eta_\xi - \bar{w}_{1\phi} = 0 \,.$$

Taking BC (164) into account, which is now at $\phi = 0$, we get

$$\bar{w}_1 = \eta_\xi \phi + V\eta_\xi + \eta_\tau + \eta\eta_\xi + \frac{1}{\pi^{1/2}} \frac{d}{d\xi} \int_\xi^\infty \frac{c(\xi')}{(\xi' - \xi)^{1/2}} \left\{1 - \exp\left[-\frac{\phi^2}{4(\xi' - \xi)}\right]\right\} d\xi' . \tag{176}$$

Finding $\bar{p}_1$ with the help of Eq.(156) and BC (164), and using the matching procedure, we get

$$\text{at } z = 0: \quad p_1 = -\eta_{\xi\xi}/B \,. \tag{177}$$

BC (174,177) are already sufficient to find the potential solution in the main bulk. Eqs.(154–156) become

$$u_{1\xi} + w_{1z} = 0 \,, \tag{178}$$

$$-u_{1\xi} + \eta_\tau + \eta\eta_\xi = -p_{1\xi} \,, \tag{179}$$

$$\eta_{\xi\xi} = -p_{1z} \,. \tag{180}$$

The solution of the problem (174,177–178) is

$$w_1 = -\eta_\tau z - \eta\eta_\xi z - \left(\frac{1}{2} - \frac{1}{B}\right) \eta_{\xi\xi\xi} + \eta_{\xi\xi\xi} + \frac{1}{\pi^{1/2}} \frac{d}{d\xi} \int_\xi^\infty \frac{\eta(\xi')}{(\xi' - \xi)^{1/2}} d\xi' \tag{181}$$

(p_1 and u_1 are not needed later on). Compared to the inviscid liquid case, an additional integral term appears in Eq.(181), which is due to a delay of the flow in the bottom boundary layer.

The matching condition between $w_0+\epsilon w_1$ and $\bar{w}_0+\epsilon\bar{w}_1$ yields an implicit equation for $c(\xi)$:

$$\frac{1}{\pi^{1/2}}\frac{d}{d\xi}\int_\xi^\infty \frac{c(\xi')}{(\xi'-\xi)^{1/2}}d\xi' = \left[2\eta_\tau + 3\eta\eta_\xi + \left(\frac{1}{3}-\frac{1}{B}\right)\eta_{\xi\xi\xi}\right] + \frac{1}{\pi^{1/2}}\frac{d}{d\xi}\int_\xi^\infty \frac{\eta(\xi')}{(\xi'-\xi)^{1/2}}d\xi' \,. \tag{182}$$

Let us now consider the temperature field. In the surface boundary layer Eqs.(157,173) and BC (167), here reduced to $\overline{T}_{0\bar{z}}=0$, yield

$$\overline{T}_0 = -\eta\,.$$

Thus the contribution to the thermocapillary term in BC (165) is zero in this order.

For the first correction in the surface boundary layer, $\overline{T}_1$, Eq.(157) and BC (167) yield

$$\overline{T}_{0\tau} - \overline{T}_{1\xi} + \bar{u}_0\overline{T}_{0\xi} - \bar{w}_0\overline{T}_{1\bar{z}} + \bar{w}_1 = \frac{1}{\mathrm{Pr}}\overline{T}_{1\bar{z}\bar{z}}\,,$$

and

$$\text{at } \bar{z}=-\eta: \quad \overline{T}_{1\bar{z}} = -Q(\tau,\xi)\,,$$

where Q is obtained by solving the problem in the air layer.

Introducing the known functions and using ϕ instead of $\bar{z}$ we get

$$-\overline{T}_{1\xi} + \eta_\xi\phi + \frac{1}{\pi^{1/2}}\frac{d}{d\xi}\int_\xi^\infty \frac{c(\xi')}{(\xi'-\xi)^{1/2}}\left\{1-\exp\left[-\frac{\phi^2}{4(\xi'-\xi)}\right]\right\}d\xi' = \frac{1}{\mathrm{Pr}}\overline{T}_{1\phi\phi}\,,$$

and

$$\text{at } \phi=0: \quad \overline{T}_{1\phi} = -Q(\tau,\xi)\,, \tag{183}$$

whose solution is

$$\begin{aligned}\overline{T}_{1\phi} = \eta\phi &+ \frac{1}{\pi^{1/2}}\int_\xi^\infty \frac{c(\xi')}{(\xi'-\xi)^{1/2}}d\xi' \\ &+ \frac{1}{\pi^{1/2}}\frac{\mathrm{Pr}}{1-\mathrm{Pr}}\int_\xi^\infty \frac{c(\xi')}{(\xi'-\xi)^{1/2}}\exp\left[-\frac{\phi^2}{4(\xi'-\xi)}\right]d\xi' \\ &- \frac{1}{\pi^{1/2}}\frac{\mathrm{Pr}^{1/2}}{1-\mathrm{Pr}}\int_\xi^\infty \frac{c(\xi')}{(\xi'-\xi)^{1/2}}\exp\left[-\frac{\mathrm{Pr}\,\phi^2}{4(\xi'-\xi)}\right]d\xi' \\ &+ \frac{1}{\pi^{1/2}}\frac{1}{\mathrm{Pr}^{1/2}}\int_\xi^\infty \frac{\eta(\xi')+Q(\xi')}{(\xi'-\xi)^{1/2}}\exp\left[-\frac{\mathrm{Pr}\,\phi}{4(\xi'-\xi)}\right]d\xi' \,.\end{aligned} \tag{184}$$

Note that as we do not need an explicit expression for the temperature field in the bottom boundary layer, the results do not depend on the type of heat exchange at the solid support.

Finally, let us use BC (165) for the tangential stress balance. It becomes

$$\text{at } \phi=0: \quad -\bar{u}_{0\phi} + m\overline{T}_{1\xi} = 0\,, \tag{185}$$

where for convenience we now use the earlier redefined Marangoni number

$$m \equiv \frac{\mathrm{Ma}\epsilon^{10}}{\mathrm{Pr}} = \frac{\mathrm{Ma}}{\mathrm{Ga}} \tag{186}$$

This modified Marangoni number, m, is of order unity and corresponds to the most general case as viscous and thermocapillary stresses are of the same order in (185). This also means that Ma is of order of Ga.

Substituting (175,184) into BC (185), and using Eq.(182), we get

$$\begin{aligned} &2\left(1-\frac{m}{\mathrm{Pr}^{1/2}+1}\right)\left[\eta_\tau+\frac{3}{2}\eta\eta_\xi+\left(\frac{1}{6}-\frac{1}{2\mathrm{Bo}}\right)\eta_{\xi\xi\xi}\right] \\ &+\left(m\frac{2\,\mathrm{Pr}^{1/2}+1}{\mathrm{Pr}+\mathrm{Pr}^{1/2}}-1\right)\frac{1}{\pi^{1/2}}\frac{d}{d\xi}\int_\xi^\infty\frac{\eta(\xi')}{(\xi'-\xi)^{1/2}}d\xi' \\ &+\frac{m}{\mathrm{Pr}^{1/2}}\frac{1}{\pi^{1/2}}\frac{d}{d\xi}\int_\xi^\infty\frac{Q(\xi')}{(\xi'-\xi)^{1/2}}d\xi'=0\,. \end{aligned} \tag{187}$$

The first bracket in Eq. (187) when set to zero is an equation derived in 1895 by Korteweg and de Vries for weakly nonlinear and dispersive, long waves in shallow inviscid liquid layers with constant surface tension. In fact it was earlier derived by Boussinesq (at the latest in 1877). The full Eq. (187) is a dissipation-modified BKdV equation with Q yet to be determined.

The leading order problem (158–161,163–164,166) in the air layer is linear. Its solution is found in terms of the Fourier components

$$f_k=\int_{-\infty}^{+\infty}f(\xi)\exp(-ik\xi)d\xi\,,\quad f(\xi)=\frac{1}{2\pi}\int_{-\infty}^{+\infty}f_k\exp(ik\xi)dk\,.$$

Thus, Eq.(36) becomes

$$\begin{aligned} &2\left(1-\frac{m}{\mathrm{Pr}^{1/2}+1}\right)\left[\eta_\tau+\frac{3}{2}\eta\eta_\xi+\left(\frac{1}{6}-\frac{1}{2\mathrm{Bo}}\right)\eta_{\xi\xi\xi}\right] \\ &-\frac{2m}{\mathrm{Pr}^{1/2}}\frac{1}{2\pi}\int_{-\infty}^{+\infty}Q_2(\xi'-\xi)\left[\eta_\tau+\frac{3}{2}\eta\eta_\xi+\left(\frac{1}{6}-\frac{1}{2\mathrm{Bo}}\right)\eta_{\xi\xi\xi}\right]d\xi' \\ &+\left(m\frac{2\,\mathrm{Pr}^{1/2}+1}{\mathrm{Pr}+\mathrm{Pr}^{1/2}}-1\right)\frac{1}{\pi^{1/2}}\frac{d}{d\xi}\int_\xi^\infty\frac{\eta(\xi')}{(\xi'-\xi)^{1/2}}d\xi' \\ &-\frac{m}{\mathrm{Pr}^{1/2}}\frac{1}{2\pi}\frac{d}{d\xi}\int_{-\infty}^{+\infty}Q_1'(\xi'-\xi)\eta(\xi')d\xi'=0\,. \end{aligned} \tag{188}$$

where Q_1' and Q_2 are the originals of $Q_{1k}\sqrt{-ik}$ and Q_{2k}, respectively, and

$$\begin{aligned} Q_{1k}=&-c_1-\frac{P}{1-P}c_1C_1-\frac{P}{1-P}c_2S_1-\frac{P^{1/2}}{1-P}c_2\frac{1}{S_2}-\frac{\sqrt{-ik}}{a}hP^{1/2}c_1\frac{C_2}{S_2} \\ &+\frac{P^{3/2}}{1-P}c_1S_1\frac{C_2}{S_2}+\frac{P^{3/2}}{1-P}c_2C_1\frac{C_2}{S_2}+c_2P^{1/2}\frac{C_2}{S_2}+\frac{\sqrt{-ik}}{a}P^{1/2}\frac{C_2}{S_2}\,, \end{aligned}$$

$$Q_{2k} = -s_1 - \frac{P}{1-P}s_1C_1 - \frac{P}{1-P}s_2S_1 - \frac{P^{1/2}}{1-P}s_2\frac{1}{S_2} + \frac{\sqrt{-ik}}{a}hP^{1/2}s_1\frac{C_2}{S_2}$$
$$+ \frac{P^{3/2}}{1-P}s_1S_1\frac{C_2}{S_2} + \frac{P^{3/2}}{1-P}s_2C_1\frac{C_2}{S_2} + s_2P^{1/2}\frac{C_2}{S_2},$$

$$c_1 = \frac{2\sqrt{-ik}a(C_1-1) - ikS_1}{2\sqrt{-ik}a(C_1-1) + ikhS_1}, \qquad s_1 = \sqrt{-ik}a\frac{C_1-1}{2\sqrt{-ik}(C_1-1)+ikhS_1},$$
$$c_2 = c_1\frac{1-C_1}{S_1}, \qquad s_2 = s_1\frac{1-C_1}{S_1} - \frac{1}{S_1},$$

$$S_1 = \sinh\left(\frac{\sqrt{-ik}}{a}h\right), \qquad C_1 = \cosh\left(\frac{\sqrt{-ik}}{a}h\right),$$
$$S_2 = \sinh\left(\frac{\sqrt{-ik}}{a}P^{1/2}h\right), \qquad C_1 = \cosh\left(\frac{\sqrt{-ik}}{a}P^{1/2}h\right).$$

Eq. (188) is the corresponding reduced dissipation-modified BKdV equation describing surface tension gradient (Marangoni) driven "long" waves in a Bénard layer. Under appropriate limiting conditions reduces to the (standard) BKdV equation for shallow inviscid liquid layers. On the other hand another dissipation modified BKdV equation with no Marangoni effect and just viscous damping follows from Eq.(187) (or Eq. (188)) by setting $m = 0$.

Eq. (188) is easier solved in Fourier space:

$$2\left(1 - \frac{m}{\mathrm{Pr}^{1/2}+1} - \frac{m}{\mathrm{Pr}^{1/2}}Q_{2k}\right)\left[\eta_{k\tau} + \frac{3}{4}ik\int_{-\infty}^{+\infty}\eta_{k-k'}\eta_{k'}dk' - ik^3\left(\frac{1}{6} - \frac{1}{2\mathrm{Bo}}\right)\eta_k\right]$$
$$-\left(m\frac{2\,\mathrm{Pr}^{1/2}+1}{\mathrm{Pr}+\mathrm{Pr}^{1/2}} + \frac{m}{\mathrm{Pr}^{1/2}}Q_{1k} - 1\right)\sqrt{ik}\eta_k = 0. \qquad (189)$$

The dissipation-modified BKdV equation, (188) or (189), possesses the necessary ingredients to have solutions in the form of stationary propagating waves: there is an unstable wavenumber interval, where the energy is brought by the Marangoni effect and dissipation occurs on the wavenumbers belonging to the stability interval. The convective nonlinearity redistributes the energy from long to short waves, making possible the dynamic equilibrium and the appropriate energy balance for the dissipative wave. We may expect solutions such as sustained dissipative periodic wave trains or solitary waves.

5.3 Predictions amenable to experimental test

For illustration, let us consider the case of a thin air gap above the liquid layer ($h \ll 1$, however h should remain much larger than the surface deformation). In this case, the air motion is dissipation-dominated. Then $\theta = \eta(z-h)/h$, and $Q = \eta/h$. Using this in Eq.(187), we get

$$2\left(1 - \frac{m}{\mathrm{Pr}^{1/2}+1}\right)\left[\eta_\tau + \frac{3}{2}\eta\eta_\xi\left(\frac{1}{6} - \frac{1}{2\mathrm{Bo}}\right) + \eta_{\xi\xi\xi}\right]$$
$$+\left(m\frac{2\,\mathrm{Pr}^{1/2}+1}{\mathrm{Pr}+\mathrm{Pr}^{1/2}} + M\frac{1}{h\,\mathrm{Pr}^{1/2}} - 1\right)\frac{1}{\pi^{1/2}}\frac{d}{d\xi}\int_\xi^\infty \frac{\eta(\xi')}{(\xi'-\xi)^{1/2}}d\xi' = 0. \qquad (190)$$

The critical (modified) Marangoni number is now

$$m_b = \left(\frac{2\,\mathrm{Pr}^{1/2} + 1}{\mathrm{Pr} + \mathrm{Pr}^{1/2}} + \frac{1}{h\,\mathrm{Pr}^{1/2}} \right)^{-1} . \tag{191}$$

An important fact is that m_b considerably decreases with decreasing h. Thus to observe sustained capillary-gravity waves as the result of Marangoni-driven instability, the thinner the air gap the better. Take, for example, a water-like liquid. For illustration choose $h_1 = 0.1$, $h_2 = 0.01$ ($h = 0.1$), $d\sigma/dT = -0.15$, $g = 10^3$ and $\mathrm{Pr} = 6$ (for dimensional quantities the CGS system is used). Then, according to Eq. (191), the temperature difference applied to the liquid layer needed to excite and sustain Marangoni-driven "long" capillary-gravity waves is 15 K, hence a 150 K/cm gradient. However, in the supercritical case, with Eq. (190) all wavenumbers are unstable. This is related to the fact that, from Eq. (189), in the limit of high dissipation in the air layer ($h \ll 1$ or $a \gg 1$), the band of unstable wavenumbers shifts to higher and higher k, where our long wave approximation ceases to be valid. In this case, to obtain a suitable energy balance to maintain the waves we must proceed to a higher order approximation (ϵ^2 in Eq. (190)), hence $m - m_b \approx \epsilon^2$).

Finally, let us mention that the vanishing of the coefficient of the ideal BKdV terms in Eq. (190), at $m = \mathrm{Pr}^{1/2} + 1$, corresponds to the resonance between capillary-gravity and dilational-longitudinal waves. Our approach is not valid in the vicinity of this resonance point. However, note that the resonance value is always higher than m_b.

6 Dissipative solitons. Discussion of the simplest (1+1)D model-case

To further simplify the problem relative to the analysis given in Sect. 5 one can consider that at the bottom there is free slip and at the top just the simplest of either heat transfer or mass adsorption and otherwise passive air. We cannot expect quantitative relevance to experiment although from earlier experience we feel such drastic simplification would allow to capture the essence of the problem and hence a useful qualitative picture of the problem. For a more detailed study of the question the reader is adviced to proceed to the recent monograph by Nepomnyashchy *et al* (2001).

Nepomnyashchy and Velarde (1994) found as solvability conditions for the problem, first the linear (phase) wave velocity and the instability threshold Ma_c=-12. This particular value is not relevant to experiment save the negative sign corresponding here to heating the liquid layer from the air side, hence from above (β was taken positive when heating from, as traditionally used for the study of Benard layers). Indeed, starting with the Navier-Stokes and the heat equations earlier given and using the longwave shallow layer approximation, a slippery bottom with the appropriate surface deformability, weak amplitude and dispersion and a 2D geometry and restricting consideration to left-to-right steadily propagating long waves, past an instability threshold the following dissipation-modified BKdV equation is obtained

$$\xi_t + C_{\exp}\xi_x + \alpha_1\xi\xi_x + \alpha_2\xi_{xxx} + \alpha_3\xi_{xx} + \alpha_4\xi_{xxxx} + \alpha_5(\xi\xi_x)_x + \alpha_6\xi = 0, \tag{192}$$

where, as earlier, ξ denotes the suitably scaled dimensionless surface deformation and the coefficients $\alpha_i (i = 1, ..., 6)$ are functions of Ma, Ra, Pr, and Ga. For completeness we have added the term $\alpha_6\xi$ that allows to account for bottom friction in *ad hoc* way. A suitable transformation of variables and choice of the reference frame, brings Eq. (192) to the more compact form

$$\eta_T + (\eta^2)_x + \eta_{xxx} + \delta[\eta_{xx} + \eta_{xxxx} + D(\eta^2)_{xx} + \alpha\eta] = 0, \tag{193}$$

where $\eta(x, \tau)$ is once more the suitably scaled surface deformation. The coefficient D can be either positive or negative, while α and δ are semidefinite positive. In particular, α_3 and, consequently, δ is a function of (Ma- Ma_c). When δ vanishes, the BKdV equation for inviscid liquid layers is recovered.

Equation (192), or (193), contains a fourth-order derivative dissipating energy at short scales and the nonlinear term $D(\eta^2)_{xx}$ which helps redistributing, supercritically, the energy supplied by the Burgers term ($\delta\eta_{xx}$ with the Marangoni effect). The factor C_{exp} in Eq. (192) is a function of the actual Marangoni number in an experiment thus providing the measurable nonlinear wave velocity in the laboratory frame of reference. Its linear approximation is C_0^2 =Pr(Ga-Ma), which to the lowest order reduces to $C_0^2 = C_{\text{BKdV}}^2(1 + 12/\text{Ga})$ where $C_{\text{BKdV}}^2 = gd$ is the linear velocity of the ideal BKdV solitary wave. For the linear approximation, the correction due to the thermal constraint is about 30 for a silicone oil layer of d = 0.01mm. Subsequently, corrections are induced by the distance of the Marangoni number from the threshold value.

Note that the linear terms with second and fourth derivatives, on the one hand, provide a (phase) wave velocity selection for Eq.(192) and, on the other hand, they are expected to add a wavy forerunner to the ideal bump or bell-shaped (sech^2) BKdV wave. Such a wavy forerunner/tail, found indeed with the numerical integration of (192), acts like an infinitely long ranged albeit exponentially weak attractive "potential", which at variance with the ideal, dissipation-free BKdV equation, suggests the easiness in forming *bound* states and the possibility of inelastic wave or particle-like collisions as in the imperfect (van der Waals) gas. For particular combinations of the coefficients in Eq. (192), or (193), some exact analytical, nonperturbative solutions in the form of solitary and periodic waves propagating with constant velocity have been found.

Let us see how Eq. (192), or (193) accommodates the energy balance at and past the instability threshold. By multiplying Eq. (193) by η, and integrating over the full espace or one wavelentgh in the variable x, the energy $E = \frac{1}{2}\int \eta^2$ is governed by the balance

$$\frac{dE}{dT} = \delta\left(\int \eta_x^2 dx - \int \eta_{xx}^2 dx + 2D\int \eta\eta_x^2 dx - \alpha\int \eta^2 dx\right), \tag{194}$$

whose value vanishes at the steady state. Clearly, it now appears that the first term on the right hand side of (194) describes the energy input at rather long wavelengths due to instability (Burgers term) the second and fourth terms describe energy dissipation on short and long wave lengths, respectively; and the third term accounts for the convective nonlinearity redistributing energy as a (feedback) correction to long wave energy input (for η positive, positive if D is positive and negative otherwise). Recalling that the BKdV equation for inviscid liquid layers possesses a one-parameter family of solitary waves or cnoidal waves (periodic wave trains) thanks to the dispersion-nonlinearity balance, also existing in Eq. (193), we note that here the input-output energy balance (194) selects indeed either a single wave or periodic wavetrains or a bound state or a spatially chaotic wavetrains.

Under the same limiting conditions but a 3-D geometry Eq. (193) has been generalized. Then, when considering the three-dimensional problem phase shifts following wave collisions or reflections at walls can be obtained. It was shown that they depend upon the incident angle, α_i [e.g., measured front to front or twice the value front to wall, i.e., by $(\pi/2) - \alpha_i$; apparently, a reflection is like a collision with a mirror image wave]. At the approximate value of $\pi/2$ no phase shift is expected while for lower collision angles the phase shift has the sign of phase shifts upon head-on collisions. Higher values than $\pi/2$ (or $\alpha_i < \pi/4$) lead to a change of sign in the phase shift and the formation of a *third* wave, or *stem*, evolving *phase locked* with the post collision or reflected front. This result for surface tension gradient-driven waves generalizes the result obtained for shallow inviscid liquid layer. These phenomena were discovered a century ago by Russell (1885; plate VII of the Appendix from his 1844 Report) for water waves in a wave tank, where viscosity plays negligible role, and by Mach and collaborators for shocks in gases, where dissipation is essential for the existence of the shock. It is known that surface waves in liquids (including hydraulic pumps) share some common kinematic features with shocks in compressible gases. The phase shift in overtaking collisions was discovered by Zabusky and Kruskal (1965)in their seminal paper where they introduced the *soliton* concept. Finally, starting with e.g., an initial condition of two nearby "solitary" pulses the system is expected to evolve, according to Eq. (192), or (193), to a *bound* state, i.e., a wavetrain with unequally spaced maxima but equal amplitude, and hence wave crests traveling with the same velocity dictated by the energy balance (194).

7 Concluding remarks

In these lecture notes I have tried to provide a succint account of phenomena, theoretical predictions, and methodology dealing with the role of heat or mass transfer and surface stresses, and hence the Marangoni effect, in triggering instability in a shallow, horizontal liquid layer subject to a transverse, vertical thermal gradient. For this paradigmatic model-problem, first, I recalled the basic evolution equations and boundary conditions. Then I have discussed scalings, asymptotics and simplifications allowing drastic albeit significant approximations to the original 3D problem. For the steady case originated in a monotonic instability I have considered the problem in (1+1)D and I have discussed how to describe the evolution of surface deformations. I have dealt in detail with the ground-based and the microgravity cases by using appropriate scalings of the Galileo number. I have presented the corresponding evolution equations and have discussed some of their salient features, in particular how the input-output energy balance arising in the relationship between the Marangoni effect and viscous and heat dissipation allows exciting and sustaining steady patterned convection, a dissipative structure. Due to the (1+1)D approximation these steady patterns are reduced to a mere wavelength selected at the onset of instability.

In the particular case of waves originated in overstability or (linear) oscillatory instability, I have shown how two types of surface waves, transverse and longitudinal (or, better, capillary-gravity and dilational, respectively) can be excited by the Marangoni effect and how they are related to each other thus leading to their combined (symbiotic-like) survival and to resonance effects. I have also given comments about the excitation of internal waves in the layer heated above and hence stably stratified and how in turn these waves are, in particular, related to the surface dilational waves.

For the particular case of capillary-gravity, transverse waves, I have provided with suitable scalings and asymptotics, again in (1+1)D, the evolution equation of nonlinear long waves, and eventually solitons. Furthermore, in a much drastically simplified approximation, albeit capturing the qualitative features of the phenomena, I have shown how again the above mentioned input-output energy balance permits exciting and sustaining these localized traveling structures, that in the moving frame are steady ones. Accordingly, they are dissipative structures very much like the steady convective cellular patterns.

In all cases I have itemized the significant dimensionless groups and have described their role in the dynamics. Heuristic arguments have been presented to foresee or justify results about instability theresholds or physical mechanisms leading to instability.

I have tried to show the field attractive for newcomers, young and old, although I have not, however, tried to provide a comprehensive account of what has been so far achieved, in theory and experiment.Thus I strongly suggest to the reader to proceed to the list of references where plenty of details and questions are provided. Indeed, the references given below, with full details, are to help the reader going deeper into the problems treated in these lecture notes, including more general cases that those here described, e. g., 3D (analytical and numerical) descriptions of both steady, cellular (Benard) patterns when the heating the liquid layer from below and waves when the heating is done from above, i. e., from the air side.

References

Bar, D. and A. A. Nepomnyashchy (1995) Stability of periodic waves governed by the modified Kawahara equation. *Physica D* 86: 586-602.

Bazin, H. (1865) Recherches hydrauliques experimentales sur la propagation des ondes entreposes par Darcy et continuees par M. Bazin. *Mem. presentes par divers Savants Etrangers a l'Acad. Sci. Inst. France* 19: 495-644

Benard, H. (1900) Les tourbillons cellulaires dans une nappe liquide. Premiere partie: description generale des phenomenes. *Rev. Gen. Sci. Pures Appl.* 11: 1261-1271.

Benard, H. (1901) Les tourbillons cellulaires dans une nappe liquide transportant de le chaleur par convection en regime permanent. *Ann. Chim. Phys.* 23: 62-143.

Benjamin, T. B. (1982) The solitary wave with surface tension. *Quart. Appl. Maths* 40: 231-234.

Berg, J.C., and Acrivos, A. (1965) The effect of surface active agents on convection cells induced by surface tension. *Chem. Engng. Sci.* 20: 737-745.

Berg, J.C., Acrivos, A., and Boudart, M. (1966) Evaporative convection. *Adv. Chem. Eng.* 6: 61-123.

Bestehorn, M. (1996) Square patterns in Benard-Marangoni convection. *Phys. Rev. Lett.* 76: 46-49.

Birikh, R. V., Briskman, V. A., Cherepanov, A. A., and Velarde, M. G. (2000) Faraday ripples, parametric resonance, and the Marangoni effect. *J. Colloid Interface Sci.* 238: 16-23.

Block, M.J. (1956) Surface tension as the cause of Benard cells and surface deformation in a liquid film. *Nature* (Lond.) 178: 650-651.

Bouasse, H. (1924) *Houle, Rides, Seiches et Marees.* Delagrave (Paris). 291-292.

Boussinesq, J.V. (1871) Theorie de l'intumescence liquide appelee onde solitaire ou de translation se propageant dans un canal rectangulaire. *C. R. Hebd. Seances Acad. Sci.* (Paris) 72: 755-759.

Boussinesq, J.V. (1872) Theorie des ondes et des remous qui se propagent le long d'un canal rectangulaire horisontal en communiquant au liquid contenu dans ce canal des vitesses sensiblement pareilles de la surface au fond. *J. Math. Pures Appl.* 17: 55-108.

Boussinesq, J. V. (1877) Essai sur la theorie des eaux courantes. *Mem. presentes par divers savants ã l'Acad. Sci. Inst. France* 32: 1-680.

Bragard, J., and Velarde, M. G. (1997) Benard convection flows. *J. Non-Equilib. Thermodyn.* 22: 1-19.

Bragard, J., and Velarde, M. G. (1998) Benard-Marangoni convection: Theoretical predictions about planforms and their relative stability. *J. Fluid Mech.* 368: 165-194.

Busse, F. H. (1978) Non-linear properties of thermal convection. *Rep. Prog. Phys.* 41: 1929-1967.

Castillo, J. L., Garcia-Ybarra, P. L., and Velarde, M. G. (1988) in *Synergetics and Dynamic Instabilities*, edited by Caglioti, G., Haken, H., and Lugiato, L. A. (North-Holland, Amsterdam), 219-243.

Chandrasekhar, S. (1961) *Hydrodynamic and Hydromagnetic Stability*, Clarendon Press, Oxford.

Christov, C. I., and Velarde, M. G. (1993) On localized solutions in an equation in Benard convection. *Appl. Math. Model.* 17: 311-318.

Christov, C. I., and Velarde, M. G. (1995) Dissipative solitons. *Physica D* 86: 323-347.

Christov, C.I., Maugin, G. A., and Velarde, M. G. (1996) Well-posed Boussinesq Paradigm with purely spatial higher-order derivatives. *Phys. Rev. E* 54: 3621-3638

Chu, X.-L., and Velarde, M. G. (1988) Sustained transverse and longitudinal waves at the open surface of a liquid. *Physicochem. Hydrodyn.* 10: 727-737.

Chu, X.-L., and Velarde, M. G. (1989) Transverse and longitudinal waves induced and sustained by surfactant gradients at liquid-liquid interfaces. *J. Colloid Interface Sci.* 131: 471-484.

Chu, X.-L., and Velarde, M. G. (1991) Korteweg-de Vries soliton excitation in Benard-Marangoni convection. *Phys. Rev. A* 43: 1094-1096.

Colinet, P., Legros, J. C., and Velarde, M. G. (2001) *Nonlinear Dynamic of Surface Tension Driven Instabilities.* Wiley-VCH, Weinheim.

Courant, R., and Friedrichs, K. O. (1948)*Supersonic flow and shock waves.* Interscience, N.Y.

Dauzere C. (1908) Recherches sur la solidification cellulaire. *J. Phys.* (Paris) 7: 930-934.

Dauzere C. (1912) Sur les changements qu'eprouvent les tourbillons cellulaires lorsque la temperature s'eleve. *C. R. Hebd. Seances Acad. Sci.* (Paris) 155: 394-398.

Davis, S. H. (1987) Thermocapillary instabilities. *Ann. Rev. Fluid Mech.* 19: 403-435.

Davis S. H., and Segel, L. A. (1968) Effects of surface curvature & property variation on cellular convection. *Phys. Fluids* 11: 470-476.

de Boer, P. C. T. (1984) Thermally driven motion of strongly heated fluids. *Int. J. Heat Mass Transfer.* 27: 2239-2251.

de Boer, P. C. T. (1986) Thermally driven motion of highly viscous fluids. *Int. J. Heat Mass Transfer.* 29: 681-688.

Drazin, P. G., and Reid, W. H. (1981)*Hydrodynamic Stability.* University Press, Cambridge.

Drazin, P. G., and Johnson, R. S. (1989) *Solitons. An Introduction.* University Press, Cambridge.

Eckert, K., Besterhorn, M., and Thess, A. (1998) Square cells in surface-tension-driven Benard convection: experiment and theory. *J. Fluid Mech.* 356: 155-197.

Garazo A. N., and Velarde, M. G. (1991) Dissipative Korteweg-de Vries description of Marangoni-Benard convection. *Phys. Fluids A* 3: 2295-2300.

Garcia-Ybarra, P. L., and Velarde, M. G. (1987) Oscillatory Marangoni-Benard interfacial instability and capillary-gravity waves in single-and two-component liquid layers with or without Soret thermal diffusion. *Phys. Fluids* 30: 1649-1655.

Garcia-Ybarra, P.L., Castillo, J. L., and Velarde, M. G. (1987) Benard-Marangoni convection with a deformable interface and poorly conducting boundaries. *Phys. Fluids* 30: 2655-2661.

Gershuni, G. Z., and Zhukhovitsky, E. M. (1976)*Convective Stability of Incompressible Fluids.* Keter, Jerusalem.

Golovin, A. A., Nepomnyashchy, A. A., and Pismen, L. M. (1994) Interaction between short-scale Marangoni convection and long-scale deformational instabilty. *Phys. Fluids* 6: 34-48.

Hershey, A. V. (1939) Ridges in a liquid surface due to the temperature dependence of surface tension. *Phys. Rev.* 56: 204.

Hornung, H. (1988) Regular and Mach reflection of shock waves. *Ann. Rev. Fluid Mech.* 18: 33-58.

Huang, G.-H., Velarde, M. G., and Kurdiumov, V. (1998) Cylindrical solitary waves and their interaction in Benard-Marangoni layers. *Phys. Rev. E* 57: 5473-5482.

Hyman, J. M., and Nicolaenko, B. (1987) Coherence and Chaos in the Kuramoto-Velarde equation, in *Partial Differential Equations*, edited by Crandall, M., Academic, New York.

Kac, M., Uhlenbeck, G. E., and Hemmer, P. C (1963) On the van der Waals theory of the vapor-liquid equilibrium. I. Discussion of a one-dimensional model. *J. Math. Phys.* 4: 216-228.

Knobloch, E. (1990) Pattern selection in long-wavelength convection. *Physica D* 41: 450-479.

Koschmieder, E.L. (1988) in *Physicochemical Hydrodynamics: Interfacial Phenomena*, edited by Velarde, M. G., Plenum, New York, 189-198.

Koschmieder, E.L. (1993)*Benard Cells and Taylor Vortices*. University Press, Cambridge.

Krehl, P., and M. van der Geest (1991) The discovery of the Mach reflection effect and its demonstration in an auditorium. *Shock Waves* 1: 3-15.

Linde, H., Chu, X.-L., and Velarde, M. G. (1993a) Oblique and head-on collisions of solitary waves in Marangoni-Benard convection. *Phys. Fluids A* 5. 1068-1070.

Linde, H., Chu, X.-L., Velarde, M. G., and Waldhelm, W. (1993b) Wall reflection of solitary waves in Marangoni-Benard convection. *Phys. Fluids A* 5: 3162-3166.

Linde, H., Chu, X.-L., and Velarde, M. G. (1993c) Solitary waves driven by Marangoni stresses. *Adv. Space Res.* 13: 109-117.

Linde, H., Velarde, M. G., Wierschem, A., Waldhelm, W., Loeschcke, K., and Rednikov, A. Ye. (1997) Interfacial wave motions due to Marangoni instability. I. Traveling periodic wave trains in square and annular containers. *J. Colloid Interface Sci.* 188: 16-26.

Linde, H., Velarde, M. G., Waldhelm, W., and Wierschem, A. (2001) Interfacial wave motions due to Marangoni instability. III. Solitary waves and (periodic) wavetrains and their collisions and reflections leading to dynamic network patterns. *J. Colloid Interface Sci.* 236:214-224.

Linde, H., Velarde, M. G., Waldhelm, W., Loeschcke, K., and Wierschem, A. (2002) Interfacial wave motion due to Marangoni instability. IV. Waves with anomalous dispersion or normal dispersion and dispesion-free waves and transitions among them with decreasing driving forcein instationary experiments. *J. Colloid Interface Sci*

Lucassen, J. (1968) Longitudinal capillary waves, Part 1. Theory. *Trans. Faraday Soc.* 64: 2221-2229, Part 2. Experiments. *ibidem* 64: 2230-2235.

Lucassen-Reynders, E. H. and Lucassen, J. (1969) Properties of capillary waves. *Adv. Colloid Interface Sci.* 2: 347-395.

Maxworthy, T. (1976) Experiments on collisions between solitary waves. *J. Fluid Mech.* 76: 177-185.

Melville, W.K. (1980) On the Mach reflexion of a solitary wave. *J. Fluid Mech.* 98: 285-297

Mihaljan, J. (1962) A rigorous exposition of the Boussinesq approximations applicable to a thin layer of fluid. *Astrophys. J.* 136: 1126-1133.

Miles, J. W. (1976) Korteweg-de Vries equation modified by viscosity. *Phys. Fluids* 19: 1063

Miles, J. W. (1977) Obliquely interacting solitary waves. *J. Fluid Mech.* 79. 157-169.

Miles, J. W. (1980) Solitary waves. *Ann. Rev. Fluid Mech.* 12: 11-43.

Nekorkin, V. I., and Velarde, M. G. (1994) Solitary waves of a dissipative Korteweg-de Vries equation describing Marangoni-Benard convection and other thermoconvective instabilities. *Int. J. Bifurc. Chaos* 4: 1135-1146.

Nepomnyashchy, A. A., and Velarde, M. G. (1994) A three-dimensional description of solitary waves and their interaction in Marangoni-Benard layers. *Phys. Fluids* 6: 187-198.

Nepomnyashchy, A. A., Velarde, M. G. and Colinet, P. (2001) *Interfacial Phenomena and Convection.* Chapman and Hall/CRC, London.

Nield, D. A. (1964) Surface tension and buoyancy effects in cellular convetion. *J. Fluid Mech.* 19: 341-352.

Normand, C., Pomeau, Y., and Velarde, M. G. (1977) Convective instability: A Physicist's Approach. *Rev. Mod. Phys.* 49: 581-624.

Ostrach, S. (1982) Low-gravity fluid flows. *Ann. Rev. Fluid Mech.* 14: 313-345.

Pearson, J. R. A. (1958) On convection cells induced by surface tension. *J. Fluid Mech.* 4: 489-500.

Perez-Cordon, R. and Velarde, M. G. (1975) On the (non linear) foundations of Boussinesq approximation applicable to a thin layer of fluid. *J. Phys.* (Paris) 36: 591-601.

Pontes, J., Christov, C. I., and Velarde, M. G. (1996) Numerical study of patterns and their evolution in finite geometries. *Int. J. Bifurc. Chaos* 6: 1883-1890.

Pontes, J., Christov, C. I., and Velarde, M. G. (2000) Numerical approach to pattern selection in a model problem for Benard convection in a finite fluid layer Prandtl number. *Annu. Univ.* Sofia 93: 157-175.

Rayleigh, Lord (1876) On Waves. *Phil. Mag.* 1: 257-279.

Rayleigh, Lord (1916) On convection currents in a horizontal layer of fluid, when the higher temperature is on the under side. *Phil. Mag.* 32: 529-536.

Rednikov, A. Ye., Colinet, P., Velarde, M. G., and Legros, J. C. (1998) Two-layer Benard-Marangoni instability and the limit of transverse and longitudinal waves. *Phys. Rev. E* 57: 2872-2884.

Rednikov, A. Ye., Colinet, P., Velarde, M. G., and Legros, J. C. (2000) Rayleigh-Marangoni oscillatory instability in a horizontal liquid layer heated from above: coupling between internal and surface waves. *J. Fluid Mech.* 405: 57-77.

Rednikov, A. Ye., Colinet, P., Velarde, M. G., and Legros, J. C. (2000) Oscillatory thermocapillary instability in a liquid layer with deformable open surface: capillary-gravity waves, longitudinal waves and mode-mixing. *J. Non-Equilib. Therm.* 25:381-405.

Rednikov, A. Ye., Velarde, M. G., Ryazantsev, Yu. S., Nepomnyashchy A. A., and Kurdyumov, V. (1995), Cnoidal wave trains and solitary waves in a dissipation-modified Korteweg-de Vries equation. *Acta Appl. Math.* 39: 457-475.

Rodriguez- Bernal, A. (1992) Initial value problem and asymptotic low dimensional behavior in the Kuramoto-Velarde equation. *Nonl. Anal. Theory Meth. Appl.* 19: 643-685.

Russell, J. S. (1844), Report on waves. *Rep. 14th Meet. British Ass. Adv. Sci.*, York, 311-390, J. Murray, London; reprinted as Appendix in Russell (1885).

Russell, J. S. (1885) *The Wave of Translation in the Oceans of Water, Air and Ether.* Trubner, London.

Sanfeld, A., Steinchen, A., Hennenberg, M., Bisch, P. M., D. Van Lamsweerde-Gallez, and Dalle-Vedove, W. in *Dynamics and Instability of Fluid Interfaces* edited by Sorensen, T. S., Springer-Verlag, Berlin 1979, 168-204. See also Henenberg, M., Bisch, P. M., Vignes-Adler, M. and Sanfeld, A., *ibidem* 229-259.

Santiago-Rosanne, M., Vignes-Adler, M., and Velarde, M. G. (1997) Dissolution of a drop on a liquid surface leading to surface waves and interfacial turbulence. *J. Colloid Interface Sci.* 191: 65-80.

Schatz, M. F., Van Hook, S., Mc Comick, W., Swift. J. B., and Swinney, H. D. (1995) Onset of surface-tension-driven Benard convection. *Phys. Rev. Lett.* 75: 1938-1940.

Scriven , L. E.,and Sternling C. V. (1960) The Marangoni effects. *Nature* 187: 186-188.

Simanovsky, I. B., and Nepomnyaschchy, A. A. (1993) *Convective Instabilities in Systems with Interfaces.* Gordon and Breach, N.Y.

Smith, K. A. (1966) On convective instability induced by surface-tension gradients. *J. Fluid Mech.* 24: 401-414.

Sternling, C. V., and Scriven, L. E. (1959) Interfacial turbulence: Hydrodynamic instability and the Marangoni effect. *A. I. Ch. E. Journal* 5: 514-523.

Tanford, Ch. (1989)*Ben Franklin Stilled the Waves. An Informal History of Pouring Oil on Water With reflections on the ups and downs of scientific life in general.* Duke Univ. Press, London.

Ursell, F. (1953) The long-wave paradox in the theory of gravity waves. *Proc. Cambridge Phil. Soc.* 49: 685-694.

Van Hook, S., Schatz, M. S., Mc Cornick, W., Swift, J. B., and Swinney, H. D. (1995) Long-wavelength instability in surface-tension-driven Benard convection. *Phys. Rev. Lett.* 75: 4397-4400.

Van Hook, S., Schatz, M. F., Mc Cornick, W., Swift, J. B., and Swinney,H. D. (1997) Long-wavelength surface-tension-driven Benard convection: experiment and theory. *J. Fluid Mech.* 345: 45-78.

Velarde M. G., and Normand, C. (1980) Convection. *Sci. American* 243(1): 92-108.

Velarde, M. G., and Chu, X.-L. (1988) The harmonic oscillator approximation to sustained gravity-capillary (Laplace) waves at liquid interfaces. *Phys. Lett. A* 131: 430-432.

Velarde, M. G., and Chu, X.-L. (1989a) Waves and turbulence at interfaces. *Phys. Scr.* T25: 231-237.

Velarde, M. G., and Chu, X.-L. (1989b) Dissipative hydrodynamic oscillators. I. Marangoni effect and sustained longitudinal waves at the interface of two liquids. *Il Nuovo Cimento* D11: 707-716.

Velarde, M. G., and Chu, X.-L. (1992) Dissipative thermohydrodynamic oscillators. *Adv. Thermodyn.* 6: 110-145.

Velarde, M. G., and Perez-Cordon, R. (1976) On the (non-linear) foundations of Boussinesq approximation applicable to a thin layer of fluid. II. Viscous dissipation and large cell gap effects. *J. Phys.* (Paris) 37: 177.

Velarde, M. G., Garcia-Ybarra, P. L., and Castillo, J. L. (1987) Interfacial oscillations in Benard-Marangoni layers. *Physicochem. Hydrodyn.* 9: 387-392.

Velarde, M. G., Nekorkin, V. I., and Maksimov, A. G. (1995) Further results on the evolution of solitary waves and their bound states of a dissipative Korteweg-de Vries equation. *Int. J. Bifurc. Chaos* 5: 831-839.

Velarde, M. G., Linde, H., Nepomnyashchy, A. A. and Waldhelm, W. (1995) Further evidence of solitonic behavior in Benard-Marangoni convection: periodic wave trains, in *Fluid Physics*, edited by Velarde, M. G. and Christov, C. I., World Scientific, Singapore, 433-441.

Velarde, M. G., and Rednikov, A. Ye. (1998) Time-dependent Benard-Marangoni instability and waves, in *Time-Dependent Nonlinear Convection*, edited by P.A. Tyvand, Computational Mechanics Publications, Southampton, 177-218.

Velarde, M. G., Rednikov, A. Ye., and Linde, H. (1999) Waves generated by surface tension gradients and Instability, in *Fluid Dynamics at Interfaces*, edited by W. Shyy, University Press, Cambridge, 43-56.

Velarde, M. G., Nepomnyashchy, A. A., and Hennenberg, M. (2000) Onset of oscillatory interfacial instability and wave motions in Benard layers. *Adv. Appl. Mech.* 37: 167-237.

Vidal, A. and Acrivos, A. (1966) Nature of the neutral state in surface-tension driven convection. *Phys. Fluids* 9: 615-616.

Weidman, P. and Maxworthy, T. (1978) Experiments on strong interactions between solitary waves. *J. Fluid Mech.* 85: 417-431.

Weidman, P., Linde, H., and Velarde, M. G. (1992) Evidence for solitary waves in Marangoni-driven unstable liquid layers. *Phys. Fluids A* 4: 921-926.

Wierschem, A., Velarde, m. G., Linde, H., and Waldhelm, W. (1999) Interfacial wave motions due to Marangoni instability. II. Three dimensional characteristics of surface waves. *J. Colloid Interface Sci.* 212: 365-383.

Wierschem, A., Linde, H., and Velarde, M. G. (2000) Internal waves excited by the Marangoni effect. *Phys. Rev. E* 62: 6522-6530.

Wierschem, A., Linde, H., and Velarde, M. G. (2001) Properties of surface wave trains excited by mass transfer through aliquid surface. *Phys. Rev. E* 64: 22601-1-4.

Zabusky, N., and Kruskal, M. D. (1965) Interaction of "solitons" in collisionless plasma and the recurrence of initial states. *Phys. Rev. Lett.* 15: 57-62.

THEORETICAL ASPECTS OF INTERFACIAL PHENOMENA AND MARANGONI EFFECT

MODELLING AND STABILITY

R.Kh. Zeytounian*

12, Rue Saint-Fiacre, Paris, France

Abstract. In a liquid layer, wavy motion of the free surface open to ambiant passive air (at constant temperature T_a and constant pressure p_a) represents one of the most important cases where capillary forces are displayed. The motion induced by tangential gradients of variable surface tension is customarily called the Marangoni effect (after one of the first scientists to give an explanation of this effect), and in the present Notes we consider mainly thermocapillarity effects and pose an equation of state : $\sigma = \sigma(T)$, for the surface tension, as (only) function of the temperature T. Indeed, these Notes are devoted to a Theoretical Fluid Dynamics aspects and modelling of wave dynamics and thermocapillary instabilities processes in falling thin liquid films. The starting equations and the boundary conditions on the interface between two immiscible fluids, are the full Navier-Stokes-Fourier (NS-F) equations for a viscous, thermally conducting, compressible (expansible) Newtonian fluid (liquid) and associated jump conditions for the stress tensor, the heat flux and the temperature across the interface (free surface). The simplified case we examine involve a simple geometry - the one-layer system - in which there is a liquid (weakly expansible) layer, whose lower boundary is a heated rigid plate and whose upper boundary is a deformable free surface with a passive gas (having negligible viscosity and density) - the so-called "Bénard thermal convection problem ".

1 Introduction

Consider an infinite horizontal layer of viscous thermally conducting expansible liquid of density ρ. On bounding surface at $x_3 = 0$ this horizontal liquid layer is in contact with a solid wall of constant temperaure T_B and at the level $x_3 = d$ is free to the atmosphere of constant temperature T_a and constant atmospheric pressure p_a, having negligible viscosity and density. At the free surface Newton's law of heat transfer is invoqued (see, for instance, Davis 1987, Joseph 1976), and the surface tension $\sigma(T)$ is decreasing linearly with temperature T, namely :

$$\sigma(T) = \sigma(T_0) - \gamma\,(T - T_0), \tag{1.1a}$$

where

$$\gamma = -\,d\sigma/dT \tag{1.1b}$$

is the constant rate of change of surface tension with temperature which is positive for most liquids.

* Honorary Professor of the University of Lille, Villeneuve d'Ascq Cedex, France

In (1,1a) T_0 is the constant temperature of the free surface in the purely static basic state, when

$$T_S = T_S(x_3) = T_B - \beta x_3, \qquad (1.2)$$

since a constant vertical temperature gradient : $-dT_S/dx_3 = \beta > 0$, is imposed on the expansible liquid layer. Obviously : $T_0 = T_B - \beta d$, and, according to (1.7) :

$$T_0 - T_a = \beta k(T_0)/q^\circ = \beta d/Bi, \qquad (1.3)$$

where

$$Bi = dq^\circ/k(T_0) \qquad (1.4)$$

is the Biot number which accounts for the heat transfer at the interface. The value of Bi for a very thin layer (whose depth d is at most 0.1 cm, and in a such case the buoyancy effect can be neglected - i.e. ρ = constant) is at most 0.1 for most liquids and such a small value of Bi does not affect appreciably the results, relative to thermocapillary instability, for Bi = 0. The basic temperature state (1.2) describes a pure conduction in the liquid at rest, when the free surface (non-deformable) is $x_3 \equiv d$.

We assume that the perturbed deformable free surface is represented, in a (O, x_i), with i = 1,2,3, Cartesian coordinates system (in wich $\mathbf{g} = -g\,\mathbf{k}$ acts in the negative x_3 direction) by the equation :

$$x_3 = d + a\,\eta(t, x_1, x_2) \equiv h(t, x_1, x_2)\,. \qquad (1.5)$$

The expansible viscous liquid, with the equation of state :

$$\rho = \rho(T), \qquad (1.6)$$

is a Newtonian fluid with the viscosities $\lambda(T)$ and $\mu(T)$, the specific heat C(T) and the thermal conductivity k(T); $\kappa = k/\rho C$ is the thermal diffusivity and $\nu = \mu/\rho$ is the kinematic viscosity. In (1.3) and (1.4), the constant q° is the thermal conductance between the free surface and air. At the free surface, (1.5), Newton's law of heat transfer is :

$$k(T)\,\partial T/\partial n + q^\circ (T - T_a) = 0, \qquad (1.7)$$

where n is the distance in the direction of the outward normal **n** to the free surface. This condition (1.7) is, in fact, a so-called 'third- mixed type' (or Robin) condition, embracing the Dirichlet (corresponding to $q^\circ \to \infty$) and Neumann (when $q^\circ \to 0$) conditions. The justification, for such a condition (1.7), relies on the assumption that heat conduction, within the liquid, is so much faster than within air, that the heat flux on the free surface, considered from the inside of the fluid, may be approximated by such a difference in temperatures.

2 Navier - Stokes - Fourier (NS-F) Model Problem

2.1 Governing Equations

The NS-F equations governing the expansible fluid flow for the velocity vector **u**, pressure p and temperature T, are (see, for example, Zeytounian 1994) :

$$D\rho/Dt + \rho\, \nabla \cdot \mathbf{u} = 0, \tag{2.1a}$$
$$\rho\, D\mathbf{u}/Dt + \nabla p + \rho\, g\, \mathbf{k} = \nabla \cdot \mathbf{S}, \tag{2.1b}$$
$$\rho\, C(T)\, DT/Dt + p\, \nabla \cdot \mathbf{u} = \nabla \cdot (k\, \nabla T) + \Phi(\mathbf{u}), \tag{2.1c}$$

where **S** is the viscous stress (symmetric) tensor of the liquid, namely :

$$\mathbf{T} = - p\mathbf{I} + \mathbf{S}$$

with **I** the unit second order tenso rand $\Phi(\mathbf{u})$ is the dissipation function.
In equations (2.1a) - (2.1b), the notations are standard: $\mathbf{u}(t, \mathbf{x}) = (u_i(t, x_i))$; $D/Dt = \partial/\partial t + \mathbf{u} \cdot \nabla$; $\nabla = (\partial/\partial x_i)$, $\mathbf{x} = (x_i)$, $i = 1, 2, 3$ .For a Newtonian compressible fluid :

$$\mathbf{S} = \lambda(T)\, (\nabla \cdot \mathbf{u})\, \mathbf{I} + 2\, \mu(T)\, \mathbf{D}, \tag{2.2}$$
$$\Phi(\mathbf{u}) = 2\, \mu(T)\, \mathbf{D} : \mathbf{D} + \lambda(T)\, (\nabla \cdot \mathbf{u})^2\, ;\ \mathbf{D} : \mathbf{D} = (d_{ij})^2, \tag{2.3}$$

Since, for the specific internal energy we have: $e = e(T)$, according to (1.6), then : $De/dt = C(T)\, DT/Dt$, with $C(T) = De/DT$, the specific heat of the expansible liquid

2.2. Boundary Conditions

For the equations (2.1a,b,c) the relevant boundary conditions are :

$$\mathbf{u} = 0,\ T = T_B, \text{ at } x_3 = 0, \tag{2.4}$$

$$- (p - p_a)\, \mathbf{n} + \mathbf{S} \cdot \mathbf{n} = 2\, \sigma(T)\, H\, \mathbf{n} + \nabla_{||}\sigma(T), \quad \text{on } x_3 = h(t, x_1, x_2), \tag{2.5a}$$
$$- k(T) \nabla T \cdot \mathbf{n} = q^\circ\, (T - T_a), \text{ on } x_3 = h(t, x_1, x_2). \tag{2.5b}$$

The location of the deformable free-surface, $x_3 = h(t, x_1, x_2)$, is determined via the kinematic condition :

$$u_3 = \partial h/\partial t + u_1\, \partial h/\partial x_1 + u_2\, \partial h/\partial x_2, \text{ on } x_3 = h(t, x_1, x_2). \tag{2.6}$$

We note that in the above boudary conditions on the free surface, we don't take into account the surface viscosities

2.3. Dimensionless Dominant Equations

For a weakly expansible liquid, when the dimensionless parameter :

$$\mathrm{B} = \beta\, d\, \alpha_T, \text{ with } \alpha_T = -\,[d(\log\rho)/dT]_{T=T_0}, \qquad (2.7)$$

is a *small parameter,* from the above formulated full NS-F model problem [(2.1a,b,c), (2.4)-(2.5a,b), (2.6)], it is possible to derive a simplified dimensionless dominant problem. Let the coefficients with the subscript "$_0$" be references values for these coefficients at $T = T_0$. We introduce the dimensionless pressure and temperature perturbations by

$$\theta = (T - T_0)/\, \beta\, d, \qquad (2.8a)$$

$$\pi = (1/Fr^2)\{[(p - p_a)/gd\rho_0] + [(x_3/d) - 1]\}, \qquad (2.8b)$$

where Fr^2 is the square of the Froude number; namely :

$$Fr^2 = (\nu_0/d)^2\, /gd. \qquad (2.9)$$

We note that the introduction of the pressure perturbation π, according to (2.8b), is deduced from a carefully dimensionless analysis of the NS-F model problem written with dimensionless variables :

$$v_i = u_i\, /(\nu_0/d),\ \ \theta \text{ and } \pi\ ;\ \ x'_i = x_i\, /d \ \text{ and } t' = t/(d^2/\nu_0\, , \qquad (2.10)$$

and gives the possibility to derive asymptoticaly, when B tends to zero, the Oberbeck-Boussinesq approximate equations. With (2.10) we derive the following dominant dimensionless equations, when we neglect the term proportional to B^2 :

$$\partial v_k/\partial x'_k = \mathrm{B}\, D\theta/Dt', \qquad (2.11a)$$

$$(1 - \mathrm{B}\theta)\, Dv_i/Dt' + \partial\pi\, /\partial x'_i - (\mathrm{B}\, /Fr^2)\, \theta\, \mathbf{k} - \nabla'^2 v_i = \mathrm{B}\, L(v_i\, , \theta), \qquad (2.11b)$$

$$[1 - \mathrm{B}\theta\, (\, 1 + \Gamma^\circ)]\, D\theta\, /Dt' - (1/Pr)\, \nabla'^2\theta$$
$$- 2\, Bq\, Fr^2\, (d'_{ij})^2 = \mathrm{B}N(v_i\, , \theta). \qquad (2.11c)$$

We precise that in (2.11a,b,c) we have :

$$d'_{ij} = (1/2)\, [\partial v_i/\partial x'_j + \partial v_j/\partial x'_i],\ \ D/Dt' = \partial/\partial t' + v_i\, \partial/\partial x'_i,$$
$$\nabla'^2 = \partial^2/\partial(x'_1)^2 + \partial^2/\partial(x'_2)^2 + \partial^2/\partial(x'_3)^2 \text{ and } p'_a = p_a/gd\rho_0, \qquad (2.12)$$
$$L(v_i\, , \theta) = [1 + (\lambda_0/\mu_0)]\, \partial/\partial x'_i\, (D\theta/Dt') - M^\circ\, \partial/\partial x'_j\, (2\, \theta\, d'_{ij}\,),$$
$$N(v_i\, , \theta) = Bq\, [p'_a\, + 1 - x'_3 + Fr^2\, \pi\,]\, D\theta\, /Dt' - 2\, Bq\, Fr^2\, M^\circ\, \theta\, (d'_{ij})^2$$
$$-\ (K^\circ/Pr)\, \partial/\partial x'_i\, [\theta(\partial\theta/\partial x'_i)],$$

where

$$\Gamma^\circ = [d\log C/dT/d\log\rho/dT]_0,\ M^\circ = [d\log\mu/dT/d\log\rho/dT]_0,$$
$$K^\circ = [d\log k/dT/d\log\rho/dT]_0. \qquad (2.13)$$

Of course, it is assumed that Γ°, M° and K° are $O(1)$, when $\beta \to 0$. For the derivation of the dominant equations (2.11a,b,c) we have also taking into account the following approximate relations, valid with an error of $O(\beta^2)$,

$$\rho(T) \cong \rho_0 [1 - \beta\,\theta],\ \mu(T) \cong \mu_0 [1 - \beta\, M^\circ\theta],\ k(T) \cong k_0 [1 - \beta\, K^\circ\theta],$$
$$C(T) \cong C_0 [1 - \beta\Gamma^\circ\theta]. \qquad (2.14)$$

In the above dimensionless equations and relations,

$$Pr = \nu_0/\kappa_0, \text{ and } Bq = g/\beta\, C_0, \qquad (2.15)$$

are the Prandtl and the analogous of a Boussinesq numbers respectively

2.4. Dimensionless Dominant Boundary Conditions

For these dominant equations (2.11a,b,c) it is necessary to derive a set of dominant dimensionless boundary conditions, from (2.4), (2.5a,b) and (2.6), for v_i, θ and π. We precise that, in the free surface boundary condition (2.5b), $\nabla_{||}$ is the surface gradient and H is the mean curvature :

$$\nabla_{||} = \nabla - \mathbf{n}\,(\mathbf{n}\cdot\nabla) \text{ and } H = -(1/2)\,(\nabla_{||}\cdot\mathbf{n}).$$

First, we write the free surface equation, (1.5), in the dimensionless form:

$$x'_3 = 1 + \delta\,\eta\,(t', x'_1, x'_2) \equiv h', \qquad (2.16)$$

where

$$\delta = a/d, \qquad (2.17)$$

is an amplitude dimensionless parameter for the free surface dimensionless deformation $\eta(t', x'_1, x'_2)$ relative to plane $x'_3 = 1$.

With (2.10) and (2.8a,b), from the boundary conditions (2.5a,b) and (2.6), we derive, first, the following five dominant dimensionless boundary conditions, at the deformable free surface, $x'_3 = 1 + \delta\,\eta\,(t', x'_1, x'_2) \equiv h'$:

$$\pi = (\delta/Fr^2)\eta + 2\, d'_{ij}\, n'_i\, n'_j + [We - Ma\,\theta](\nabla'_{||}\cdot\mathbf{n}') + \beta P(v_i, \theta), \quad (2.18a)$$

$$d'_{ij}\, t'^{(s)}_i\, n'_j + (1/2)\, Ma\, t'^{(s)}_i\, \partial\theta/\partial x'_i = \beta\, Q(v_i, \theta),\ s = 1, 2, \qquad (2.18b)$$

$$\nabla'\theta\cdot\mathbf{n}' + Bi\,\theta + 1 = \beta\, K^\circ\theta\, \nabla'\theta\cdot\mathbf{n}', \qquad (2.18c)$$

$$v_3 = \delta\,[\partial\eta/\partial t' + v_1\,\partial\eta/\partial x'_1 + v_2\,\partial\eta/\partial x'_2]. \qquad (2.18d)$$

In the two (s = 1 and 2) dimensionless dominant boundary conditions (2.18b), $t'^{(s)}_i$ and $t'^{(s)}_i$ are the dimensionless components of two orthonormal tangent vectors to the free surface $x'_3 = h'$ ($\equiv 1 + \delta\,\eta$), and n'_i are the dimensionless components of the outward normal $\mathbf{n}'$ to the free surface ; namely :

$$\mathbf{t}'^{(1)} = [1/\surd N'_1]\,(1;\ \ 0;\ \ \partial h'/\partial x'_1)\,;$$
$$\mathbf{t}'^{(2)} = [1/\surd N'_1 N']\,(-(\partial h'/\partial x'_1)(\partial h'/\partial x'_2);\ \ 1 + (\partial h'/\partial x'_1)^2;\ \ \partial h'/\partial x'_2); \qquad (2.19)$$
$$\mathbf{n}' = [1/\surd N']\,(-\partial h'/\partial x'_1;\ \ -\partial h'/\partial x'_2;\ \ 1).$$

According to Pavithran and Redeekopp (1994), tangential and normal unit vectors to the deformed free surface : $x'_3 = h'(t', x'_1, x'_2)$, are written (see (2.19)) in terms of the (x'_1, x'_2, x'_3) Cartesian system of coordinates. In (2.18a), for $(\nabla'_{||} \cdot \mathbf{n}')$ we have the following explicit formulae :

$$\nabla'_{||} \cdot \mathbf{n}' = -(N')^{-3/2}\{N'_2\,(\partial^2 h'/\partial x'_1\partial x'_1) + N'_1\,(\partial^2 h'/\partial x'_2\partial x'_2)$$
$$- 2\,(\partial^2 h'/\partial x'_1\partial x'_2)\,(\partial h'/\partial x'_1)\,(\partial h'/\partial x'_2)\}, \qquad (2.20a)$$

where

$$N'_1 = 1 + (\partial h'/\partial x'_1)^2;\ \ N'_2 = 1 + (\partial h'/\partial x'_2)^2;$$
$$N' = 1 + (\partial h'/\partial x'_1)^2 + (\partial h'/\partial x'_2)^2. \qquad (2.20b)$$

In (2.18a,b), for the terms $P(v_i\,,\theta)$ and $Q(v_i\,,\theta,)$ we have the following formulae:

$$P(v_i\,,\theta) = (\lambda_0/\mu_0)D\theta/Dt' - 2M^\circ\theta\,d'_{ij}\,n'_i\,n'_j\,; \qquad (2.21a)$$
$$Q(v_i\,,\theta) = M^\circ\theta\,d'_{ij}\,t'^{(s)}_i\,n'_j\,. \qquad (2.21b)$$

Finally, instead of (2.4) we write :

$$v_i = 0 \text{ and } \theta = 1, \text{ at } x'_3 = 0, \qquad (2.22)$$

if we take into account (2.8a) and (2.10).
In the dimensionless boundary conditions, (2.18a,b,c,d), the dimensionless parameters:

$$We = \sigma_0\,d/\rho_0(v_0)^2,\ \ Bi = q^\circ d/k_0,\ \ Ma = \gamma\,d^2\,\beta/\rho_0\,(v_0)^2 \qquad (2.23)$$

are, respectively, the Weber, Biot and Marangoni numbers (when we take into account (1.1a)). We note also that :

$$Cr = 1/We \text{ and } Bo = Cr/Fr^2, \qquad (2.24)$$

are, respectively, the crispation (or capillary) and the Bond numbers.

3 From NS-F Model Problem to Rayleigh-Bénard (R-B) Problem

In the above formulation of the classical Bénard thermal-free surface problem, via the NS-F model problem, there are two mechanism responsible for driving the instability : the first one is the density variation generated by the thermal expansion of the liquid, the second cause of the instability results from the surface tension gradients due to temperature fluctuations at the upper free surface of the liquid layer.
It is usual in the literature (see, for instance, Drazin and Reid 1981), to call *Rayleigh-Bénard* (R-B) shallow convection the instability problem produced mainly by *buoyancy* (with eventually a Marangoni and Biot effects in a *non-deformable free surface).*
Naturally, in the above full mathematical formulation of the Bénard thermal-free sueface problem - dominanant equations, (2.11a,b,c), with dominant boundary conditions, (2.18a,b,c,d), (2.22), for a weakly ($\beta \ll 1$) expansible viscous and heat conducting liquid - both, buoyancy and surface-gradient effects are operative, so it is important (from our point of view!) to ask :

How the two effects are coupled when $\beta \to 0$ *?*

Indeed, in order to extract a tractable problem from the complicated dimensionless system of dominanant equations, (2.11a,b,c), and dominant boundary conditions, (2.18a,b,c,d), (2.22), which we have just presented above, we will now undertake an asymptotic analysis based mainly upon the small parameter $\beta \to 0$!

More precisely, from equation (2.11b), when $\beta \to 0$, it is obvious that if we want to take into account the buoyancy term : $(\beta / Fr^2)\, \theta\, \mathbf{k}$, then it is necessary to impose (according to Zeytounian 1989) the following "*Boussinesq limit process*"

$$\beta \to 0 \text{ and } Fr^2 \to 0, \text{ with } \beta / Fr^2 \equiv Gr = O(1), \tag{3.1}$$

where

$$Gr = \beta\, \alpha_T\, d^4\, g/(\nu_0)^2 \tag{3.2a}$$

is the Grashof number and the associated Rayleigh (Ra) number is

$$Ra = Pr\, Gr = \beta\, \alpha_T\, d^4\, g/\, \nu_0\, \kappa_0\, . \tag{3.2b}$$

Now, if we consider the free surface boundary condition (2.18a) for π then, as consequence of $Fr^2 \ll 1$, the first term, $(\delta / F^2)\, \eta$, in the right hand side of (2.18a) is unbounded when $Fr^2 \to 0$!

Thus, when $Fr^2 << 1$:

The free surface pressure condition (2.18a) is (asymptotically) consistent with the Boussinesq limit process (3.1), only for a small free surface amplitude parameter δ, *such that:*

$$\delta / Fr^2 = \delta^* = O(1), \text{ when } \delta \to 0 \text{ and } Fr^2 \to 0. \qquad (3.3)$$

The above Boussinesq limit process, (3.1), is valid only for the *thickness* d of the liquid layer such that :

$$d >> [(\nu_0)^2 /g]^{1/3} \approx 1\text{mm}. \qquad (3.4)$$

3.1 Boussinesq Approximate Equations

If now all the dependent and independent variables, the parameters Pr and Bq, the coefficient $\Gamma°$ in (2.11c), and the terms $L(v_i , \theta)$ and $N(v_i , \theta)$ in (2.11b,c) are fixed and of the order of O(1), we derive asymptotically the following classical shallow convection dimensionless (Boussinesq) equations (written without the primes) for the limit functions v_{i0}, π_0, θ_0 :

$$\partial v_{k0}/\partial x'_k = 0, \qquad (3.5a)$$

$$Dv_{i0}/Dt + \partial\pi_0 /\partial x_i - Gr\ \theta_0\ \mathbf{k} = \nabla^2 v_{i0}, \qquad (3.5b)$$

$$D\theta_0/Dt = (1/Pr)\ \nabla^2\theta_0. \qquad (3.5c)$$

In these Boussinesq equations (3.5a,b,c) we have :

$$[v_{i0} , \pi_0 , \theta_0] = \lim\ [v_i , \pi , \theta], \text{ when } \beta\to 0, Fr^2 \to 0, \text{ with } \beta/F^2 \equiv Gr = O(1). \qquad (3.6)$$

As : Bo = O(1) the above Boussinesq model equations (3.5a,b,c) are valid when

$$\beta \cong g/C_0. \qquad (3.7)$$

3.2 Formulation of the R-B Rigid - Free Model Problem with the Buoyancy, Marangoni and Biot Effects

If now we take into account (3.3) and consider the case of the large Weber numbers, We >> 1, such that :

$$\delta\ We = We^* = O(1), \qquad (3.8)$$

then, as consequence of $\delta << 1$, we derive the following R-B (rigid-free) problem, which take into account at the non-deformable free surface $x_3 = 1$, the buoyancy, Marangoni and Biot effects :

$$\partial v_{k0}/\partial x'_k = 0,$$
$$Dv_{i0}/Dt + \partial\pi_0/\partial x_i - Gr\ \theta_0\ \mathbf{k} = \nabla^2 v_{i0},\ i = 1,2,3, \tag{3.9a}$$
$$D\theta_0/Dt = (1/Pr)\ \nabla^2\theta_0,$$

$$v_{10} = v_{20} = v_{30} = 0, \text{ and } \theta = 1, \text{ at } x_3 = 0,$$
$$v_{30} = 0, \text{ and } \partial^2 v_{30}/\partial x_3^2 = Ma\ [\partial^2\theta/\partial x_1^2 + \partial^2\theta/\partial x_2^2], \text{ at } x_3 = 1, \tag{3.9b}$$
$$\partial\theta/\partial x_3 + Bi\ \theta + 1 = 0, \text{ at } x_3 = 1.$$

The deformation of the free surface $\eta(t, x_1, x_2)$ is determined, when the perturbation of the pressure $\pi_0(t, x_1, x_2, x_3 = 1)$ is known at $x_3 = 1$(after the resolution of the R-B problem (3.9a,b)), by the equation :

$$\partial^2\eta/\partial x_1^2 + \partial^2\eta/\partial x_2^2 - (\delta^*/We^*)\ \eta = -(1/We^*)\ \pi_0\ (t, x_1, x_2, 1). \tag{3.10}$$

It is important to note that the R-B problem (3.9a,b) is asymptotically derived from the dominant (relative to $\beta >> 1$) problem formulated in the §2, when :

$$\beta \to 0,\ Fr^2 \to 0,\ \delta \to 0 \text{ and } Cr \to 0,$$

with the following three similarity relations:

$$\beta/Fr^2 = Gr = O(1),\ \delta/Fr^2 = \delta^* = O(1),\ \delta/Cr = We^* = O(1) \tag{3.11}$$

aand in a such case

$$Bo = Cr/Fr^2 = O(1). \tag{3.12}$$

When:

$$\delta^* = We^* = Bi = Ma \equiv 0,$$

we obtain the classical R-B rigid-free shallow convection problem. The R-B problem (3.9a,b) has been recently considered by Dauby and Lebon (1996).

Finally, it is necessary to note that, it is *not consistent* (from an asymptotic point of view) *to take into account simultaneously* in the R-B thermal convection model problem (for a weakly expansible liquid) the *buoyancy* effect and the *deformation of the free surface* (according to Zeytounian 1997, 1998).

4 From NS-F Model Problem to Bénard -Marangoni (B-M) Problem

Indeed, the Bénard (1900), famous *convective cells* are primarily induced by the surface tension gradients resulting from temperature variations across the free surface (the so-called, Marangoni effect). For this, it seems justified to use the term of Bénard-Marangoni (B-M) thermocapillary instability problem when, as in Bénard's experiments, the dominant acting driving force is the surface tension gradient on the deformable free surface (without the influence of the buoyancy force).

From (2.18a), obviously, for the derivation of the full B-M model thin film problem, when the free surface deformation [$\delta = O(1)$ in equation $h' = 1 + \delta\,\eta$] plays an essential role, it is necessary to assume that the Froude number Fr is O(1). As consequence we must consider the following *incompressible limit process* :

$$\beta \to 0 \text{ and } Fr^2 = O(1) . \tag{4.1}$$

But in this case, for the most liquids : $Bq \ll 1$, since

$$C_0\,(\beta d)/g \gg d \approx (\nu_0^2/g)^{1/3}. \tag{4.2}$$

4.1 Formulation of the Full B-M Model Problem

As consequence of (4.1) the buoyancy term in equation (2.11b) is negligible for a weakly expansible liquid, and as consequence, in place of Boussinesq equations (3.5a,b,c), we derive from the dominant dimensionless equations (2.11a,b,c) the following "incompressible model equations" :

$$\begin{aligned} \partial v_{k0}/\partial x_k &= 0, \\ Dv_{i0}/Dt + \partial \pi_0/\partial x_i &= \nabla^2 v_{i0}, \\ D\theta_0/Dt &= (1/Pr)\,\nabla^2 \theta_0, \end{aligned} \tag{4.3a,b,c}$$

where $D/Dt = \partial/\partial t + v_{i0}\,\partial/\partial x_i$.

For these above equations (4.3a,b,c), since $Fr^2 = O(1)$ and $\delta = O(1)$, we must write the *deformable* free - surface conditions. These below boundary conditions (4.5a,b,c,d) are written for the unknowns : v_{i0}, π_0, θ_0, and as consequence we obtain a coupled B-M problem for the limit functions:

$$[v_{i0}, \pi_0, \theta_0] = \lim_{\beta \to 0,\, Fr^2 = O(1)} [v_i, \pi, \theta], \tag{4.4}$$

and also $h(t, x_1, x_2)$ - the thickness of the unknown film free surface.
According to the dominant dimensionless boundary conditions (2.18a,b,c,d) we derive a set of dimensionless very complicated conditions written at the deformable free surface :$x_3 = h(t, x_1, x_2)$, namely :

$$\pi_0 = (h - 1)/Fr^2 + (2/N)\{\partial v_{10}/\partial x_1 (\partial h/\partial x_1)^2 + \partial v_{02}/\partial x_2 (\partial h_{1/}\partial x_2)^2$$
$$+ \partial v_{30}/\partial x_3 + [\partial v_{10}/\partial x_2 + \partial v_{20}/\partial x_1](\partial h/\partial x_1)(\partial h/\partial x_2)$$
$$- [\partial v_{10}/\partial x_3 + \partial v_{30}/\partial x_1](\partial h/\partial x_1) - [\partial v_{20}/\partial x_3 + \partial v_{30}/\partial x_2](\partial h/\partial x_2)\}$$
$$- [(We - Ma\,\theta_0)/N^{3/2}]\{N_2 (\partial^2 h/\partial x_1 \partial x_1) + N_1 (\partial^2 h/\partial x_2 \partial x_2)$$
$$- 2 (\partial^2 h/\partial x_1 \partial x_2)(\partial h/\partial x_1)(\partial h/\partial x_2)\}; \quad (4.5a)$$

$$[\partial v_{10}/\partial x_1 - \partial v_{30}/\partial x_3](\partial h/\partial x_1) + (1/2)[\partial v_{10}/\partial x_2 + \partial v_{20}/\partial x_1](\partial h/\partial x_2)$$
$$+ (1/2)[\partial v_{20}/\partial x_3 + \partial v_{30}/\partial x_2](\partial h/\partial x_1)(\partial h/\partial x_2) - (1/2)[1 - (\partial h/\partial x_1)^2][\partial v_{30}/\partial x_1$$
$$+ \partial v_{10}/\partial x_3] = (Ma/2)[\partial\theta_0/\partial x_1 + (\partial h/\partial x_1)\partial\theta_0/\partial x_3] N^{1/2}; \quad (4.5b)$$

$$[\partial v_{10}/\partial x_1 - \partial v_{20}/\partial x_2](\partial h/\partial x_2)(\partial h/\partial x_1)^2 + [\partial v_{10}/\partial x_3 + \partial v_{30}/\partial x_1](\partial h/\partial x_1)(\partial h/\partial x_2)$$
$$+ [\partial v_{20}/\partial x_2 - \partial v_{30}/\partial x_3](\partial h/\partial x_2)$$
$$+ (1/2)[1 + (\partial h/\partial x_1)^2 - (\partial h/\partial x_2)^2][\partial v_{10}/\partial x_2 + \partial v_{20}/\partial x_1](\partial h/\partial x_1)$$
$$- (1/2)[1 + (\partial h/\partial x_1)^2 - (\partial h/\partial x_2)^2][\partial v_{20}/\partial x_3 + \partial v_{30}/\partial x_2]$$
$$= (Ma/2)\{-(\partial h/\partial x_1)(\partial h/\partial x_2)\partial\theta_0/\partial x_1 + [1 + (\partial h/\partial x_1)^2]\partial\theta_0/\partial x_2$$
$$+ (\partial h/\partial x_2)\partial\theta_0/\partial x_3\} N^{1/2}; \quad (4.5c)$$

$$\partial\theta_0/\partial x_3 = (\partial h/\partial x_1)\partial\theta_0/\partial x_1 + (\partial h/\partial x_2)\partial\theta_0/\partial x_2 - (Bi\,\theta_0 + 1) N^{1/2}. \quad (4.5d)$$

We note that, according to (2.20b),

$N_1 = 1 + (\partial h/\partial x_1)^2$; $N_2 = 1 + (\partial h/\partial x_2)^2$; $N = 1 + (\partial h/\partial x_1)^2 + (\partial h/\partial x_2)^2$.

Finally, in place of (2.18d) and (2.22) we write :

$$v_{30} = \partial h/\partial t + v_{10}\, \partial h/\partial x_1 + v_{20}\, \partial h/\partial x_2, \text{ at } x_3 = h(t, x_1, x_2), \tag{4.6}$$

$$v_{10} = v_{20} = v_{30} = 0\,, \; \theta_0 = 1, \text{ at } x_3 = 0. \tag{4.7}$$

A similar to [(4.3a,b,c), (4.5a,b,c,d), (4.6), (4.7)] B-M problem was considered recently by Nepomnyaschy and Velarde (1994), but the above boundary conditions (4.5a,b,c) are somewhat different [these conditions (4.5a,b,c) are consistent with exact starting boundary condition (2.5a) or boundary conditions (2.18a,b) if we take into account the definition (2.19), with (2.20b), of the unit vectors tangential ($\mathbf{t}'^{(1)}$ and $\mathbf{t}'^{(2)}$) and normal ($\mathbf{n}$) to the deformed free surface, under the incompressible limit (4.1).

The full B-M model problem, [(4.3a,b,c), (4.5a,b,c,d), (4.6), (4.7)], is a very complicated problem, but for a thin film flow we can apply the long wave approximation. In this case, we derive (as in next Section 4.2) a simplified Bénard-Marangoni Boundary Layer (B-MBL) model problem for the high Reynolds numbers.

4.2 The B-MBL Long Wave Thin Film Model Problem

According to long waves approximation, we assume that the characteristic value of the horizontal (in the x and y directions) wave-length $\lambda \gg d$. In this case, instead of dimensionless variables (t', x'_1, x'_2, x'_3), it is judicious to introduce the following news dimensionless variables :

$$x = \varepsilon\, x'_1,\; y = \varepsilon\, x'_2,\; z \equiv x'_3,\; \tau = \varepsilon \,\mathrm{Re}\, t', \tag{4.8a}$$

and news functions:

$$u = v_{10}/\mathrm{Re},\; v = v_{20}/\mathrm{Re},\; w = v_{30}/\varepsilon\, \mathrm{Re},\; \Pi = \pi_0/(\mathrm{Re})^2, \tag{4.8b}$$

in the above, Section 4.1, formulated dimensionless full B-M problem . In (4.8a,b)

$$\varepsilon = d/\lambda, \tag{4.9a}$$

is the "long wave dimensionless parameter and

$$\mathrm{Re} = U^\circ\, d/\nu_0. \tag{4.9b}$$

the Reynolds number. The choice of the characteristic velocity U° is such that :

$$\varepsilon\, \mathrm{Re}/F^2 = 1 \Rightarrow U^\circ = g\, d^3/\lambda\, \nu_0, \tag{4.10}$$

where F is a modified Froude number, namely :

$$F = U^\circ/(g\, d)^{1/2}. \tag{4.11}$$

In place of dimensionless parameters We and Ma we introduce corresponding modified Weber and Marangoni numbers based on the chatacteristic velocity U° :

$$W = \sigma_0 / \rho_0 \, d U^{\circ 2}, \quad M = \gamma \beta / \rho_0 U^{\circ 2} . \tag{4.12}$$

When

$$\varepsilon \to 0, \ \mathrm{Re} \to \infty \text{ and } We \to \infty \tag{4.13a}$$

with the following similarity relations

$$\varepsilon \, \mathrm{Re} = \mathrm{Re}^* = O(1) \text{ and } \varepsilon^2 W = W^* = O(1), \tag{4.3b}$$

in place of the full B-M model problem formulated in the Section 4.1, we derive for the functions :

$$\mathbf{V} = \big(u(\tau, x, y, z), v(\tau, x, y, z) \big), \ w(\tau, x, y, z), \ \Pi(\tau, x, y, z),$$
$$\theta_0 = \Theta(\tau, x, y, z), \ h' = H(\tau, x, y),$$

the following approximate B-M_{BL} model thin film problem :

$$\begin{gathered} \mathbf{D} \cdot \mathbf{V} + \partial w / \partial z = 0, \\ D\mathbf{V}/D\tau + \mathbf{D}\Pi = (1/\mathrm{Re}^*) \partial^2 \mathbf{V} / \partial z^2, \\ \partial \Pi / \partial z = 0 \Rightarrow \Pi = (H - 1)/\mathrm{Re}^* - W^* \mathbf{D}^2 H, \\ \Pr D\Theta/D\tau = (1/\mathrm{Re}^*) \partial^2 \Theta / \partial z^2, \end{gathered} \tag{4.14a}$$

at $z = 0$:

$$u = v = w = 0 \text{ and } \Theta = 1, \tag{4.14b}$$

at $z = H(\tau, x, y)$:

$$\begin{gathered} \partial \mathbf{V} / \partial z = - \mathrm{Re}^* M \, [\mathbf{D}\Theta + (\mathbf{D}H) \, \partial \Theta / \partial z], \\ \partial \Theta / \partial z + 1 + \mathrm{Bi} \, \Theta = 0, \\ w = \partial H / \partial \tau + \mathbf{V} \cdot \mathbf{D}H. \end{gathered} \tag{4.14c}$$

In these above BM_{BL} model equations we have the following operators :

$$D/D\tau = \partial / \partial \tau + \mathbf{V} \cdot \mathbf{D} + w \, \partial / \partial z \ , \quad \mathbf{D} = (\partial / \partial x, \partial / \partial y),$$

and

$$\mathbf{D}^2 = \partial^2 / \partial x^2 + \partial^2 / \partial y^2 .$$

In the reduced B-M_{BL} problem (4.14a,b,c) the parameters M, Pr, Bi, Re* and W* are O(1). The above BM_{BL} problem is a very sinificant approximate model for the investigation of the thin, slightly viscous, film instability problem and this model deserves further consideration.

4.3 The B- M_{BL} Problem for a vanishing Prandtl Number

In order to derive, from the above problem (4.14a,b,c), a more simple model problem for $\mathbf{V}(\tau, x, y, z)$ and $H(\tau, x, y)$, we consider, below, the case when :

$$\Pr \to 0, \text{ in the equation (4.14a), for } \Theta.$$

In this case the for the temperature perturbation Θ we obtain the following reduced linear problem :

$\partial^2\Theta/\partial z^2 = 0$; $\Theta = 1$, at $z = 0$, and $\partial\Theta/\partial z + 1 + \mathrm{Bi}\,\Theta = 0$, at $z = H(\tau, x, y)$,
and the solution is very simple :

$$\Theta = 1 - [(1 + \mathrm{Bi})/(1 + \mathrm{Bi}\,H)]z. \tag{4.15}$$

On the other hand (for all Pr and Re* fixed) from the continuity equation of the above problem (II,14), with the conditions :

$$w = 0, \text{ at } z = 0, \text{ and } w = \partial H/\partial\tau + \mathbf{V}\cdot\mathbf{D}H, \text{ at } z = H,$$

we derive the following averaged (evolution) equation :

$$\partial H/\partial\tau + \mathbf{D}\cdot\int_0^H \mathbf{V}\,dz = 0. \tag{4.16a}$$

As consequence, in this case, we derive for the two functions $\mathbf{V}(\tau, x, y, z)$, and $H(\tau, x, y)$, a system of two nonlinear evolution equations ; namely :(4.16a) and

$$D\mathbf{V}/D\tau - (1/\mathrm{Re}^*)\,\partial^2\mathbf{V}/\partial z^2 = -\,(1/\mathrm{Re}^*)\;\mathbf{D}H + W^*\,\mathbf{D}\,[\mathbf{D}^2H], \tag{4.16b}$$

since

$$w = -\int_0^z (\mathbf{D}\cdot\mathbf{V})\,dz.$$

For these two equations (4.16a,b) we have as boundary conditions (in z) :

$$\mathbf{V} = 0, \text{ at } z = 0, \tag{4.17a}$$

and

$$\partial\mathbf{V}/\partial z = M\,\mathrm{Re}^*\,(1 + \mathrm{Bi})\,[\mathbf{D}H/(1 + \mathrm{Bi}\,H)^2], \text{ at } z = H. \tag{4.17b}$$

The above problem (4.16a,b), (4.17a,b) is again a complicated problem for the two functions $\mathbf{V}(\tau, x, y, z)$, and $H(\tau, x, y)$, since in (4.16b),

$$D\mathbf{V}/D\tau = \partial\mathbf{V}/\partial\tau + (\mathbf{V}\cdot\mathbf{D})\,\mathbf{V} + w\,\partial\mathbf{V}/\partial z, \text{ with } w = -\int_0^z (\mathrm{D}\cdot\mathbf{V})\,dz!$$

But, it seems that, a numerical computation is more easy, than for the preceding, (4.14a,b,c) , B-MBL long-wave thin film model problem.

4.4 A Lubrication Evolution Equation for the Thickness of the Film

But, if we assume that:

$$Re^* << 1 \text{ in the equation (4.16b),}$$

then, in limit :

$$Re^* \to 0 \text{ and for large } W^* \text{ and } M, \tag{4.18}$$

from (4.16b), we obtain, with (4.17a,b), the following limit solution for the horizontal velocity :

$$\mathbf{V}(z; H) = B(1 + Bi)[1/(1 + Bi\,H)^2](\mathbf{D}H)\,z + (1/2)\{\mathbf{D}H - A\,\mathbf{D}\,[\mathbf{D}^2H\}[z^2 - 2H\,z], \tag{4.19}$$

where

$$B = \gamma\,\beta/\rho_0\,g\,d, \text{ and } A = \sigma_0/\rho_0\,g\,\lambda^2, \tag{4.20}$$

if we assume that

$$Re^*\,W^* = O(1) \quad \text{and } Re^*M = O(1),$$

and utilize the definition of U° (see (4.10)). The two coeffeicients A and B are very significant and it is important to note that the Marangoni effect (proportional to the dimernsionless parameter B) is different to zero for a vanishing Biot effect, when $Bi \to 0$.

As consequence of (4.16a) and (4.19), in place of the problem (4.16a,b) - (4.17a,b), when $Re^* \to 0$, we derive a single evolution equation for the thickness of the film $H(\tau, x, y)$:

$$\partial H/\partial\tau + (1/3)\,\mathbf{D}\cdot\{H^3[A\,\mathbf{D}\,(\mathbf{D}^2H) - \mathbf{D}H] + B(1 + Bi)[1/(1 + Bi\,H)^2]\,H^2(\mathbf{D}H)\} = 0. \tag{4.21}$$

The above, (4.21), lubrication equation, is very similar to the one derived by Oron and Rosenau (1992). Unfortunately, in the equation derived by these above authors, the term responsible of the Marangoni effect is vanishing with a vanishing Biot number ($Bi \to 0$)! This is clearly a paradoxical conclusion?

The term proportional to A, has a stabilizing effect while the term proportional to B (which take into account the Marangoni effect), has, on the contrary, a destabilizing impact.

It should be noted that the gravity (term : $-(1/3)\,\mathbf{D}\cdot[H^3\,\mathbf{D}H]$), in (4.21) stabilizes the evolution of the interface when the film is supported from below.

4.5 The B-M Problem for a Free-Falling Vertical Film

In what follows we consider the thermocapillary instabilities on a free-falling vertical two-dimensional (2D) film, since most experiments and theories have focused on the latter - in fact, wave dynamics on an inclined plane is quite analogous.
Using dimensionless variables and functions, the governing equations and boundary conditions for the (incompressible but thermally conducting) liquid motion down vertical plane are as follows :

$$\partial u/\partial x + \partial w/\partial z = 0, \tag{4.22a}$$

$$Du/D\tau + \partial\Pi/\partial x - 1/\varepsilon F^2 = (1/\varepsilon \, Re)\,[\partial^2 u/\partial z^2 + \varepsilon^2\, \partial^2 u/\partial x^2], \tag{4.22b}$$

$$\varepsilon^2\, Dw/D\tau + \partial\Pi/\partial z = (\varepsilon/Re)\,[\partial^2 w/\partial z^2 + \varepsilon^2\, \partial^2 w/\partial x^2], \tag{4.22c}$$

$$Pr\, D\Theta/D\tau = (1/\varepsilon\, Re)\,[\partial^2\Theta/\partial z^2 + \varepsilon^2\, \partial^2\Theta/\partial x^2], \tag{4.22d}$$

with $D/D\tau = \partial/\partial\tau + u\partial/\partial x + w\partial/\partial z$;

at $z = 0$:

$$u = w = 0 \text{ and } \Theta = 1; \tag{4.23}$$

at $z = H(\tau, x)$:

$$\begin{aligned}\partial u/\partial z = &- \varepsilon M\,[\partial\Theta/\partial x + (\partial H/\partial x)\, \partial\Theta/\partial z] \\ &- \varepsilon^2\,[\partial w/\partial x + 4(\partial H/\partial x)\, \partial w/\partial z] \\ &- (3/2)\,\varepsilon^3 M\,(\partial H/\partial x)^2\,[\partial\Theta/\partial x + (\partial H/\partial x)\,\partial\Theta/\partial z]\,;\end{aligned} \tag{4.24a}$$

$$\begin{aligned}\Pi = &\ 2(\varepsilon/Re)\,[\partial w/\partial z - (\partial H/\partial x)\, \partial u/\partial z] \\ &- \varepsilon^2\, W\, \partial^2 H/\partial x^2 + \varepsilon^2\,(M/Re)\,(\partial^2 H/\partial x^2)\,\Theta \\ &+ 2(\varepsilon^3/Re)\,[(\partial H/\partial x)^3\, \partial u/\partial z - 2\,(\partial H/\partial x)^2\, \partial w/\partial z \\ &- (\partial H/\partial x)\, \partial w/\partial x]\,;\end{aligned} \tag{4.24b}$$

$$\begin{aligned}\partial\Theta/\partial z = &- (1 + Bi\,\Theta) + \varepsilon^2\,[(\partial H/\partial x)\,\partial\Theta/\partial x \\ &- (1/2)(\partial H/\partial x)^2\,(1 + Bi\,\Theta)]\,;\end{aligned} \tag{4.24c}$$

$$w = \partial H/\partial\tau + u\, \partial H/\partial x. \tag{4.24d}$$

We precise that the deformable free surface boundary conditions (at $z = H(\tau, x)$), (4.24a,b,c), are written with an error of $O(\varepsilon^4)$.

5 Integral Boudary Layer B-M Model for the Free-Falling Vertical Film Problem

We consider the B-M problem for a free-falling vertical film, formulated in above Section 4.5. First, in Section 5.1, we reduce the problem [(4.22a,b,c,d), (4.23), (4.24a,b,c,d)] to a boundary layer form, according to the following limit process:

$$\varepsilon << 1,\ Re >> 1,\ \text{but}\ \ \varepsilon\, Re = Re^* = O(1). \tag{5.1}$$

Obviously, from the continuity equation (4.22a), the kinematic condition (4.24d)) and condition $w = 0$ at $z = 0$, we derive easily the following averaged continuity equation :

$$\partial H/\partial \tau + \partial q/\partial x = 0, \tag{5.2}$$

where

$$q(\tau, x) = \int_0^{H(\tau, x)} u(\tau, x, z)\, dz. \tag{5.3}$$

The function $q(\tau, x)$ is the average over the film thickness of the (parallel to the solid vertical plate $z = 0$) component u of the velocity. Due to the fact that $q(\tau, x)$, as a general rule, cannot be expressed in terms of $H(\tau, x)$ or some of its spatial derivatives, equation (5.2) is not a closed form evolution equation for the thickness $H(\tau, x)$. Our main goal, in this §5 is to examine a particular situation, arising mainly from averaging process, from which a closed system of equations may be obtained.

5.1 Boundary Layer B-M Model Problem for a Free-Falling Vertical Film

We consider, again, as in Section 4.2, the situation corresponding to long-waves ($\varepsilon << 1$) and large Reynolds numbers ($Re >> 1$), according to (5.1). With (5.1), in place of the equation (4.22c) for w, we obtain the limit equation $\partial \Pi/\partial z = 0$ and in this case, when ε tends to zero, according to boundary conditions (4.24b), we derive the following relation between Π and $H(\tau, x)$:

$$\Pi = \Pi(H) \equiv -W^*\, \partial^2 H/\partial x^2, \tag{5.4}$$

if we assume that (large Weber numbers) :

$$\varepsilon^2\, W = W^* = O(1). \tag{5.5}$$

As consequence of (5.4), the following leading order limit, boundary-layer (BL), equations for u, w, Θ and h are derived :

$$\partial u/\partial x + \partial w/\partial z = 0, \tag{5.6a}$$

$$Re^* \, Du/D\tau - \partial^2 u/\partial z^2 = 1 + (1/K^*) \, \partial^3 H/\partial x^3, \tag{5.6b}$$
$$Re^* \, Pr \, D\Theta/D\tau = \partial^2\Theta/\partial z^2, \tag{5.6c}$$

with

$$1/K^* = Re^* \, W^* = O(1) \text{ and if we assume that : } Re/F^2 = 1.$$

For the equations (5.6a,b,c)) we have the following reduced boundary conditions

$$u = w = 0 \text{ and } \Theta = 1, \text{ at } z = 0, \tag{5.7a}$$
$$\partial u/\partial z = - M^* \, [\partial\Theta/\partial x + (\partial H/\partial x) \, \partial\Theta/\partial z], \quad \text{at } z = H(\tau, x), \tag{5.7b}$$
$$\partial\Theta/\partial z = - (1 + Bi \, \Theta), \text{ at } z = H(\tau, x), \tag{5.7c}$$
$$w = \partial H/\partial\tau + u \, \partial H/\partial x., \text{ at } z = H(\tau, x), \tag{5.7d}$$

if we assume that (large Marangoni numbers) :

$$\varepsilon \, M = M^* = O(1). \tag{5.8}$$

The above reduced boundary layer problem [(5.6a,b,c), (5.7a,b,c,d)] remains a complicated boundary value problem. This problem is, in fact, very similar to problem (4.14a,b,c)), but it is formulated for a two-dimensional free-falling vertical film. When $M^* = 0$ (in this case the thermal field is decoupled to dynamical one), Shkadov (1967), using the integral method have reduced the problem to a system of two averaged equations for $H(\tau, x)$ and $q(\tau, x)$, to use the self-similarity assumption of the horizontal velocity component u.
Indeed, using the integral method, we can write in place of (5.6b) and (5.6c) the following two, averaged, integral boundary layer (IBL), equations :

$$Re^*\left\{\partial q/\partial\tau + \partial/\partial x \left[\int_0^{H(\tau, x)} u^2 \, dz\right]\right\} + (\partial u/\partial z)_{z=0} = - M^* \, [\partial\Theta/\partial x + (\partial H/\partial x) \, \partial\Theta/\partial z]$$
$$+ (1/K^*) \, H\partial^3 H/\partial x^3 + H, \tag{5.9}$$

and

$$Pr \, Re^*\left\{\partial\chi/\partial\tau + \partial/\partial x \left[\int_0^{H(\tau, x)} u \, [\Theta - (1 - z)] \, dz\right] - \int_0^{H(\tau, x)} w \, dz]\right\}$$
$$+ (\partial\Theta/\partial z)_{z=0} + 1 = 0, \tag{5.10}$$

where

$$\chi(\tau, x) = \int_0^{H(\tau, x)} [\Theta(\tau, x, z) - (1 - z)] \, dz, \tag{5.11}$$

when we use the boundary conditions (5.7b) and (5.7c) with Bi = 0.

5.2 The Shkadov Averaged Classical IBL Model ($M^* = 0$)

When $M^* = 0$, the term proportional to M^* in (5.9) disappears and the thermal field is decoupled to dynamical one. In a such case (when $M^* = 0$):

$$u = 0, \text{ at } z = 0 \text{ and } \partial u/\partial z = 0, \text{ at } z = H(\tau, x),$$

and we can use as self-similarity assumption for u(t, x, z) :

$$u(\tau, x, z) = [U(\tau, x)/H(\tau, x)]\, [z - (1/2h(t, x))\, z^2], \tag{5.12}$$

where $U(\tau, x)$ is an arbitrary unknown function. With (5.12) we easily derive, in place of (5.9) the folowing averaged equation betwwen $H(\tau,x)$ and $q(\tau, x)$:

$$Re^*\{\partial q/\partial\tau + (6/5)\, \partial/\partial x\, [q^2/H\,]\} + (3/H^2)\, q = (1/K^*)\, H\, \partial^3 H/\partial x^3 + H, \tag{5.13}$$

when we assume that the capillary number $K^* = O(1)$. We note that with (5.3) and (5.12) we have the following relation between $q(\tau, x)$ and $U(\tau, x)$:

$$U(\tau, x) = [3/H(\tau, x)]\, q(\tau, x)\, . \tag{5.14a}$$

The system of two evolution equations, (5.2) and (5.13), for $H(\tau, x)$ and $q(\tau, x)$, forms the *Shkadov averaged, IBL model.*

5.3 Averaged IBL Model with Marangoni Effect, but Pr = 0

When $M^* \neq 0$ and $Pr = 0$, the horizontal velocity u satisfy the following two boundary conditions :

$$u = 0 \text{ at } z = 0 \text{ and } \partial u/\partial z = M^*\, (1 + Bi)\, [\partial H/\partial x/(1 + BiH)^2] \text{ at } z = H(\tau, x),$$

when we take into account the solution (4.15) for Θ. As consequence, in place of (5.12) we write :

$$u(\tau, x, z) = [U(\tau, x)/H(\tau, x)]\, [z - (1/2H(\tau, x))\, z^2] + M^*\, (1 + Bi)\, [\partial H/\partial x/(1 + BiH)^2]\, z, \tag{5.15}$$

which satisfy the above boundary conditions for $u(\tau, x, z)$ at $z = 0$ and $z = H(\tau, x)$. In this case, we obtain (in place of (5.13))

$$U(\tau, x) = (3/H)\, q - (3/2)\, M^*[(1 + Bi)/(1 + BiH)^2]\, H\, (\partial H/\partial x), \tag{5.14b}$$

and from (5.9) we derive a second (more complicated) averaged equation between $H(\tau, x)$ and $q(\tau, x)$:

$$\begin{aligned}&\mathrm{Re}^*\{\partial q/\partial\tau + (6/5)\,\partial/\partial x[q^2/H]\\
&+ (1/20)M^*\,(1 + Bi)\,\partial/\partial x\,[H(\partial H/\partial x)\,q/(1 + BiH)^2\,]\} + (3/H^2)\,q\\
&= (1/K^*)\,H\,\partial^3H/\partial x^3 + H + (3/2)\,M^*(1 + Bi)[\partial H/\partial x/(1 + BiH)^2]\\
&- (1/120)\,\mathrm{Re}^*M^{*2}\,(1 + Bi)^2\,\partial/\partial x\,[(\partial H/\partial x)^2\,H^3/(1 + BiH)^4].\end{aligned} \tag{5.16}$$

The above system of two equations (5.2) and (5.16), for two functions $H(\tau, x)$ and $q(\tau, x)$, generalizes the Shkadov classical system for the case when the Marangoni effect is important and Biot effect is taken into account, but $Pr = 0$

5.4 Averaged IBL Model with Marangoni and Prandtl Effects, but Bi = 0

This case is more complicated, since for $Pr \neq 0$ it is necessary to derive a third averaged equation from (5.10), which is valid only for $Bi = 0$. At the present time, it is not clear if it is possible to derive a such averaged equation for the case of $Bi.\neq 0$! When $Bi = 0$, we can write the condition :

$$\partial u/\partial z = M^*\,[\partial H/\partial x - \partial Q\,/\partial x],\ \text{at}\ z = H(\tau, x), \tag{5.,17}$$

with $Q(\tau, x) \equiv \Theta - 1 + z$, and we can assume the following self-similarity for the solution $Q(\tau, x, z)$:

$$Q(\tau, x, z) = 2\,[1 - \Sigma(\tau, x)/H(\tau, x)]\,(z - (1/2H(\tau, x))\,z^2). \tag{5.18}$$

Our problem is now to derive two averaged equations for three unknown functions $H(\tau, x)$, $q(\tau,x)$ and $\Sigma(\tau, x)$. First, we obtain easily:

$$\text{at}\ z = H(\tau, x) : \partial Q/\partial x = \partial H/\partial x - \partial\Sigma/\partial x\ \ \text{and}\ \partial u/\partial z = M^*\,\partial\Sigma/\partial x. \tag{5.19}$$

As consequence of (5.19) we write for $u(\tau, x, z)$ [by analogy with (5.15), but with $Bi = 0$]:

$$\begin{aligned}u(\tau, x, z) = [U(t, x)/H(\tau, x)]\,(z - (1/2H(\tau, x))\,z^2)&\\
+\ M^*\,(\partial\Sigma/\partial x)\,z,&\end{aligned} \tag{5.20}$$

and we obtain

$$q(\tau, x)\ = (1/3)\,UH + (1/2)\,M^*\,H^2(\partial\Sigma/\partial x), \tag{5.21}$$

and (since $Bi = 0$)

$$U(t, x) = (3/H)\,q - (3/2)\,M^*\,H(\partial\Sigma/\partial x). \tag{5.22}$$

By analogy with (5.16) we derive the following averaged equation (in terms proportional to M^* we replace $\partial H/\partial x$ by $\partial \Sigma/\partial x$ and we assume that $Bi = 0$) :

$$\begin{aligned} Re^*\{\partial q/\partial t + (6/5)\,\partial/\partial x\,[q^2/h] + (1/20)M^*\,\partial/\partial x\,[h\,(\partial\Sigma/\partial x)\,q]\} \\ + (3/h^2)\,q = (1/K^*)\,h\,\partial^3 h/\partial x^3 + h \\ + (3/2)\,M^*\,\partial\Sigma/\partial x - (1/120)\,Re^* M^{*2}\,\partial/\partial x\,[(\partial\Sigma/\partial x)^2 h^3]. \end{aligned} \quad (5.23)$$

Now from (5.11), and (5.18) we obtain the relation :

$$\chi = (2/3)\,H\,(H - \Sigma), \quad (5.24)$$

and, in (5.10) for the term $\partial\chi/\partial\tau$, we find :

$$-(2/3)\,H\{\partial\Sigma/\partial\tau + [2 - (\Sigma/H)]\,\partial q/\partial x\}. \quad (5.25)$$

On the other hand the term :

$$\int_0^{H(\tau,x)} w\,dz = -(1/48)\,M^*\,\partial/\partial x\,[H^3(\partial\Sigma/\partial x)] + q\,\partial H/\partial x - (3/8)\,\partial/\partial x\,(Hq), \quad (5.26)$$

and the term

$$\int_0^{H(\tau,x)} u\,Q\,dz = (12/15)\,q\,(H - \Sigma) + (1/60)\,M^*\,(H - \Sigma)\,H^2\,(\partial\Sigma/\partial x). \quad (5.27)$$

Finally, for $\Sigma(\tau, x)$we derive the following averaged equation, in place of (5.10),

$$\begin{aligned} \partial\Sigma/\partial\tau + [2 - (\Sigma/H)]\,\partial q/\partial x + (6/H)\,\partial/\partial x\,[q\,(\Sigma - H)] \\ - (9/16H)\,\partial(q\,H)/\partial x + (3/2)\,(q/H)\,\partial H/\partial x \\ + (M^*/40H)\,\partial/\partial x\,[H^2(\Sigma - H)(\partial\Sigma/\partial x) + (M^*/32H)\,\partial/\partial x\,[H^3\,(\partial\Sigma/\partial x)] \\ + (3/Re^*Pr)\,[(\Sigma - H)/H^2] = 0. \end{aligned} \quad (5.28)$$

Finally, for our three unknown functions, $H(\tau, x)$, $q(\tau, x)$ and $\Sigma(\tau, x)$, we derive an averaged IBL system of three equations - (5.2), (5.23) and (5.28).
This averaged system of three equations has been first derived in Zeytounian (1998).

6 The KS and KS-KdV Model Equations

6.1 KS Equation

We start again with the equations (4.22a,b,c,d), and boundary conditions (4.23) and (4.24a,b,c,d), governing the B-M problem for a free-falling vertical film.

We assume that

$$\mathrm{Re} = O(1),$$

and we get that ε is the main small parameter occuring within equations and conditions , for the full starting two-dimensional B-M problem related with a free-falling vertical film, formulated in the Section 4.5. As characteristic velocity $U°$ we choose the interface velocity : $U° = g\, d^2/\nu_0$, and in this case : $\mathrm{Re}/F^2 = 1$. We assume that the Weber number W is large and we introduce again a Weber similarity parameter : $W^* = \varepsilon^2 W = O(1)$.

Concerning the Prandtl, Marangoni and Biot numbers in (4.22d) and in boundary conditions (4.24a,b,c) we suppose that : $\mathrm{Pr} = O(1)$, $M = O(1)$ and $\mathrm{Bi} = O(1)$.

We look for the solution of (4.22a,b,c,d)) with (4.23) and (4.24,a,b,c,d), taking into account the above relations for Re, F^2, W, Pr, M and Bi, an expansion in the form proposed by Benney (1966); namely :

$$U = [u, w, \Pi, \Theta]^T = U_0 + \varepsilon U_1 + O(\varepsilon^2), \text{ when } \varepsilon \to 0, \qquad (6.1)$$

but we do not expand the thickness of the film $H(\tau, x)$ for the moment. The solution U_0 is :

$$u_0 = -z\,[(1/2)\,z - H];\ w_0 = -(1/2)\,(\partial H/\partial x)\,z^2; \qquad (6.2a)$$

$$\Pi_0 = -W^*\,(\partial^2 H/\partial x^2); \qquad (6.2b)$$

$$\Theta_0 = 1 - (1 + \mathrm{Bi})\,[z/(1 + \mathrm{Bi}\,H)]. \qquad (6.2c)$$

From (5.2), with (5.3), we get also :

$$\partial H/\partial \tau + H^2\,(\partial H/\partial x) = O(\varepsilon). \qquad (6.3)$$

Now, writing out set of equations and boundary conditions for U_1, and again, assuming that $H(\tau, x)$ is not expanded, it is easy to get an analytic expression for u_1, as a function of z. Using u_0 and u_1, we may compute q_0 and q_1 within the expansion of :

$$q(\tau, x) = \int_0^{H(\tau, x)} u(\tau, x, z)\,dz = q_0 + \varepsilon\, q_1 + O(\varepsilon^2). \qquad (6.4)$$

according to (5.3). Concerning $q_0 = (1/3)\, H^3$, this has already been taken into account with (6.3), while we get for q_1 the following expression :

$$q_1(\tau, x) = (1/3)\, Re\, W^*\, H^3\, (\partial^3 H/\partial x^3) + (1/2)\, M'(1 + Bi)\, H^2\, (\partial H/\partial x)\, /[1/(1 + BiH)^2] + (2/15)\, Re\, H^6\, (\partial H/\partial x). \tag{6.5}$$

Now, from equation (5.2), we may get an equation (with an error of $O(\varepsilon^2)$) for the thickness of the film $H(\tau, x)$, involving the $O(\varepsilon)$ term occuring in (6.3) :

$$\partial H/\partial\tau + H^2\, (\partial H/\partial x) + \varepsilon\, \partial/\partial x \{(1/3)\, Re\, W^*\, H^3\, (\partial^3 H/\partial x^3) + (2/15)\, Re\, H^6\, (\partial H/\partial x) + (1/2)\, M(1 + Bi)\, H^2\, (\partial H/\partial x)\, /\, [1/(1 + BiH)^2]\} = 0. \tag{6.6}$$

This evolution equation (6.6), of the "Benney type", contains the small parameter ε, a fact which is due to that $H(\tau, x)$ has not been expanded at it should in a fully consistent asymptotic approach through an expansion with respect to ε. Of course we may expand $H(\tau, x)$ in different ways and we shall investigate, here only the same kind of phenomenon that the one which led the the Kuramoto - Sivashinsky (so-called, 'KS') equation. In order to achieve this, we put in (6.6) :

$$\theta = \varepsilon\, \tau,\ \xi = x - \tau,\ H(\tau, x) = 1 + \delta\, \eta(\theta, \xi) + ..., \tag{6.7a}$$

and we assume that

$$\varepsilon = \delta. \tag{6.7b}$$

Since:

$$\partial H/\partial\tau = -\ \varepsilon\, \partial\eta/\partial\xi + \varepsilon^2\, \partial\eta/\partial\theta \ \text{ and } \ \partial H/\partial x = \varepsilon\, \partial\eta/\partial\xi,$$

if we let $\varepsilon \to 0$ within the transformed version of equation (6.6) we derive the following KS equation :

$$\partial\eta/\partial\theta + 2\, \eta\, \partial\eta/\partial\xi + \alpha\, \partial^2\eta/\partial\xi^2 + \gamma\, \partial^4\eta/\partial\xi^4 = 0, \tag{6.8}$$

where

$$\alpha = (2/15)\, Re + (1/2)\, M/(1 + Bi); \quad \gamma = (1/3)\, ReW^*. \tag{6.9}$$

The KS equation (6.8) is asymptotically consistent, when $\varepsilon = \delta \to 0$, and is an approximate equation valid with an error of the order $O(\varepsilon)$.

It is interesting to note that the Benney type equation (6.6) is also derived from the averaged "Shkadov type" equations (5.2), (5.16), when $Re^* \to 0$, if we assume that K^* and M^* remains $O(1)$ in this limit. In this last case we derive again a KS equation, similar to (6.8), but (see (6.9)) the first term in the coefficient α is absent and in place of M we have M*. Respectively, in coefficient γ, in place of ReW*, we have the coefficient 1/K*.

Now, if we introduce the amplitude : $A(\theta, \xi) = 2\eta(\theta, \xi)$, then we obtain in place of (6.8) the following KS, canonical, equation

$$\partial A/\partial\theta + A\, \partial A/\partial\xi + \alpha\, \partial^2 A/\partial\xi^2 + \gamma\, \partial^4 A/\partial\xi^4 = 0. \tag{6.10}$$

When $\alpha = 0$ and $\gamma = 0$, we obtain the well - known equation :

$$\partial A/\partial\theta + A\, \partial A/\partial\xi = 0,$$

and along characteristics (defined by $d\xi/d\theta = A(\theta, \xi)$) the solution $A(\theta, \xi(\theta))$ is constant.

When $\gamma = 0$ ($W^* = 0$), the surface tension term is removed and (6.10) reduces to Burger's equation. In this case, the Cole - Hopf transformation further reduces it to heat equation. Since $\alpha > 0$ the Cole - Hopf transformation produces a heat equation backward in time and initial disturbance will then grow without limit. Hence, we shall include the surface tension term and discuss equation (6.10) when: $\alpha > 0$ and $\gamma > 0$. The full KS equation (6.10) is capable of generating solutions in the form of irregularity fluctuating quasi-periodic waves. This KS model equation provides a mechanism for the saturation of an instability, in which the energy in long-wave instabilities is transferred to short-wave modes which are then damped by surface tension.

In full KS equation (6.10), the terms: $\partial A/\partial\tau + A\, \partial A/\partial\xi$ leads to steepening and wave breaking in the absence of stabilizing terms. The term: $\alpha\, \partial^2 A/\partial\xi^2$ destabilizes shorter wavelength modes preferentially and therefore aggravates wave steepening (since M and Bi are both positive; for small Bi the Marangoni effect aggravates this destabilization). Finally, the term : $\gamma\, \partial^4 A/\partial\xi^4$ is required for saturation. Unfortunately, explicit analytic solution of KS equation are not available!

6.2 KS - KdV Equation

The problem considered below is interesting when we look at (6.6). As a matter of fact this equation (6.6) is a singular perturbation of the hyperbolic equation

$$\partial h/\partial\tau + H^2\, (\partial H/\partial x) = 0.$$

Curiously, we get again a singular pertubation of this same equation, but of an another type, if we make the assumption (the low Reynolds number case) :

$$Re \ll 1,\ M \ll 1, \tag{6.11a}$$

such that

$$Re/\varepsilon = R^\circ \quad \text{and } M/\varepsilon = M^\circ, \tag{611b}$$

and again assume that :

$$\varepsilon^2 W = W^* \text{ and } Re/F^2 = 1,$$

in the full starting problem (4.22a,b,c,d), (4.23) and (4.24a,b,c,d).
In this case, in place of this starting problem, we obtain the following problem :

$$\partial u/\partial x + \partial w/\partial z = 0, \tag{6.12a}$$
$$\partial^2 u/\partial z^2 + 1 = \varepsilon^2 R^\circ [\, Du/D\tau + \partial\Pi/\partial x \, - (1/R^\circ)\partial^2 u/\partial x^2 \,], \tag{6.12b}$$
$$\partial\Pi/\partial z - (1/R^\circ)\partial^2 w/\partial z^2 = \varepsilon^2 \, [(1/R^\circ)\partial^2 w/\partial x^2 - Dw/D\tau \,], \tag{6.12c}$$
$$\partial^2\Theta/\partial z^2 = \varepsilon^2 \, [\Pr R^\circ D\Theta/D\tau \, - \partial^2\Theta/\partial x^2]. \tag{6.12d}$$

At $z = 0$:

$$u = w = 0 \text{ and } \Theta = 1. \tag{6.13}$$

At $z = H(\tau, x)$:

$$\partial u/\partial z = -\, \varepsilon^2 M^\circ \, [\partial\Theta/\partial x + (\partial H/\partial x)\, \partial\Theta/\partial z] - \varepsilon^2 \, [\partial w/\partial x + 4(\partial H/\partial x)\, \partial w/\partial z] + O(\varepsilon^4), \tag{6.14a}$$
$$\Pi = -\, W^* \, \partial^2 H/\partial x^2 + (2/R^\circ)[\partial w/\partial z - (\partial H/\partial x)\, \partial u/\partial z] + O(\varepsilon^2), \tag{6.14b}$$
$$\partial\Theta/\partial z = -\,(1 + Bi\, \Theta) + O(\varepsilon^2), \tag{6.14c}$$
$$w = \partial H/\partial t + u\, \partial H/\partial x. \tag{6.14d}$$

Obviously in this case, the formal Benney expansion in ε is modified. Namely :

$$U = (u, w, \Pi, \Theta)^T = U_0 + \varepsilon^2 U_2 + ..., \text{ when } \varepsilon \to 0. \tag{6.15}$$

The solution U_0 is got in a straightforward way. In this case, when $\varepsilon \to 0$ in the above problem [(6.12a,b,c,d), (6.13), (6.14a,b,c,d) we obtain the following leading order solution :

$$u_0 = -\,(1/2)\, z^2 + H\, z, \quad w_0 = -\,(1/2)\, (\partial H/\partial x)\, z^2 , \tag{6.16a}$$
$$\Pi_0 = -\, W^* \, \partial^2 H/\partial x^2 - (1/R^\circ)\, (\partial H/\partial x)\, (H + z), \tag{6.16b}$$
$$\Theta_0 = 1 - (1 + Bi)\, [z/(1 + Bi\, H)]. \tag{6.16c}$$

From (5.2), with (5.3) and the first relation of (6.16a), we get, this time, again:

$$\partial H/\partial\tau + H^2\, \partial H/\partial x = O(\varepsilon^2), \tag{6.17}$$

since $q_0 = (1/3)\, H^3$ - but valid with an error of $O(\varepsilon^2)$.
Writting out the set of equations and boundary conditions at the order ε^2 (from (6.12b) and (6.14a), (6.13), for u_2, and assuming, yet, that $H(\tau, x)$ is not

expanded, we may get an awkward expression for u_2 that may be integrated with respect to z in order to obtain an explicit expression for q_2 in (as consequence of (5.2)) equation :

$$\partial H/\partial \tau + H^2\, \partial H/\partial x + \varepsilon^2\, \partial q_2/\partial x = O(\varepsilon^4).$$

The final result is analogous to (6.6), but with some additional terms, which reads

$$\begin{aligned}\partial H/\partial \tau + H^2\, \partial H/\partial x + \varepsilon^2\, \partial/\partial x\, \{(1/3)\, H^3\, [R°W^*\, \partial^3 H/\partial x^3 + 7\, (\partial H/\partial x)^2] \\ + H^4\, \partial^2 H/\partial x^2 + (2/15)\, H^6\, \partial H/\partial x \\ + (1/2)\, M°\, (1 + Bi)\, [H^2\, (\partial H/\partial x)/(1 + Bi\, H)^2]\} = 0. \qquad (6.18)\end{aligned}$$

This above evolution equation (6.18), for $H(\tau, x)$, is valid with an error of $O(\varepsilon^4)$. Now, we intend to play, with this last evolution equation (6.18), the same game as the one considered for the reduction of (6.6) in a KS equation (6.8).
We use, namely :

$$\theta = \delta\, \tau, \quad \xi = x - \tau, \quad H = 1 + (1/\phi)\varepsilon^2\, \eta(\theta, \xi) + ..., \qquad (6.19a)$$

and

$$\delta = (1/\phi)\varepsilon^2, \qquad (6.19b)$$

where ϕ is the dispersive similarity parameter.
Carrying out, again the limiting process $\varepsilon \to 0$, we find, in place of (6.18), an equation which cumulates the features of KdV on the one hand and KS on the other hand :

$$\partial\eta/\partial\theta + 2\, \eta\, \partial\eta/\partial\xi + \alpha\, \partial^2\eta/\partial\xi^2 + \phi\, \partial^3\eta/\partial\xi^3 + \gamma\, \partial^4\eta/\partial\xi^4\, ,$$

(6.20)
where

$$\alpha = \phi\, [(2/15)\, R° + (1/2)\, M°/(1 + Bi)], \quad \gamma = \phi\, (1/3)\, R°W^*. \qquad (6.21)$$

The evolution KS-KdV equation (6.20) is again a significant model equation valid for large time with an error of $O(\delta)$. The coefficients α, γ and ϕ are all positive constants characterizing *instability*, *dissipation* and *dispersion.*
As consequence of the derivation of the KS-KdV equation (6.20), valid for the *low Reynolds numbers,* we conclude that the features of the thin film for a strongly viscous liquid is quite different: The dispersive term, $\phi\, (\partial^3\eta/\partial\xi^3)$, changes the behavior of the thickness $\eta(\theta, \xi)$ in space and in time.
The above derivation of the KS-KdV model equation has been first published in Zeytounian (1995). For $\phi = 0$, equation (6.20) reduces to a self-exciting dissipative KS equation (6.8) which exhibits a turbulent (chaotic) behavior. On the other hand, for $\alpha = \gamma = 0$ (limiting case when : $R° = 0$ and $M° = 0$ - non -

viscous liquid film, without the Marangoni effect), the equation (6.20) reduces to the KdV equation which is known to admit soliton solutions instead of chaos! Thus, in the general case of non zero α, γ and ϕ, the increasing value of ϕ is expected to change the character of the solution of the equation (6.20) from an irregular wave train to a regular row of solitons (a row of pulses of equal amplitude). The trend is more amplified at larger values of ϕ , and the asymptotic state of the solution for large ϕ takes the form of a row of the KdV solitons. There seems to exist a critical value of about unity for the adimensional parameter :

$$\mu = \phi/(\gamma\alpha)^{1/2}$$

which represents the relative importance of the dispersion corresponding to the transition from an irregular wave train to a regular row of solitons. We precise that, the complicated evolution of solutions of (6.20) is described by the weak interaction of pulses, each of which is a steady solution of (6.20). When the dispersion is strong, pulse interactions become repulsive, and the solutions tend, in fact, to form stable lattices of pulses.

7 Some Aspects of the Linear Stability Analysis of the B-M Problem

In the investigations of Takashima (1981a), the effect of a free surface deformation on the onset of surface tension driven instability in a horizontal thin liquid layer subjected to a vertical temperature gradient is examined using linear stability theory. Assuming that the neutral state is a stationary one, the conditions under which instability sets in are determined in detail. In Takashima (1981b), the above linear theory is extended to include the possibility that surface tension driven instability in a horizontal thin liquid layer confined between a solid wall and a deformable free surface can set in a purely oscillatory motions. More recently, linear thermocapillary instabilities with surface deformation in a fluid layer heated from below are studied by Regnier and Lebon (1995).
In order to consider the linear stability of thin film wavy motions governed by the B-M problem, we must first determine the steady-state solutions of the evolution equations presented. Then we can perturb about the steady-states using the normal modes. Assuming that the perturbing quantities are small, we linearize the evolution equations about the steady-state solutions to obtain an eigenvalue system for the considered B-M problem.

7.1 The Eigenvalue Linear System for the B-M Problem

For the B-M problem, (4.3a,b,c), (4.5a,b,c,d), (4.6) and (4.7), the steady-state solutions are :

$$v_{10} = v_{20} = v_{30} = 0\,,\ \theta_0 = 1 - z\,,\ \pi_0 = 0, \text{ and } h = 1. \tag{7.1}$$

In this case, if we consider only the two-dimensional case (x, z), for the perturbed quantities : (u,' w', π', θ'), as function of t, x, and z, and $\eta(t, x)$, we obtain, first, from (4.3a,b,c) the following linear equations :

$$\begin{aligned}
&\partial u'/\partial x + \partial w'/\partial z = 0,\\
&\partial u'/\partial t + \partial \pi'/\partial x = \partial^2 u'/\partial x^2 + \partial^2 u'/\partial z^2,\\
&\partial w'/\partial t + \partial \pi'/\partial z = \partial^2 w'/\partial x^2 + \partial^2 w'/\partial z^2,\\
&Pr\,(\partial \theta'/\partial t - w') = \partial^2 \theta'/\partial x^2 + \partial^2 \theta'/\partial z^2 .
\end{aligned} \tag{7.2a}$$

Next, for these linear equations (7.2a) we derive from (4.7), (4.5a,b,c,d) and (4.6), the following linear boundary conditions :

$$\text{at } z = 0: \qquad u' = w' = 0 \text{ and } \theta' = 0. \tag{7.2b}$$

and

$$\text{at } z = 1: \quad \begin{aligned} &\pi' = (1/F^2)\eta - \mathrm{We}\,\partial^2\eta/\partial x^2 + 2\,\partial w'/\partial z, \\ &\partial u'/\partial z + \partial w'/\partial x = -\,\mathrm{Ma}\,(\partial\theta'/\partial x - \partial\eta/\partial x), \\ &\partial\theta'/\partial z + \mathrm{Bi}\,(\theta' - \eta) = 0, \\ &w' = \partial\eta/\partial t. \end{aligned} \tag{7.2c}$$

Now, the linear equations (7.2a) with the linear boundary conditions (7.2b,c) are then simplified in the usual manner by decomposing the solution in terms of normal modes, so that :

$$[u', w', \pi', \theta', \eta] = [U(z), W(z), P(z), T(z), H^\circ]\exp[\sigma t + ikx], \tag{7.3}$$

where σ is the time constant, which is in general complex, k is the (real) wave number and H° = constant.
With (7.3) from (7.2a) we derive for W(z) and T(z) the following two ODE :

$$\{\sigma - [\,d^2/dz^2) - k^2]\}\,[(d^2W/dz^2) - k^2\,W] = 0, \tag{7.4a}$$

$$\{\Pr\sigma - [\,d^2/dz^2) - k^2]\}T = \Pr W. \tag{7.4b}$$

For W(z) and T(z), from (7.2b,c) we derive the following boundary conditions:

$$\begin{aligned} &dW/dz|_{z=0} = 0,\ W(0) = T(0) = 0, \\ &d^2W/dz^2|_{z=1} + k^2\,W(1) + k^2\,\mathrm{Ma}\,[T(1) - H^\circ] = 0, \\ &[\sigma + 3\,k^2]\,dW/dz|_{z=1} - d^3W/dz^3|_{z=1} + k^2\,[(1/\mathrm{Fr}^2) + k^2\,\mathrm{We}]H^\circ = 0, \\ &dT/dz|_{z=1} + \mathrm{Bi}\,[T(1) - H^\circ] = 0; \end{aligned} \tag{7.4c}$$

$$W(1) = \sigma\,H^\circ. \tag{7.4d}$$

Equations (7.4a,b) and the boundary conditions (7.4c,d) contitute an eigenvalue (linear) system for the B-M problem formulated in the § 4.
We note that, if in place of Froude (Fr) and Weber (We) numbers we introduce the crispation (Cr*) and Bond (Bo) numbers, according to ($\mu_0 = \rho_0 \nu_0$) :

$$\mathrm{Cr}^* = 1/\Pr\mathrm{We} = \mu_0\,\kappa_0/\sigma_0 d, \text{ and } \mathrm{Bo} \equiv \mathrm{Bo}^* = \Pr\mathrm{Cr}^*/\mathrm{Fr}^2 = \rho_0\,g\,d^2/\sigma_0, \tag{7.5}$$

then we obtain from the above eigenvalue system (7.4a,b,c,d), the eigenvalue system considered in Takashima (1981a)and also in Regnier and Lebon (1995), with

$$\omega = \Pr\sigma \text{ and } w = \Pr W, \tag{7.6}$$

in place of σ and W.

7.2 The Case when the Neutral State is a Stationary One

In a such case, in the above eigenvalue (linear) system (7.4a,b,c,d),

$$\sigma = 0. \tag{7.7}$$

Then, if we take into account (7.5) and (7.6), we obtain the following steady problem :

$$[(d^2w/dz^2) - k^2 w] = 0, \tag{7.8a}$$

$$[d^2/dz^2) - k^2] T + w = 0, \tag{7.8b}$$

$$dw/dz|_{z=0} = 0, \; w(0) = T(0) = 0, \tag{7.8c}$$

$$(d^2w/dz^2)|_{z=1} + k^2 w(1) + k^2 \, Ma \, [T(1) - H^\circ] = 0, \tag{7.8d}$$

$$Cr^* \left[3 k^2 (dw/dz)|_{z=1} - (d^3w/dz^3)|_{z=1}\right] + k^2 [Bo + k^2] H^\circ = 0 \tag{7.8e}$$

$$(dT/dz)|_{z=1} + Bi \, [T(1) - H^\circ] = 0. \tag{7.8f}$$

The general solutions of the two equations (7.8a,b), for w(z) and T(z), can easily be obtained through the sinh(kz) and cosh(kz). When these solutions are substituted into the boundary conditions, (7.8c)-(7.8f), then we derive the following eigenvalue relationship :

$$Ma = \frac{8k\,[\sinh(k)\cosh(k) - k]\,[k\cosh(k) + Bi\,\sinh(k)]\,(Bo + k^2)}{8\,Cr^*\,k^5\cosh(k) + (Bo + k^2)\,[\sinh^3(k) - k^3\cosh(k)]} \tag{7.9}$$

The result (7.9) was derived a few years ago by M. Takashima (1981a) and see, also, Regnier and Lebon (1995) paper, where the growth rate of disturbances for the non-zero mode is studied. For fixed values of Bi, Bo and Cr*, relation (7.9) enables us to plot a neutral stability curve in the (k, Ma)-plane; see, the Fig. 7.1 below.

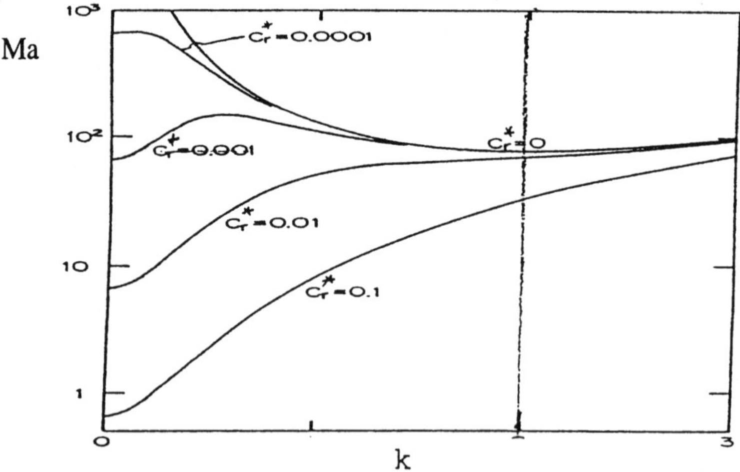

Fig. 7.1. Neutral stability curves for various values of Cr* (Bi = 0 and Bo = 0.1). The region below each curve represents stable state

Indeed, the case when Bi = 0 is still representative since the value of Bi for a thin layer with which we are concerned is at most 0.1 for most liquids, and such a small value of Bi does not affect appreciably the results for Bi = 0. Concerning Bo, for most liquids in contact with air, if the depth of the layer d is assumed to vary from 0.01 cm to 0.1 cm, we have : $10^{-3} \sim |Bo| \sim 1$.

The neutral stability curves, when Bi = 0 and Bo = 0.1, are shown in above Fig. 7.1 as a sample case. The values of Cr* are given in the figure. Since the region below each curve represents stable state, the lowest point of each curve gives the critical Marangoni number Ma_c and the corresponding critical wave number k_c.

When $Cr^* \rightarrow 0$ (i.e. in the case of a non-deformable free surface, and in a such case $Fr^2 \rightarrow 0$ also but $Bo \neq 0$), the result of Pearson's (1958) :

$$Ma_c \approx 80 \text{ and } k_c \approx 2.0, \qquad (7.10)$$

is again obtained. When Cr* exceeds 0.00083, k_c changes abruptly from 2.0 to 0 and Ma_c decreases rapidly as Cr* increases. In order to examine analytically the behavior of Ma in the vicinity of k = 0, we expand the right-hand side of relation (7.9) in power of k. It is of the form :

$$\begin{aligned} Ma = &(2Bo/3Cr^*)(1 + Bi) \\ &+ (2/3Cr^*)\left\{(1 + Bi)\,[1 - (2/15)\,Bo] + (Bo/3) - (1 + Bi)\,[Bo^2/120\,Cr^*]\right\}k^2 \\ &+ O(k^4). \end{aligned} \qquad (7.11)$$

This above result (7.11), of Takashima (1981a), is also derived by Regnier and Lebon (1995). It follows from relation (7.11) that when

$$Cr^* > [(1 + Bi)\,Bo^2]/120\left\{(1 + Bi)\,[1 - (2/15)\,Bo] + (Bo/3)\right\}, \qquad (7.12)$$

the coefficient of k^2 is positive and therefore Ma has a minimum value at k = 0. It is interesting to note that each curve for Ma_c has a turning point at

$$Cr^* = Bo/120 \qquad (7.13)$$

and this turning point is accompanied by a discontinuous change in k_c.

When Cr* < Bo/120 the free surface deformation is not important, but, on the contrary, when Cr *> Bo/120, the value of Ma_c is strongly dependent of Bo and Cr* and the value of k_c is zero - this means that when Cr* > Bo/120, the free surface deformation is important. According to Takashima (1981), the condition under which the free surface deformation becomes important can also be expressed as

$$d < d^* = (120\,\nu_o\,\kappa_o)/g)^{1/3}, \qquad (7.14)$$

and for water the value of d* is 0.015 cm. In fact, when $Pr \approx 1$ ($\nu_o = \kappa_o$), the condition (7.14) is equivalent to our condition (4.2)

7.3 The Case when $\omega \neq 0$: Overstability

In this case, in place of the problem (7.8a,b)-(7.8c,d,e,f), we have the following more complicated problem for w(z) and T(z) :

$$[(d^2/dz^2) - k^2]\,[(d^2w/dz^2) - c^2 w] = 0, \quad c^2 = k^2 + (1/Pr)\,\omega, \qquad (7.15a)$$
$$[d^2/dz^2) - b^2]\,T + w = 0, \quad b^2 = k^2 + \omega, \qquad (7.15b)$$

$$dw/dz|_{z=0} = 0, \; w(0) = T(0) = 0,$$
$$(d^2w/dz^2)|_{z=1} + k^2 w(1) + k^2 Ma\,[T(1) - H°] = 0,$$
$$Cr^* \left[(2k^2 + c^2)\,(dw/dz)|_{z=1} - (d^3w/dz^3)|_{z=1}\right] \qquad (7.15c)$$
$$+ k^2 [Bo + k^2] H° = 0,$$
$$(dT/dz)|_{z=1} + Bi\,[T(1) - H°] = 0,$$

$$w(1) - \omega H° = 0. \qquad (7.15d)$$

Again, the general solutions of the above equations (7.15a,b) can easily be obtained and these, when substituted into the boundary conditions (7.15c,d), yield a time dependent eigenvalue relationship of the form :

$$Ma = L\,(Bi, Bo, Cr^*, Pr, k, \omega)\,, \qquad (7.16)$$

where L is a real-valued function of parameters in parentheses. The explicit espression for relation (7.16) when $b \neq c$ (i.e; $Pr \neq 1$) is given by Takashima (1981b). Since the neutral state for an oscillatory mode is characterized by:

$$\omega = i\,\omega_i, \text{ with } \omega_i \text{ real},$$

it is clear from relation (7.16) that for arbitrarily assigned values of Bi, Bo, Cr*, Pr, k and ω_i, the value of Ma will be complex! But the physical meaning of Ma requires it to be real. Hence, a numerical search must be conducted to find the value of ω_i which makes the imaginary part of Ma vanish. If such a value of ω_i is not found for wide ranges of Bi, Bo, Cr*, Pr and k, it may be concluded that the neutral state is indeed a stationary one. If such a value of ω_i is found, this value is substituted for ω_i in the real part of Ma. The repetition of this procedure for various values of k enables us to plot a neutral stability curve for the onset of overstability in the (k, Ma)-plane. It was found from numerical computations by M. Takashima (1981b), that the neutral stability curves for the onset of overstability lie only in the negative region of Ma and, contrary to the case of a stationary mode (see the above Section 7.2), the region above each curve, in the below Fig. 7.2, represents stable state.

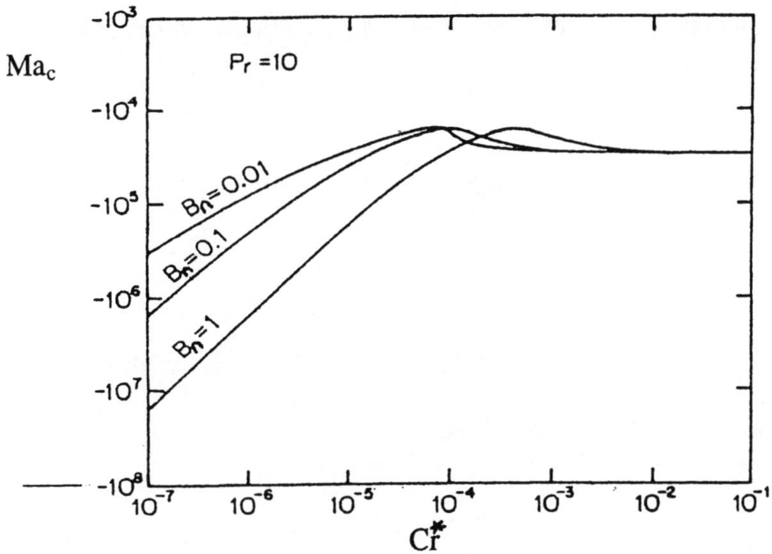

Fig. 7.2. The critical Marangoni number Ma_c for the onset of overstability as a function of Cr* for various values of Bo > 0, Bi = 0 and Pr = 10

It follows from this latter fact that the highest point of each curve gives the critical Marangoni number Ma_c, for the onset of overstability and the corresponding critical wave number k_c. When, according to (2.23), $Ma = \gamma d^2 \beta/\rho (\nu_0)^2 < 0$, and upside of the layer is a free surface (Bo > 0), then it is necessary that (see (1.3)) :

$$\beta < 0 \Rightarrow T_0 < T_a,$$

and the liquid layer can become unstable via a marginal state of purely oscillatory motions. Therefore, when the upside of the liquid layer is a free surface instability is possible for heating in either direction, but the manner of the onset of instability depends on the direction of heating. It is finally concluded from the investigation of M. Takashima (1981b), that the *free surface deformation must be taken into account* when :

(1) $Ma \geq 0$ and $Bo < 0$ (in this case the underside of the layer is a free surface);

(2) $Ma \leq 0$ (when $Ma < 0$ we assume that $T_0 < T_a$) and $Bo < 0$;

(3) $Ma > 0$ and $0 < Bo < 120\, Cr^*$;

(4) $Ma < 0$ and $Bo > 0$.

It is either in the case in the trivial case when $Ma > 0$ and $Bo > 120\, Cr^*$ or in the trivial case when $Ma = 0$ and $Bo > 0$ that the assumption of a *non-deformable free surface is valid.* Concerning, more precisely the growth rate around k = 0, we precise that in cited above paper by Regnier and Lebon (1995), the relation (7.11) is derived from an approximate expression for the growth rate of the perturbations for the zero mode; namely :

$$\omega = \{ [\mathrm{Ma}/2(1+\mathrm{Bi})] - (G/3)] \} k^2 + \{(G^2/360) - G\,[(3-2\mathrm{Bi})/45\,(1+\mathrm{Bi})] - (1/3\mathrm{Cr}^*)\} k^4, \tag{7.17}$$

where

$$G = \mathrm{Bo}/\mathrm{Cr}^* = \mathrm{Pr}/\mathrm{Fr}^2 = gd^3/\nu\kappa. \tag{7.18}$$

Setting $\omega = 0$ in (7.17) yields the expression (7.11) of the marginal curve Ma versus k in the vicinity of $k = 0$ with an error od $O(k^4)$.
In the particular case of a zero G (zero Bond, Bo, number), the critical Marangoni number is zero and relation (7.17) yields :

$$\omega = [\mathrm{Ma}/2(1+\mathrm{Bi})]\,k^2 - (1/3\mathrm{Cr}^*)\,k^4, \tag{7.19}$$

and for a fixed Marangoni number the growth rate is positive at the condition that

$$k^2 \leq [3\mathrm{Ma}\,\mathrm{Cr}^*/2(1+\mathrm{Bi})] = (k_{crit})^2, \tag{7.20}$$

which was the result derived by Scriven and Sterling (1964).
We precise that the dispersion relation for the zero mode was found, by Regnier and Lebon (1995), analytically by using an appropriate scaling (horizontal characteristic length of waves significantly exceeds the thickness of the liquid layer).
As a main result of this analysis, it is shown that the presence of the zero mode can only be detected in very large aspect ratio boxes and for very thin fluid layers.
It should be realized that the results for $k = 0$ obtained above provide a first step towards a (weakly) nonlinear analysis.
It it is also interesting to note, that for the KS equation, (6.8), according to (6.10) we have the following dispersion relation :

$$\omega = [(2/15)\,\mathrm{Re} + (1/2)\,M/(1+\mathrm{Bi})]\,k^2 - (1/3)\,\mathrm{Re}\,W^*\,k^4, \tag{7.21}$$

which is very similar to (7.19), when $\mathrm{Re} \rightarrow 0$, but $\mathrm{Re}\,W^* = O(1)$.

8 Some Stability Results for the Lubrication Equation (4.21) and Averaged System : (5.2), (5.23) and (5.28)

8.1 A Stability Criterion for the Lubrication Equation (4.21)

In the case when $H \longrightarrow h(\tau, x)$, in place of equation (4.21), derived in the Section 4.4, we obtain the following one-dimensional nonlinear evolution equation for $h(\tau, x)$:

$$\partial h/\partial\tau - \partial/\partial x\{[(h^3/3) - (B/3)(1+\mathrm{Bi})\ h^2/(1+\mathrm{Bi}\ h)^2]\partial h/\partial x\} + (A/3)\ \partial/\partial x\ [h^3\ \partial^3 h/\partial x^3] = 0, \quad (8.1)$$

The full one-dimensional evolution equation (8.1) can be obviously rewritten in the conservative form :

$$\partial h/\partial t = -\partial/\partial x\ \{\ h^3\ \partial/\partial x\ [Q(h) + (A/3)\ \partial^2 h/\partial x^2]\}, \quad (8.2)$$

with

$$Q(h) = -h/3 + (B/3)(1+\mathrm{Bi})\{\ln[h/(1+\mathrm{Bi}h)] + [1/(1+\mathrm{Bi}h)] + \text{const.} \quad (8.3)$$

This above equation (8.2) can be also rewritten in a so-called “Cahn-Hilliard” type form, namely :

$$\partial h/\partial\tau = \partial/\partial x\ \{\ h^3\ \partial/\partial x(d\mathcal{I}/dh)\}, \quad (8.4)$$

with

$$\mathcal{I} = \int_\Omega [U(h) + (A/6)\ (\partial h/\partial x)^2] d\Omega\ , \quad (8.5)$$

playing the role of free energy and h^3 of mobility coefficient.
In (8.5)

$$dU(h)/dh = Q(h), \quad (8.6)$$

and Ω is a periodic domain in x and we have integrated over Ω by parts using periodic boundary conditions in x.
On the other hand, the right-hand side of the equation :

$$d\mathcal{I}/d\tau = -\int_\Omega h^3\ \{\partial/\partial x\ [Q(h) + (A/3)\ \partial^2 h/\partial x^2]\}^2\ d\Omega\ , \quad (8.7)$$

derived from (8.2), after multiplication by :

$$H(h) = Q(h) + (A/3)\ \partial^2 h/\partial x^2]$$

and integration over Ω by parts using periodic boundary conditions in x, is always non-negative :

$$d\mathcal{I}/d\tau \leq 0, \quad (8.8)$$

and Л is a decreasing function along any trajectory.
Therefore Л is a *Liapounov functional* for the evolution equation (8.1). Steady states of the system, therefore, if exist, are solutions of the equation :

$$H(h) = \text{Const.}$$

Relations (8.2) - (8.7), with H(h) = Const, provide a start for any meaning full stability analysis of equation (8.1). We recall that if a system has a *Liapounov functional bounded from below* (a free energy functional decreasing along all trajectories) then, *any initial data evolves into a steady state.* The above short discussion is directly inspired by the paper of Oron and Rosenau (1992).
Note that equation (4.21), and its descendants as well, equations (8.1) and (when $B = 0$) equation :

$$\partial h/\partial \tau + \partial/\partial x \left\{ (h^3/3)\, \partial/\partial x\, [A\, \partial^2 h/\partial x^2 - h] \right\} = 0, \qquad (8.9)$$

preserve the total mass, i.e.: $\int_\Omega h \, d\Omega$ = const, where Ω is either infinite or periodic domain. Indeed, for equation (8.9) another conservation law is available, namely :

$$d/d\tau \int_\Omega (1/h) d\Omega = -(2/3) \int_\Omega \left[(\partial h/\partial x)^2 + A\, (\partial^2 h/\partial x^2)^2\right] d\Omega. \qquad (8.10)$$

8.2 The Associated to: (5.2), (5.23) and (5.28), Linear Averaged System

A basic (constant) solution of the IBL averaged system,(5.2), (5.23) and (5.28), derived in the § 5, for : $h(\tau, x)$, $q(\tau, x)$ and $\Sigma(\tau, x)$, is :

$$h = 1,\ q = 1/3 \text{ and } \Sigma = 1. \qquad (8.11)$$

Since :

$$h(\tau, x) = 1 + \delta\, \eta(\tau, x), \qquad (8.12a)$$

we can write

$$q = (1/3) + \delta\, \varphi(\tau, x) \text{ and } \Sigma = 1 + \delta\, \zeta(\tau, x). \qquad (8.12b)$$

When $\delta \ll 1$, with (8.12a,b), we derive, from (5.2), (5.23) and (5.28),the following system of two linear evolution equations for $\eta(\tau, x)$ and $\zeta(\tau, x)$:

$$\begin{aligned} &\partial^2\eta/\partial\tau^2 + (4/5)\partial^2\eta/\partial x\partial\tau + (2/15)\partial^2\eta/\partial x^2 \\ &+ (3/2)\,(M^*/Re^*)\partial^2\zeta/\partial x^2 - (M^*/60)\,\partial^3\zeta/\partial x^3 \\ &+ W^*\,\partial^4\eta/\partial x^4 + (3/Re^*)\,[\partial\eta/\partial x + \partial\eta/\partial\tau] = 0, \end{aligned} \qquad (8.13a)$$

$$\begin{aligned} &\partial\zeta/\partial\tau - (7/16)\partial\eta/\partial\tau - (7/80)\partial\eta/\partial x + (2/5)\partial\zeta/\partial x \\ &+ (M^*/32)\partial^2\zeta/\partial x^2 + (3/Pr\, Re^*)\,[\zeta - \eta] = 0. \end{aligned} \qquad (8.13b)$$

Since

$$\partial\varphi/\partial x = - \partial\eta/\partial\tau \,. \tag{8.13c}$$

and Re* W* = 1/K*.

This closed linear system (8.13a,b) can be resolved numerically (with appropriate initial conditions and periodicity relative to x) for the investigation of the stability of the (constant) Nusselt type smooth basic regime (8.11). Here we consider this linear problem for the infinitesimal disturbances of the form :

$$\eta(\tau, x) = A^\circ \exp[ik(x - c\tau)] \text{ and } \zeta(t, x) = B^\circ \exp[ik(x - c\tau)], \tag{8.14}$$

and in such case we derive the following dispersion relation

$$\begin{aligned}(3/Re^*)\,[c - 1] - i\,k\,[c^2 - (4/5)\,c + (2/15)] + i\,k^3\,W^* \\ = (B^\circ/A^\circ)\,M^*\,[k^2/60 + (3/2)i\,(1/Re^*)\,k],\end{aligned} \tag{8.15}$$

with

$$\begin{aligned}B^\circ\,[3/Pr\,Re^* - (1/32)\,M^*k^2 - \,i\,k\,[c - (2/5)]] \\ - A^\circ\,[3/Pr\,Re^* + (7/16)\,i\,k\,[c - (1/5)]] = 0.\end{aligned} \tag{8.16}$$

From (8.16) we see that the ratio B°/A° is a complex number function of k and c, and as consequence, when $Pr \neq 0$, the dispersion relation (8.15) is very complicated! For this, below we consider only the case when :

$Pr = 0$, but $M^* \neq 0$, and in this case : $A^\circ = B^\circ$.

As consequence, in place of characteristic equation (8.15) we obtain the following reduced dispersion relation (with the Marangoni effect) :

$$\begin{aligned}(3/Re^*)\,[c - 1] - i\,k\,[c^2 - (4/5)\,c + (2/15)] + i\,k^3\,W^* \\ = (1/2)\,M^*[(k^2/30) + 3i\,k/Re^*].\end{aligned} \tag{8.17}$$

8.3 Analysis of the Reduced Dispersion Relation (8.17)

Since the stability with respect to disturbances limited throughout the space at any moment of time is here of particular interest, the wavenumber k, in (8.17), is assumed to be real. Then the values of complex phase velocity, $c = c_r + i\,c_i$, may be found from real and imaginary part of (8.17), respectively :

$$(3/Re^*)\,[1 - c_r] + k\,c_i\,[(4/5) - 2\,c_r\,] + (1/60)\,M^*\,k^2 = 0, \tag{8.18a}$$

and

$$\begin{aligned}(3/Re^*)\,c_i - k\,[c_r^2 - \,c_i^2 - (4/5)\,c_r + (2/15) + (3/2)(M^*/Re^*)] \\ + k^3\,W^* = 0.\end{aligned} \tag{8.18b}$$

If $c_i > 0$ the disturbances is amplified and if $c_i < 0$ it disappears. From (8.18a), when $c_i = 0$, we derive for the phase celerity of a neutral disturbance the following relation

$$c_r{}^* = c^* = 1 + (M^*Re^*/180)k^2, \qquad (8.19)$$

and infinitesimal disturbances thus appears to be dispersive - the coefficient

$$\beta^* = M^*Re^*/180 \qquad (8.20)$$

is the rate of the dispersive term and is a explicit function of M^*.
The imaginary part, (8.18b), when $c_i = 0$ and according to (8.19), give for neutral wavemunber k^* ($\neq 0$) a biquadratic algebraic equation

$$\beta^* (k^{*2})^2 + [\,(6/5)\,\beta^* - W^*]k^{*2} + (1/3) + (3/2)\,(M^*/Re^*) = 0. \qquad (8.21)$$

It is obvious that for (when $M^* \neq 0$ or $\beta^* \neq 0$):

$$W^* \geq (6/5)\,\beta^* + 2\,\beta^* \left[(1/3) + (3/2)\,(M^*/Re^*)\right]^{1/2} \qquad (8.22)$$

we obtain two or one values for k^{*2} and these values are always positive.
A particular case corresponds to the relation [in (W^*, M^*, Re^*) space] :

$$30\ W^*/M^*\ Re^* = (1/5) + (1/3)\left[(1/3) + (3/2)\,(M^*/Re^*)\right]^{1/2}, \qquad (8.23)$$

between the dimensionless parameters W^*, M^* and Re^*, and in this case a single neutral wavenumber k^* (associated with the neutral curve of stability, corresponding to $c_i = 0$) exists such that :

$$k^{*2} = (180/M^*Re^*)\left[(1/3) + (3/2)\,(M^*/Re^*)\right]^{1/2} \qquad (8.24)$$

All the disturbances with $k < k^*$ are unstable and those with $k > k^*$ disappear. Finally, from (8.21), we derive for M^* the following relation :

$$M^* = [k^{*2}\,W^* - (1/3)]/\{(3/2)(1/Re^*) + k^{*2}\,(Re^*/180)\,[k^{*2} + (6/5)]\}, \qquad (8.25)$$

if we take into account (8.20). For fixed values of Re^* and W^*, relation (8.25) for M^*, enables us to plot a neutral stability curve in the (k^*, M^*) - plane.
In fact, M^* is positive only for : $k^* > (1/3W^*)^{1/2}$.
When $M^* = 0$ ($\beta^* = 0$), from (8.18b), the condition for neutral stability ($c_i = 0$) is

$$-(1/3) + k^2\,W^* = 0, \qquad (8.26)$$

since, according to (8.18a), $c = c_r = 1$.

The condition (8.26) gives us a condition on W^*, i. e., we will have linear stability for values of $W^* > W^*_c$ where:

$$W^*_c = 1/3\ k^2\ , \tag{8.27}$$

or on k, i. e., we will have linear stability for wavenumber $k > k_c$, where :

$$k_c^2 = 1/3\ W^*_c\ . \tag{8.28}$$

The above results, (8.27), (8.28), are classical (see, for instance, Trifonov and Tsvelodub (1991) and also Alekseenko et al. (1985).
In above either case, we find that the parameter Re^* do not affect the linear stability of our film when $M^* = 0$.
On the contrary, for $M^* \neq 0$, we observe the existence of a cut-off wavenumber which is function of the three dimensionless parameters : W^*, M^* and Re^* and in a such case, each of these parameters can play the role of a bifurcation parameter (in contrast with (8.27), where the bifurcation parameter is only W^*).
We precise that, when $M^* = 0$, from (8.18a) we have $c_r < 1$ for $c_i < 0$ and $c_r > 1$ for $c_i > 0$, for all real $k > 0$.
On the other hand, if we determine c_i from (8.18b), then when $c_i < 0$ (linear stability), and if $c_r = 1 - a$, $1 > a > 0$, we obtain

$$(1/3) - k^2\ W^* - a\ [(6/5) - a] < 0,$$

and, since $a\ [(6/5) - a] > 0$, we will have obviously linear stability for above indicated values of $W^* > W^*_c = 1/3\ k^2$.

9 Stability and Dynamical System Approach for the KS and KS-KdV Equations

A very naive linear stability analysis show that for the KS equation (6.8) exists a cutoff wave number. Indeed, if :

$$\eta(\theta, \xi) \sim \exp [\omega\, \theta + i\, k\xi],$$

then for ω we derive the following dispersion relation

$$\omega - \alpha\, k^2 + \gamma\, k^4 = 0. \qquad (9.1)$$

The curve $\omega = 0$ determines neutral curve of the (linear) steady stability; in this case the phase velocity, $\omega / k = c = 0$, where the wavenumber k is assumed to be real, and as consequence we obtain a cutoff wavenumber k^* such that :

$$(k^*)^2 = \alpha/\gamma = \big[(2/5) + (3/2)\, M/Re(1 + Bi)\big]/W^*. \qquad (9.2)$$

Linear dispersion relation (9.1) shows that short waves are stable, and long waves are unstable. The critical wavenumber is $k^* = \sqrt{(\alpha/\gamma)}$ which ought to be small for the analysis of long waves to make sense. The maximum growth rate is $(\alpha^2/4\gamma)$ and occurs at $k^*/\sqrt{2}$. It is anticipated that the effect of the nonlinear term in canonical KS equation (6.10) will be allow energy exchange between a wave with wavenumber k and its harmonics with the end result being nonlinear saturation. The final state may be either chaotic oscillatory motion or a state involving only a few harmonics. The energy equation, corresponding to (6.10), is obtained by multiplying (6.10) by A and integrating by parts, assuming A is periodic with period 2L :

$$(1/2)\, \partial/\partial\theta\Big[\int_0^{2L} A^2\, d\xi\Big] = \int_0^{2L} [\alpha\, (\partial A/\partial\xi\,)^2 - \gamma\, (\partial A^2/\partial\xi^2)^2]d\xi\,. \qquad (9.3)$$

The minimization of the right hand side of (9.3), over all periodic functions, shows that this right hand side will be negative for : $\pi/L > k^*$; and therefore the nonlinear KS equation (6.10) is globally stable for an initial condition with a wavenumber satisfying the linear stability criterion.

In other words, if you put in an initial disturbance (e.g. $\sin(k\xi)$) with a wavenumber k' greater than k^*, then the nonlinear term in (610) creates higher harmonics, but it will not create waves with wavenumbers smaller than k', so there will be stability.

9.1 Dynamical System Approach for the KS Equation

If you want to generate a component with a wavenumber in the unstable region, you have to put in an initial condition with a wavenumber less than k*. Hence, we need to consider only the case $k < k^*$. The periodic boundary conditions allow A to be write as the Fourier series :

$$A = \sum_{-\infty}^{+\infty} A_n(\theta) \exp(ink\xi), \quad A_{-n} = A^*_n . \qquad (9.4)$$

where A^*_n is the complex conjugate of A_n.
Since A_0 = constant we may put $A_0 = 0$ and substitution of (9.4) into KS equation (6.10) gives:

$$\partial A_n/\partial\theta - \sigma_n A_n + inkB_n = 0, \qquad (9.5)$$

where

$$B_n = \sum_{r=1}^{\infty} A^*_r A_{r+n} + (1/2) \sum_{r=1}^{n-1} A_r A_{n-r}, \quad \sigma_n = \alpha (nk)^2 - \gamma (nk)^4 . \qquad (9.6)$$

The significant feature of the system of equations (9.5) is that :

For any given k, only a finite number of Fourier modes, $A_1, A_2, ..., A_n$ say, are unstable ($\sigma_n > 0$), and all higher modes are stable.

Note that the nth mode has a critical wavenumber of k*/n, and a maximum growth rate of ($\alpha^2/4\gamma$) - independent of n - at $k^*/(n\sqrt{2})$. This implies that unstable modes will be stabilized by energy transfer to higher harmonics.
The simplest case amenable to some analysis is when : $k^*/2 < k < k^*$. Only $n = 1$ is unstable and in the following it is assumed that it is sufficient to consider just the interaction between $n = 1$ and $n = 2$. The approximate version of (9.5) is then:

$$\partial A_1/\partial\theta - \sigma_1 A_1 + ikA_2A^*_1 = 0, \qquad (9.7a)$$
$$\partial A_2/\partial\theta - \sigma_2 A_2 + i\kappa (A_1)^2 = 0, \qquad (9.7b)$$

and note that A is unstable ($\sigma_1 > 0$) but A_2 ($\sigma_2 < 0$) is stable. Note that

$$\sigma_1 |A_1|^2 + \sigma_2 |A_2|^2 = 0,$$

reflecting the requires energy balance in the approximate version of

$$\partial/\partial\theta \left(\sum |A_n|^2\right) = 2 \sum \sigma_n |A_n|^2 , \ n = 1 \text{ to } \infty,$$

as consequence of (9.3) and (9.4).
Equation (9.7a) has the steady solution :

$$|A_1| = [- (1/k^2) \sigma_1 \sigma_2]^{1/2}, \qquad (9.8a)$$

since from (9.7b)

$$A_2 = (ik/\sigma_2)\,(A_1)^2. \tag{9.8b}$$

Here, A_1 is growing and A_2 is stabilizing. However, as k is decreased, the hypothesis that only two modes are involved becomes more suspect! Indeed, as k is decreased the steady solution of (9.5), given approximately by (9.8a,b), is at first modified by the presence of a small correction due to A_3 and then when :

$$k^*/3 < k < k^*/2 \text{ (i.e. } \sigma_2 > 0, \text{ but } \sigma_3 < 0),$$

is replaced by another "two-mode equilibrium" in which A_2 and A_4 are the dominant components. Further decrease in k then leads to a succession of states, alterning between "two-mode equilibria" and "bouncy states". If the steady solution for A_2 in (9.8b) is substituted into the equation (9.7a) a Landau - Stuart (LS) equation is obtained for A_1 ; namely :

$$\partial A_1/\partial\theta = \sigma_1 A_1 + (k^2/\sigma_2)\,|A_1|^2 A_1, \tag{9.9}$$

and this LS equation (9.9) is, in fact, valid only for k close to k*. If in (9.9) we assume that $A_1 = |A_1| \exp(i\phi)$, then ϕ = constante and for $|A_1|$ we derive a classical Landau equation:

$$\partial|A_1|/\partial\theta = \sigma_1 |A_1| + \lambda |A_1|^3, \tag{9.10}$$

with $\lambda = (k^2/\sigma_2) < 0$, since $\sigma_2 < 0$.

The solution of (9.10) gives :

$$|A_1| \sim A_1{}^\circ \exp(\sigma_1\theta), \text{ as } \theta \to -\infty,$$

where $A_1{}^\circ$ is the initial value at $\theta = 0$ and $\sigma_1 > 0$, which decays like the linearized theory.

However, $| A_1|^2 \to |A_1|_e{}^2 = - (2\sigma_1/\lambda)$ as $\theta \to +\infty$,

for all value of $A_1{}^\circ$ - this case is called the *supercritical stability*.

If now we introduce a small perturbation parameter, κ, defined by :

$$\kappa^2 \mu = k^2 [(\alpha/\gamma) - k^2] > 0, \tag{9.11}$$

and a slow time scale: $T = \kappa^2 \theta$, then for the slowly varying amplitude of the fundamental wave: H(T) such that $|A_1| = \kappa H$, from (9.10), with (9.6) for σ_1 and σ_2, we derive the following canonical Landau equation for H(T) :

$$\partial H/\partial T = \gamma \mu H - \lambda H^3, \tag{9.12}$$

where the (positive) Landau constant is: $\lambda = 1/16\gamma\, [k^2 - (\alpha/4\gamma)] > 0$.

9.2 Stability and Dynamical System Approach for KS-KdV Equation

The linear dispersion relation of the KS-KdV equation (6.20) for the wave :

$$\eta(\theta,\xi) \approx \exp\,[ik\xi + \sigma\theta]$$

is expressed as (9.13)

$$\sigma = \alpha\, k^2 - \gamma\, k^4 + i\,\phi\, k^3.$$

For Real(σ) > 0 we have instability and for Real(σ) < 0, stability, and Real(σ) = 0, if

$$k = k_c = (\gamma/\alpha)^{1/2}. \qquad (9.14)$$

Consequently, the cut-off wavenumber for the KS-KdV equation (6.20), satisfy the relation

$$k_c{}^2 = (2/5W^*) + (3/2)\, M°/R°W^*(1 + Bi). \qquad (9.15)$$

Thus the wave of small wavenumber are amplified while those of large wavenumbers are damped, and the maximum growth rate occurs at:

$$k_m = (\gamma/2\alpha)^{1/2}.$$

To demonstrate the competition between the stationary waves and the non-stationary (possibly chaotic) attractors of KS-KdV equation (6.20), we do so by rendering (6.20), with $\alpha = \gamma = 1$, into a dynamical system by the Galerkin projection in a periodic medium with wavelength $2\pi/k$:

$$\eta(\tau, \xi) = (1/2)\Sigma\, A_p(\tau) \cos(pk\xi) + B_p(\tau) \sin(pk\xi)\,,\ p \geq 1. \qquad (9.16)$$

For a qualitative analysis of projections of the chaotic phase trajectory onto the plane it seems suffisant (!) to consider a dynamical system truncated at the three harmonics. This system can be easily written in an explicit form.
Namely, first we make a simple linear transformation of the coordinate : $k\,\xi \to x$, $k\,\theta \to t$, the space period of the equation (6.20), with the initial : $\eta(0, \xi) = \eta°(\xi)$
and periodic boundary conditions : $\eta(\theta, \xi) = \eta(\theta, \xi + 2\pi/k)$, $\xi \in [0, 2\pi/k]$, will transit to $x \in [0, 2\pi]$.

Next, substituting (9.16) into KS-KdV equation (6.20), we derive for the amplitudes $A_1(t)$, $B_1(t)$ and $B_2(t)$, the following reduced dynamical system :

$$\begin{aligned} dA_1/dt &= \sigma_1\, A_1 + k^2\phi\, B_1 - 2A_1B_2, \\ dB_1/dt &= \sigma_1\, B_1 - k^2\phi\, A_1 + 2B_1B_2, \\ dB_2/dt &= 2\,\sigma_2\, B_2 + 2\,[(A_1)^2 - (B_1)^2\,], \end{aligned} \qquad (9.17)$$

where

$$\sigma_1 = k(1 - k^2) \text{ and } \sigma_2 = 2\,k(1 - 4k^2).$$

The phase flow of the above dynamical system (9.17) is dissipative if the following relation is satisfied:

$$\sigma_1 + \sigma_2 < 0,$$

and because this dissipative effect, the corresponding strange attractors have zero phase volume and dimensionality smaller than 3 (when t tends to infinity) for the wavenumber k such that :

$$0.58 < k < 1. \tag{9.18}$$

This three-amplitude DS (9.17) can be studied qualitatively and numerically.

An another way for the derivation of a three-amplitude DS for KS-KdV equation (6.20) is the Fourier series. In, this case we assume that :

$$\eta(\theta, \xi) = (1/2)\sum A_n(\theta) \exp[in\,(\omega_1{}^\circ\theta - k_1\,\xi)], \tag{9.19}$$

where $A_n(\theta)$ is the complex amplitude of the n-th spatial harmonic, and where $\omega_1{}^\circ$ is the linear angular frequency (in fact, the angular frequency of the fundamental harmonic, with k_1 as wavenumber, at the first stage of its growth) .
It must be stressed that, if the wavenumber $k_n = n\,k_1$ is the actual wavenumber of the n-th harmonic, on the contrary, the frequency : $n\,\omega_1{}^\circ$, cannot be considered as its actual frequency ω_n (the latter may vary a little, owing to possible small dispersive effects). The slow variation $\psi_n(\tau)$ of the phase corresponding to his small frequency shift is taken into account in the complex amplitude :

$$A_n(\theta) = |\,A_n(\theta)|\ \exp(i\psi_n(\theta)). \tag{9.20}$$

Inserting (9.19) into (6.20), for the first three harmonics we derive the following three - amplitude DS (again, with $\alpha = \gamma = 1$) :

$$dA_1/d\theta = \gamma_1 A_1 + i\,k_1\,A_1{}^*\,A_2\,, \tag{9.21a}$$
$$dA_2/d\theta = (\gamma_2 - 6i\,k_1{}^3\,\phi)\,A_2 + i\,k_1\,A_1{}^2\,, \tag{9.21b}$$
$$dA_3/d\theta = (\gamma_3 - 24i\,k_1{}^3\,\phi)\,A_3 + 3i\,k_1\,A_1A_2, \tag{9.21c}$$

where

$$\gamma_n = (nk_1)^2\,[1 - (nk_1)^2],\ n = 1, 2, 3.$$

Near criticality, where the mode A_1 is the only unstable mode, while the others are linearly strongly damped, the dynamics is controlled by the marginally unstable mode A_1, to which the other two modes are *slaved.*
As consequence, from (9.21b), we have that the dynamics of the harmonics A_2 is slaved to the dynamics of the fundamental harmonic, A_1, according to :

$$A_2 = - [i\, k_1/(\gamma_2 - 6i\, k_1^3\, \phi)]A_1^2. \qquad (9.22)$$

From (9.21a), with (9.22), the fundamental harmonic A_1 obeys the following Stuart - Landau type equation :

$$dA_1/d\theta = \gamma_1 A_1 + \lambda\, A_1^*\, A_1^2\,, \qquad (9.23)$$

where

$$\lambda = (\gamma_2\, k_1^2/a^2)\ [1 + i\, (6k_1^3\, \phi)/\gamma_2], \qquad (9.24)$$

with : $a^2 = \gamma_2^2 + 36\, k_1^6\, \phi^2$, and $\gamma_2 < 0$, is the complex Landau constant. Its real part (positive) corresponds to nonlinear dissipation; and its imaginary part to nonlinear frequency correction (due to dispersive effects).
The dispersive character of the waves plays a crucial role (via the parameter ϕ in (6.20)) in the occurence of amplitude collapses and frequency locking.
This may be understood within the framework of DS (9.21) after separation of modulus and phase of the complex amplitudes (according to (9.20)).
For the simple case, of $|A_1(\theta)|$ and $|A_2(\theta)|$ and phase difference:

$$\Theta(\tau) = \psi_2 - 2\psi_1,$$

we derive the following DS of three - equations in place of (9.21a,b,c) :

$$d|A_1|/d\theta = \gamma_1 |A_1| - k_1 |A_1||A_2| \sin\Theta, \qquad (9.25a)$$
$$d|A_2|/d\theta = \gamma_2 |A_2| + k_1 |A_1|^2 \sin\Theta, \qquad (9.25b)$$
$$d\Phi/d\theta = - 6\, (k_1)^3\, \phi\ +\ k_1\{[\,|A_1|^2 - 2|A_2|^2\,]/|A_2|\,\} \cos\Theta. \qquad (9.25c)$$

This above DS (9.25a,b,c) deserve a further carefully numerical investigation!

10 Weakly Nonlinear Stability Analysis

The linearized stability theory is based upon the main assumption of small (infinitesimal) amplitude disturbances, that is, all terms involving quadratic or higher powers in the disturbances are neglected. Consequently, the solution of the governing partial differential equations for the disturbances is simplified considerably by linearization [see, for example, the linearsystem of equations (7.2a) with the boundary conditions (7.2b,c)] and a Fourier analysis of the disturbances is used with great success to obtain the solutions. In fact, the linearized stability theory determines the conditions of a given steady flow (Nusselt smooth regime, for the case considered in the Section 8.2) which allow the growth of a small disturbance.

According to the linear theory, the amplitude of the disturbance is found to grow exponentially in time for values of certain flow parameters above a critical value [see, for example, (8.27)]. In reality, such disturbances do not grow exponentially without limit! So the validity of the process of linearization, even though it is so often used in many physical problems, has been questioned in connection with instability problem. Landau (1944), (see also, Landau 1965), first described nonlinear instability phenomena of certain classes of flows by the nonlinear amplitude equation

$$d/dt\,(|A|^2) = \nu\,|A|^2 - \lambda\,|A|^4, \tag{10.1}$$

where $A = A(t)$ is the amplitude of the leading Fourier mode, ν and λ are constants, and the latter one is called the Landau constant. In general, ν, λ and A are complex The parameter ν is the eigenvalue of the linear stability problem and if W^* is the bifurcation parameter , then

$$\mathrm{Real}\,(\nu) \sim W^* - (W^*)_c \ \text{ as } W^* \to (W^*)_c. \tag{10.2}$$

The case $\lambda = 0$ corresponds to the linear equation given by the linearized theory. The second term on the right hand side of (10.1) is due to nonlinearity, and may accelerate or decelerate the exponential growth of the disturbance depending on the signs of ν and λ. The original Stuart (1960), equation is :

$$dA/dt = \nu\,A + \mu\,A\,|A|^2\,, \tag{10.3}$$

and equivalent form of (10.3) is

$$-\,c\,dA/d\xi = \nu\,A + \mu\,A\,|A|^2\,, \tag{10.4}$$

where: $\xi = x - c\,t$ and $A = A(\xi\,)$. Below, in the Section 10.1, we derive, first, an evolution equation of the form (10.4) for the Shkadov averaged nonlinear IBL system (5.2), (5.13), for $H(\tau, x)$ and $q(\tau, x)$.

10.1 From IBL Averaged Shkadov System to Landau - Stuart Evolution NL Equation

The derivation of the amplitude, Stuart - Landau type, equation near criticality using the technique of multiple scales is now well known and the details can be found in Newell (1974) or Stewartson and Stuart (1971).

Our main small parameter is: $\delta \ll 1$, according to (2.16), (2.17). First, we write :

$$W^* = W^*_c + \sigma\, \delta^2, \qquad (10.5)$$

where σ is a scalar, and for the analysis of the instability (see, for instance, the Section 8.3) it is necessary that $\sigma < 0$.

As $\delta \ll 1$ we consider, in fact, a weakly nonlinear 'instability' theory - i. e., near-linear analysis near neutral stability. For the phase velocity we write, according to (10.5)

$$c_r = c_r^* + \delta^2 c_2\,, \text{ with } c_r^* = 1. \qquad (10.6)$$

Next, we introduce slow variables :

$$\xi_k = \delta^k \xi\,,\ k = 1, 2, 3, \ldots, \qquad (10.7)$$

where $\xi = x - c_r \tau$.

But for a weakly nonlinear case, if we want to derive the associated Landau - Stuart (LS envelope) evolution equation, it is suffisant to assume that the amplitude of the wave packet envelope is a function of $\xi_2 \equiv \eta$ only.

As consequence, we can assume that $H(\tau, x)$ and $q(\tau, x)$ in equations (5.2) and (5.13), have the following approximate form :

$$H = H^*(\xi, \eta; \delta) = 1 + \delta\, h_1 + \delta^2 h_2 + \delta^3 h_3 + \ldots, \qquad (10.8a)$$

$$q = q^*(\xi, \eta; \delta) = (1/3) + \delta\, q_1 + \delta^2 q_2 + \delta^3 q_3 + \ldots. \qquad (10.8b)$$

But, according to linear theory,

$$h_1(\xi, \eta) = A(\eta)\, E(\xi) + A^*(\eta)\, E(-\xi), \qquad (10.9)$$

with $\quad E(\pm\xi) = \exp[\pm i k^* \xi]$

where k^* is the neutral (cut-off) wave number, and A^* is the complex conjugate of the amplitude A ($A A^* = |A|^2$).

For the derivation of the LS equation for the amplitude $A(\eta)$ of the wave packet envelope it is necessary to eliminate the secular terms in the equation for h_3 and q_3 ! In others words, we assume that the asymptotic expansions (10.8a,b) are uniformly valid with respect to variable ξ.

Now, taking into account the relations

$$\partial/\partial\tau = -c_r\,(\partial/\partial\xi + \delta^2\, \partial/\partial\eta) \text{ and } \partial/\partial x = \partial/\partial\xi + \delta^2\, \partial/\partial\eta, \qquad (10.10)$$

substitute the above expansions (10.8a,b) and (10.5), (10.6), for W* and c_r, into the nonlinear system of two equations (5.2), (5.13) and identify different orders to derive a sequence of differential equations.
To obtain the amplitude, LS, equation at the lowest order, we only need to consider n = 1, 2 and 3 powers of the small parameter δ.

n = 1

For $h_1(\xi, \eta)$ and $q_1(\xi, \eta)$ we obtain the classical homogeneous linear system :

$$\partial h_1/\partial\xi = \partial q_1/\partial\xi, \quad \Lambda(h_1, q_1) = 0, \tag{10.11}$$

where

$$\Lambda(h, q) \equiv - (1/5)\, \partial q/\partial\xi - (2/15)\, \partial h/\partial\xi + (3/Re^*)\,(q - h) - (1/3k^{*2})\partial^3 h/\partial\xi^3. \tag{10.12}$$

Since $\Lambda(E(\pm\xi)\,,\, E(\pm\xi)) \equiv 0,$
the solution of (10.11) is simply :

$$q_1(\xi, \eta) = h_1(\xi, \eta) = A(\eta)\, E(\xi) + A^*(\eta)\, E(-\xi). \tag{10.13}$$

n = 2

For $h_2(\xi, \eta)$ and $q_2(\xi, \eta)$ we derive a nonhomogeneous system :

$$\partial h_2/\partial\xi = \partial q_2/\partial\xi, \tag{10.14a}$$

$$\Lambda(h_2, q_2) = - (8/15)\, \partial/\partial\xi\, (q_1{}^2) + (3/Re^*)\, q_1{}^2 + (1/3k^{*2})\, q_1 \partial^3 q_1/\partial\xi^3. \tag{10.14b}$$

If we take into account the solution (10.13) for $q_1(\xi, \eta)$ and $h_1(\xi, \eta)$, then in place of (10.14b) we obtain the following equation :

$$\Lambda(h_2, q_2) = (6/Re^*)\, |A|^2 + [(3/Re^*) - i\,(7/5)\,k^*]\, A^2\, E(2\xi) + [(3/Re^*) + i\,(7/5)\,k^*]\, A^{*2}\, E(-2\xi). \tag{10.15}$$

The two equations (1014a) and (10.15), for $q_2(\xi, \eta)$ and $h_2(\xi, \eta)$, support solutions of the following type :

$$h_2(\xi, \eta) = q_2(\xi, \eta) - \alpha\, |A|^2, \tag{10.16a}$$

$$q_2(\xi, \eta) = B(\eta)\, E(2\xi) + B^*(\eta)\, E(-2\xi), \tag{10.16b}$$

with : $E(\pm 2\xi) = \exp[\pm 2ik^*\xi]$.

But, since (according to (10.16a,b)):

$$\Lambda(h_2, q_2) \equiv (3/Re^*)\,\alpha\,|A|^2 + 2ik^*B(\eta)\,E\,(2\xi) - 2ik^*\,B^*(\eta)\,E\,(-2\xi),$$

and we see that :

$$\alpha = 2, \text{ and } B(\eta) = \beta\,[A(\eta)]^2,$$

with

$$\beta = -\,[(7/10) + i\,(3/2\,k^*Re^*)]. \tag{10.17}$$

Finally, for $q_2(\xi, \eta)$ and $h_2(\xi, \eta)$ we obtain as solution

$$h_2(\xi, \eta) = q_2(\xi, \eta) - 2\,|A|^2, \tag{10.18a}$$

$$q_2(\xi, \eta) = \beta\,[A(\eta)]^2\,E\,(2\xi) + \beta^*\,[A^*(\eta)]^2\,E\,(-2\xi), \tag{10.18b}$$

where β^* is the complex conjugate of the coefficient β (see (10.17)).

n = 3

For $h_3(\xi, \eta)$ and $q_3(\xi, \eta)$ we obtain first the following equation

$$\partial h_3/\partial\xi - \partial q_3/\partial\xi = c_2\,\partial h_1/\partial\xi,$$

since $h_1 = q_1$, and we derive the following relation between h_3 and q_3 :

$$h_3 = q_3 - c_2\,h_1 + H(\eta). \tag{10.19a}$$

The second equation between $h_3(\xi, \eta)$ and $q_3(\xi, \eta)$ is:

$$\Lambda(h_3, q_3) = \gamma\,[A(\eta)]^3\,E\,(3\xi) + S(A)\,E\,(\xi) + C.C, \tag{10.19b}$$

when we take into account the solutions (10.13) and (10.18a,b). In (10.19b) we have introduced the following operator :

$$S(A) = -\,(2/3)\,\partial A/\partial\eta + \lambda\,A - \mu\,A\,|A|^2\,. \tag{10.20}$$

For the complex coefficients γ, λ and μ we have the following expressions :

$$\gamma = -\,(55/2)\,(1/Re^*) + i\,[(507/50)\,k^* - (9/k^*)\,(1/Re^*)^2], \tag{10.21a}$$

$$\lambda = -\,i\,k^*\,[c_2\ -\ \sigma\,k^{*2}], \tag{10.21b}$$

$$\mu = (93/10Re^*) +\ i\,[\,(31/50)\,k^* + (9/k^*)\,(1/Re^*)^2]. \tag{10.21c}$$

From (10.19b) we conclude that, necessarily : $H(\eta) = 0$, and if we utilize (10.19a) (with $H(\eta) = 0$) and the expression (10.12) for the operator $\Lambda(h, q)$ with h_3 and q_3, we derive for the term q_3 the following equation:

$$\partial/\partial\xi\,[(1/k^{*2})\,\partial^2 q_3/\partial\xi^2 + q_3] = 3\,[\kappa\,c_2\,A - S(A)]\,E\,(\xi) - 3\gamma\,[A(\eta)]^3\,E\,(3\xi) + C.\,C. \tag{10.22}$$

where

$$\kappa = (3/Re^*) - (1/5)\, i\, k^* \qquad (10.23)$$

The solution of (10.22) for q_3 is of the following form

$$q_3 = D(A)E\,(3\xi) - \{(3/2)\,[\kappa\, c_2\, A - S(A)]\, E\,(\xi)\}\, \xi + C.\,C, \qquad (10.24)$$

and as consequence the term proportional to : $-(3/2)\,[\kappa\, c_2\, A - S(A)]\, E(\xi)$ is a secular term - this term is very large for the very large values of ξ, and as consequence, the term $\delta^3\, q_3$ in (10.8b) may not be small relatively to the second order term $\delta^2\, q_2$!

Finally, for the amplitude $A(\eta)$ we derive the following evolution LS equation :

$$(2/3)\, \partial A/\partial\eta + \nu\, A + \mu\, A\, |A|^2 = 0, \qquad (10.25)$$

where :

$$\nu = \kappa\, c_2 - \lambda = (3/Re^*)\, c_2 + i\,[-(6/5)\, k^* c_2 + \sigma\, k^{*3}], \qquad (10.26)$$

and for the complex Landau coefficient μ we have the formulae (10.21c).

It is not difficult to derive an analogous LS equation for the system (5.2), (5.16), when the Marangoni effect is taken into account, but this derivation is more laborious and the coefficients ν and μ, for this case, are the rather awkward-looking one! In fact, in (10.26) it is reasonable to choose :

$$[-(6/5)\, k^* c_2 + \sigma\, k^{*3}] = 0 \rightarrow c_2 = (5/6)\sigma\, k^{*2}, \qquad (10.27)$$

and in this case, the LS evolution equation is

$$-\partial A/\partial\eta = \alpha\, \sigma\, A + (3/2)\, \mu\, A\, |A|^2, \qquad (10.28)$$

vith

$$\alpha = (15/4)\, (k^{*2}/Re^*) > 0, \qquad (10.29a)$$

$$\mu = (93/10Re^*) + i\,[\,(31/50)\, k^* + (9/k^*)\, (1/Re^*)^2]. \qquad (10.29b)$$

Now, we write for the complex amplitude $A(\eta)$, solution of the evolution equation (10.28) with complex Landau constant $\mu = \mu_r + i\, \mu_i$:

$$A(\eta) = |A(\eta)|\, \exp\,[i\psi(\eta)]. \qquad (10.30)$$

In this case for $|A(\eta)|$ we derive the following Landau equation

$$d|A(\eta)|/d\eta = -\alpha\, \sigma\, |A(\eta)| - (3/2)\, \mu_r\, |A(\eta)|^3, \qquad (10.31)$$

where

$$\mu_r = 93/10Re^*.$$

For the phase $\psi(\eta)$ we have the following relation :

$$d\psi(\eta)/d\eta = -(3/2)\mu_i\, |A(\eta)|^2, \qquad (10.32)$$

where $\mu_i = (31/50)\, k^* + (9/k^*)\, (1/Re^*)^2.$

In fact, the above Landau equation (10.31) is a first order linear ODE for the ratio:

$$B(\eta) = 1/|A(\eta)|^2$$

and as consequence we may solve for $|A(\eta)|$ explicitly to obtain :

$$|A(\eta)| = |A(0)| \exp[-\alpha\sigma\eta]\{1/[1 - X]\}^{1/2} \quad (10.33)$$

where

$$X = (3\mu_r/2\alpha\sigma)|A(0)|^2 [\exp(-2\alpha\sigma\eta) - 1]. \quad (10.34)$$

Using the above result, we may estimate the *critical rupture* $\eta = \eta^*$, corresponding to $A(\eta) = \infty$.

In this case $X = 1$ and :

$$\eta^* = -(1/2\alpha\sigma) \ln\left[1 + (2\alpha\sigma/3\mu_r|A(0)|^2)\right]. \quad (10.35)$$

The positive values of σ, in (10.5), would correspond to stable solutions, and we would thus expect that σ would be negative, while negative values of σ would imply to opposite to be true : $\sigma < 0$. Ignoring the Marangoni effect and analyzing the simpler problem which results we find that this is exactly what happens.

10.2 Interaction Between Short-Scale Marangoni Convection and Long-Scale Deformational Instability

Now, it is important to note that in the case of the Bénard-Marangoni convection in a liquid layer with a deformable interface, as was previously shown by Takashima (1981a) through a linear stability analysis (see the § 7), there exist *two monotonous modes* of surface tension driven instability.

One, a *short-scale* mode, is caused by surface tension gradients alone, without surface deformation, and it leads to formation of a stationary convection with a characteristic scale of the order of the liquid layer depth.

The other, a *long-scale* mode, is influenced also by gravity and capillary (Laplace) forces, and surface deformation plays a crucial role in its development. This instability mode results in a large-scale convection and in the growth of long surface deformations which the characteristic scale is large in comparaison with the thickness of the liquid layer. As shown in the paper by Golovin, Nepomnyaschy and Pismen (1994), these two types of the Marangoni convection, having different scales, can interact with each other in the course of their nonlinear evolution. There are two mechanisms of the coupling between them. On the one hand, surface deformation change locally the Marangoni number, which depends on the depth of the liquid layer - this leads to a space-dependent growth rate of the short-scale convection and, hence, its intensity becomes also space dependent. On the other hand, the short-scale convection generates an additional mean mass/heat flux from the bottom to the free surface, which is proportional to its intensity (square of the amplitude).

When the intensity is not uniform, this leads to additional long-scale surface tension gradients affecting the evolution of the long-scale mode. Indeed, the coupling effects are most pronounced in the case when the long-and the short-scale modes have instability threshold close to each other. In the above cited paper, Golovin, Nepomnyaschy and Pismen (1994), have studied these effects analytically (via a weakly nonlinear analysis) in the vicinity of the instability thresholds.

Close to the bifurcation point, the mean long-scale flow generated by the short-scale convection is very weak, of the order of ε^3, and it will considerably effect the long-scale surface deformations only if the later are also small, of the order of ε^2, where ε is the amplitude of the deformationless convective mode - this happens when the surface tension is sufficiently large.

According to Golovin, Nepomnyaschy and Pismen (1994), near the instability threshold, the nonlinear evolution and interaction between the two modes can be described by a system of two coupled nonlinear equations, namely :

$$\partial A/\partial T = (\pm A) + \partial^2 A/\partial x^2 + A\,|A|^2 + \eta\, A, \tag{10.36a}$$

$$\partial \eta/\partial T = -(\pm m)\, \partial^2 \eta/\partial x^2 - w\, \partial^4 \eta/\partial x^4 + s\, \partial^2 |A|^2/\partial x^2, \tag{10.36b}$$

where the parameters m, w, s are all positive. The parameter m characterizes the effect of surface tension gradients and gravity, the parameter w corresponds to the Laplace pressure ans s is the interaction parameter characterizing the coupling between the two modes of Marangoni convection.

The complex amplitude of the short-scale convection, A, undergoes a long-scale evolution, described by the *Ginsburg-Landau equation* (10.36a), but the later, however, contains an additional nonlinear (quadratic) term, ηA, connected with the surface deformation η - this term adds to the linear growth rate for the amplitude A and describes, in fact, a nonuniform space-dependent supercriticality. It plays a stabilizing role when the surface is elevated ($\eta > 0$), and supresses the short-scale convection under the surface deflections.

The surfave deformation η is governed by the *nonlinear evolution equation* of the fourth order (10.36b). However, the only nonlinear term in this equation is the coupling term proportional to $\partial^2|A|^2/\partial x^2$, describing the effect of the mean flow generated by the short-scale convection - this term always plays a stabilizing role.

In two equations (10.36a, b) the various signs of the terms correspond to the four cases described by Golovin, Nepomnyaschy and Pismen (1994).

In fact :

(i)- when in (10.36a) we have +A and in (10.36b) + m, both the short-scale deformations mode and the long-scale deformational one are unstable ;

(ii)- if in (10.36a) we have +A and in (10.36b) - m, only the short-scale mode is unstable;

(iii)- when in (10.36a) we have - A and in (10.36b) + m, the dformational mode is unstable;

(iv)- if in (10.36a) we have - A and in (10.36b) - m, both modes linearly stable, but their nonlinear interaction may lead to an instability.

The typical neutral stability curve, Ma(k), represented in the Fig. 10.1 below, has two minima : first Ma_l correspondingto k = 0, and describes the long-wave instability, and then Ma_s, related with $k_c \neq 0$, and indicates the threshold of short-scale convection. In this Fig. 10.1, the dashed line corresponds to layer with undeformable interface and in this case only one minimum exist, Ma_s.

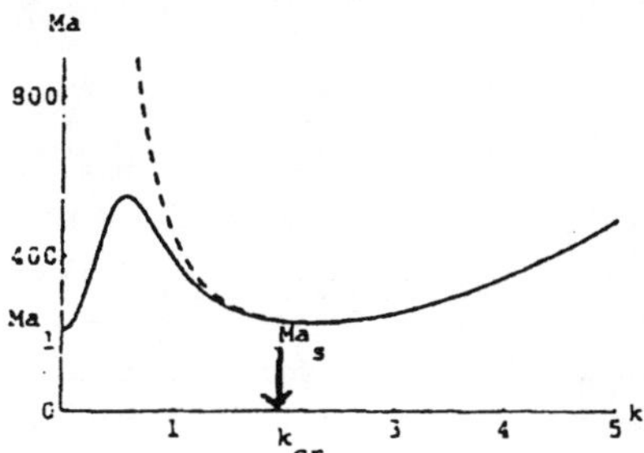

Fig. 10.1. Neutral stability curves for the Marangoni convection

The surface deformation η is a real quantity, whereas the amplitude A of the small-scale convection is complex. If we assume, for the sake of simplicity, A to be also reel, thus considering the evolution of the short-scale convective structure with the fixed wave number $k = k_c$, then we write for the derivation of a three-mode truncated model,

$$A(T) = A_0 (T) + A_1(T) \cos(k_m x) + ...; \quad \eta = B_1(T) \cos(k_m x) + ..., \quad (10.37)$$

with : $k_m = (m/2w)^{1/2}$, corresponding to the maximum linear growth rate of the first harmonic. In this case, substituting (10.37) into (10.36a, b) and consider the third case (-A, +m), after appropriate rescaling, we obtain the following dynamical system for A_0 (T), A_1(T), and B_1(T) :

$$\begin{aligned} dA_0/dT &= - A_0 [1 + A_0^2 + (3/2) A_1^2] + (1/2)A_1 B_1, \\ dA_1/dT &= - A_1 [1 + (m/2w) + 3A_0^2 + (3/4) A_1^2] + A_0 B_1, \\ dB_1/dT &= - \mu B_1 - \sigma A_0 A_1, \end{aligned} \tag{10.38}$$

where :

$$\mu(k_m) = m k_m^2 - w k_m^4 , \quad \sigma = 2 k_m^2 s`$$

The system (10.38) describes the time evolution of a periodic surface deformation (mode B_1), which can generate not only a periodic mode of the short-scale convection (A_1) following the surface deformation, but also a uniform zero mode (A_0). The below Fig. 10.2, shows projection of one of the (strange) chaotic attractor on the planes : (a) - (A_0, B_1) and (b) - (A_1 ,B_1). Thus, the coupling between the short-scale convection and large-scale deformations of the interface can lead to stochastization of the system and be one of the causes of interfacial turbulence of the thin film.

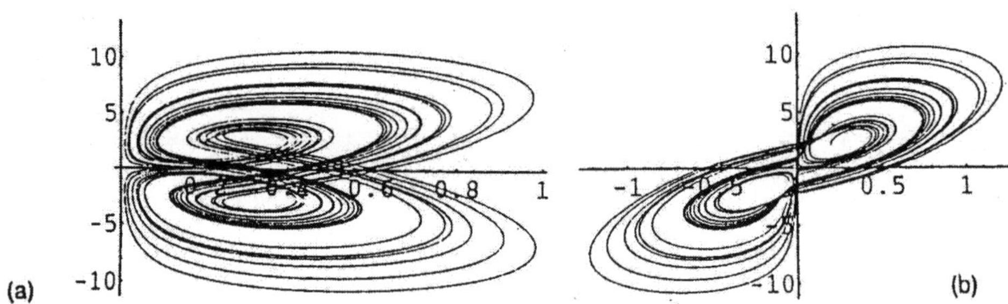

Fig. 10.2. Chaotic attractor of system (10.38).
$k_m = 1$, $\mu = 1$, $\sigma = 30$; time interval T $\in$ [0, 150]

The reader can find in Kazhdan et al. (1995), the numerical analysis of the system of two nonlinear coupled equations (10.36a,b), which confirms the predictions of weakly nonlinear analysis and shows the existence of either standing or travelling waves in the proper parametric regions, at low supercriticality. With increasing supercriticality, the waves undergo various transformations leading to the formation of pulsating travelling waves, nonharmonic standing waves as well as irregular wavy behavior resembling “interfacial turbulence”.

The below Fig. 10.3, according to numerical study of Kazhdan et al. (1995), presents long-time series for the surface deformation η and the amplitude A (dashed line) - normalized by their maximum absolute values - at a fixed location. It can be seen that oscillations of η and A are highly correlated : the amplitude of the short-scale convection follows the surface oscillations, being large beneath surface elevations and small under surface depressions.

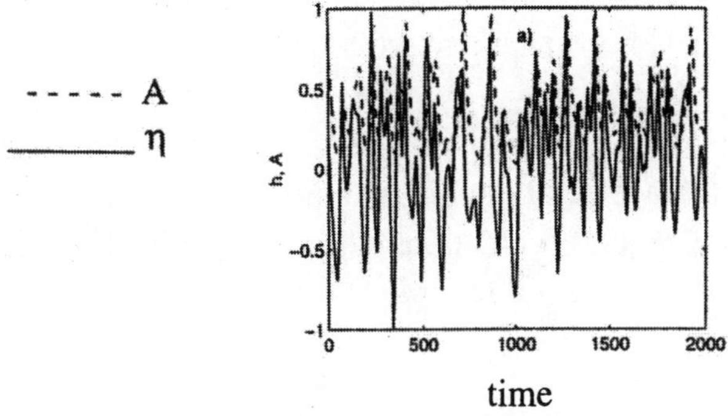

Fig. 10.3. Chaotic oscillations of surface deformation h and of the amplitude of the short-scale convection A at a fixed space point , in case : $s = 70$, $m = 4.5$, $w = 1$; $Ma_L < Ma < Ma_S$.

The Fig. 10.4, below, shows the projection of the phase portrait of the system (10.36a,b) on the plane (η, $\partial\eta/\partial T$). The motion is apparently chaotic. In the space Fourier spectrum, due to the strong damping of the higher harmonics caused by the fourth derivative in the evolution equation (10.36b) for η, only a small number of modes are excited (less than eight), and the number of modes does not change significantly with the increase of the coupling parameter. Hence the observed irregular pattern can be characterized as a low mode chaotic system

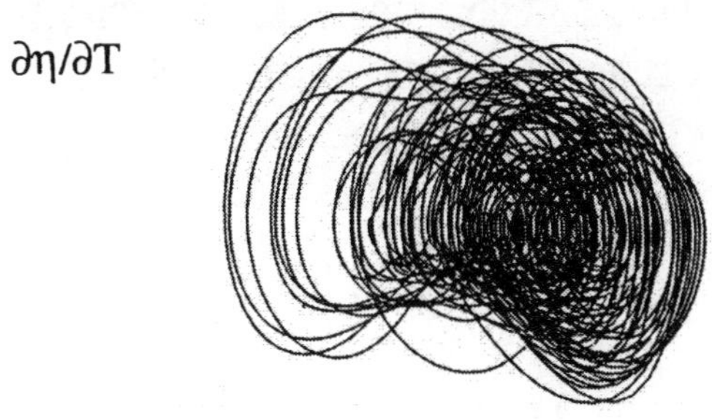

Fig. 10.4. Phase portrait of the chaotic oscillations of the surface deformation at a fixed point shown in above Fig. 10.3 ($T < 6000$)

With increasing s [coupling parameter in equations (10.36a,b)], the growth of the surface deformations is suppressed by the short-scale convection and the coupling between the two modes gives rise to long-scale standing waves modulating the short-scale roll convection pattern.

As the coupling becomes stronger, these oscillations decay and a stationary marge-scale structure appears instead. This structure consists of narrow well separated depressions, with the rest of the interface being almost flat - under surface depressions the fluid is almost quiescent while in flat regions convection has an almost constant amplitude.

With further increase of the coupling parameter s, the stationary pattern becomes unstable and both long-scale surface deformations and the amplitude of the short-scale roll convection undergo irregular oscillations as shown in Fig. 10.5. below, in the case : $s = 70$, $m = 4.5$, $w = 1$; $Ma_l < Ma < Ma_s$.

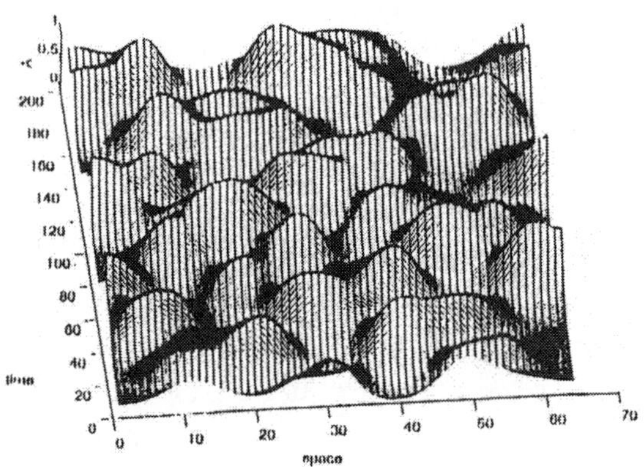

Fig. 10.5. Long-scale modulation of the amplitude of the short-scale convection generated by the surface deformation ($s = 70$, $m = 4.5$, $w = 1$; $Ma_l < Ma < Ma_s$)

11 Concluding Remarks

Although significant understanding has been achieved, yet surface tension gradient-driven (B-M) convection flows still deserve further study.
Indeed, as a paradigmatic form of a spontaneous self-organizing system, the doctrine about the original Bénard problem has not reached the degree of sophistication, in theory and experimentation, attained in buoyancy-driven (R-B) convection. There are still challenging problems like relative stability of patterns (hexagons, rolls, squares,..., labyrinthine convection flows!), higer transitions and interfacial turbulence (at low Marangoni number ?), a case of space-time chaos with high dissipation. It should be pointed out, also, that the ever increasing number of industrial applications of thin film flows and the richness of behavior of the governing equations make this area a particularly rewarding one for mathematicians, engineers, and industrialists alike.

Although Bénard was aware of the role of surface tension and surface tension gradients in his experiments, it took, however, five decades to unambiguously assess, experimentally and theoretically [see, for instance, the papers by Block (1956) and Pearson (1958)], that indeed the surface tension gradients rather than buoyancy was the cause of Bénard cells in thin liquid films. Only in 1997 this almost evident physical fact has been rigorously proved, through an asymptotic approach, by Zeytounian (1997) :

" *Either the buoyancy is taken into account and in this case the free-surface deformation effect is negligible and we have the possibility to take only partially into account the Marangoni effect, or this free-surface deformation effect is taken into account and in this case the buoyancy don't play a significant rôle in the Bénard - Marangoni full thermocapillary problem*".

It seems that the first author to explain the effect of the surface tension gradients on Bénard convection was Pearson (1958). The review articles by Normand, Pomeau and Velarde (1977) and by Davis (1987) consider the role of both buoyancy and surface tension gradients in triggering convective instability. Recently, a review article by Cross and Hohenberg (1993) was devoted to non-equilibrium pattern formation, with a sketchy section dealing with genuine Bénard cells i. e. Bénard - Marangoni convection. Koschmieder (1993) who has been for decades a key figures in the experimental investigation of the Bénard problem wrote recently a very valuable monograph. The more recent review article by Bragard and Velarde (1997), is provided of salient findings, old and recent, about Bénard convection flows in a liquid layer heated from below and open to the ambient air.

In Zeytounian (1998) I have considered, for a thin film provblem, three main situations in the relation with the magnitude of the characteristic Reynolds number and discussed various model equations - these model equations are analyzed from various point of view but the central intent of the my review paper is to elucidate the role of the Marangoni number on the evolution of the free-surface in space-time.

The Myers (1998) recent paper is a review of work on thin films when (high) surface tension is a driving mechanism. Its aim is to highlight the substantial amount of literature dealing with relevant physical models and also analytic work on the resultant equations. The recent paper of Ida and Miksis (1998a) consider also the dynamics of a general 3D thin film subject to van der Waals forces, surface tension, and surfactants. Using an asymptotic analysis based upon the thinness of the film with respect to the lateral extent, evolution equations for the leading-order film thicknesses, tangential velocities, and surfactant concentrations are obtained. The scaling has been chosen by the above authors such that the surface tension effects occur at leading order in the dynamics model of the thin film. Note that the analysis applies to the breaking of a thin liquid film off of a stable centersurface.

Unfortunately, the model equations as presented form a complicated set of evolution equations and cannot be solved until the centersurface is prescribed. In Part II of their work, Ida and Miksis (1998b), consider a series of special centersurfaces and in each case consider the linear stability and solve the resulting nonlinear equations numerically. In particular, it is showed that increasing surface tension is stabilizing, while increasing the effects of van der Waals forces is destabilizing. The effects of surfactants, although irrelevant in the determination of the neutral stability curves, is stabilizing. The results obtained by solving the full evolution equations numerically agreed with the stability results obtained analytically. Time-dependent B-M (oscillatory) instability and waves has been considered recently by Velarde and Rednikov (1998). As the liquid layer is subjected to a thermal gradient orthogonal to its open surface, the attention is focused on the role played by surface tension non-uniformity, which induces surface stresses and (Marangoni) convective motions (beyond an instability threshold). Long waves in a highly viscous Bénard layer placed on an almost stress-free support is also considered by these authors and a KdV- like equation is derived [see, for instance, Garazo and Velarde (1991)], but augmented with dissipative terms entering with a smallness parmeter - travelling periodic and solitary waves of this equation are discussed. In particular, the dissipative solitary wave has been shown to possess a nonmonotonic tail. It is important to note that the presence of the free-surface introduces additional interesting effects of surface tension and gravity, which changes the character of the instability dramatically in a parallel flow. While the instability of the parallel flow between two rigid walls takes the form of short shear waves the instability in a liquid film takes the form of long gravity-capillary waves at the relatively small Reynolds number.

Another interesting feature of the film instability is that there exists no finite critical wavelength according to the linear theory, in contrast to the case of rigid boundaries. The linear theory predicts the instability to take place in the form of

an infinitely long-wave. On the other hand, surface waves of finite wavelengths were observed, as the consequence of the instability, by Kapitza and Kapitza (1949) and Binnie (1957). This led to the conjecture that the observed waves are the most amplified waves with the wavelength λ_m predicted by the linear theory. However, referring to the case of a free-falling vertical film, Benjamin (1957) pointed out that : "one can scarcely expect waves to appear with a strictly uniform and distinct periodicity, because under all conditions infinitesimal waves with a wide range of wavelengths are unstable, and the wave with length λ_m comes into prominence only through a rather uncritical selection process depending on differences in the rates of amplification of different wavelengths. The ultimate state of the amplified waves is, of course, determined largely by nonlinear effects which remain unknown". This statement is consistent with the experiment of Kapitza and Kapitza (1949), who found that distinctively periodic waves could not be observed unless the disturbaces were introduced at precisely controlled frequencies. Thus Lin (1969, 1970) was led to investigate the nonlinear evolution of the so-called Benjamin (1957) - Yih (1963) wave of a given mode. In Lin (1974), the nonlinear instability to disturbances of a finite frequency bandwidth is studied, in the case of a layer of an incompressible viscous fluid flowing down a plane inclined at an angle β to the horizontal. Indeed, Lin consider the case of weakly nonlinear wave motion which pertubs the free-surface only slightly, and derive by a multiple scale asymptotic expansion an amplitude equation for the leading-wave envelope. This Lin (1974, eq.(18)) equation is the appropriate equation for the description of the weak nonlinear evolution of relatively short waves near the upper branch of the neutral curve where the amplification rate c_i is $O(\varepsilon^2)$ - see in Lin (1974) its amplitude equation (18) and its Figure1 [which represents in the (Re, α) plane, the stability curves for water at 15°C with $\beta = 90°$ and Weber number We = 463.3]. Near the lower branch of the neutral curve the wave number α (in a distance 2π d, d is the constant film thickness) is zero, even where $c_i < O(\alpha^2) = O(\varepsilon)$ the modal interaction is stronger, and the Lin (1974) equations, (25) or (26) - which is very similar to KS-KdV equation (6.20), but with $\gamma = 0$ - without the diffusion term - is then the governing equation of the nonlinear evolution. Farther away from the neutral curve where $c_i = O(\varepsilon)$ the diffusion term in Lin equations (25), or (26), becomes important.
Unfortunately, the Lin (1974) film stability study do not take into account the Marangoni effect. In a recent paper by Wilson and Thess (1997), explicit analytical expressions for the linear growth (and decay) rates of long-waves modes in Bénard-Marangoni convection are derived and discussed. These analytical predictions are shown to be in good agreement with experimental observations [of VanHook et al. (1995)] and are used to estimate the minimum experimental time necessary in order to observe the long-wave instability under microgravity conditions. This work is a natural extension of previous linear stability studies by Pérez - Garcia and Carneiro (1991) which concentrated on the determination of the marginal stability curves, the nonlinear analyses of Marangoni convection in a thin layer of fluid by Kopbosynov and Pukhnachev (1986) and Davis (1987) and the recent investigation of the linear growth rates of

the Marangoni problem near the onset of convection by Regnier and Lebon (1995).

It is interesting to note that long-wave instability (L-WI) occurs even in highly viscous fluids, but its growth rate becomes so small as to be experimentally undetectable! On the other hand in the limit of small Biot number (which is typical of experimental situations), neither the dimensional growth rate S and the corresponding critical temperature difference ΔT_L (for the onset of the L-WI at which S = 0), depends on the thermal diffusivity of the fluid, but S does not depend on the absolute value of the surface tension. The results which are obtained by Regnier and Lebon (1995) indicate that the influence of surface deformation on the relaxation time and the correlation length is weak for the non-zero wavenumber instability mode; in contrast, the zero mode which is inherent to the surface deformation, exhibits a high sensitivity to the crispation number Cr. The dispersion relation for the zero mode was found analytically by using an appropriate scaling, as a main result of the analysis of Regnier and Lebon (1995), it is shown that the presence of the zero mode can only be detected in very large aspect ratio boxes and for very thin fluid layers - the results for zero mode provide a first step towards a (weakly) nonlinear analysis. Just a weakly nonlinear analysis of coupled surface-tension and gravitational- driven instability in thin fluid layer (but, again, with a flat upper free surface!) is presented in Parmentier, Regnier and Lebon (1996) paper. In a weakly nonlinear analysis, it is sufficient to take into acount the modes that are critical at the linear threshold and as consequence for the critical modes, in Parmentier, Regnier and Lebon (1996) paper, a system of three coupled Ginzburg-Landau type equations for the three amplitudes A_1, A_2, A_3, is derived :

$$\tau\, \partial A_i/\partial t = \varepsilon\, A_i + a\, A_j^*\, A_k^* - b\, A_i\, [|A_j|^2 + |A_k|^2] - c\, A_i\, |A_i|^2 \qquad (11.1)$$

with, i = 1 and j = 2, k = 3; i = 2 and j = 3, k = 1; i = 3 and j = 1, k = 2, wherein the coefficients τ, a, b and c depend generally on the Prandtl and the Biot number and also on the ratio α (the percentage of buoyancy effect with regard to thermocapillary effect - $\alpha = 0$ corresponds to pure thermocapillarirty and $\alpha = 1$ to pure buoyancy). The relative distance from the threshold is :

$$\varepsilon = 1 - \lambda/\lambda_c \text{ with } \lambda = Ra/Ra^\circ + Ma/Ma^\circ, \qquad (11.2)$$

where the wave number k corresponding to λ_c is the critical wave number k_c ; Ra° is the critical Rayleigh number for pure buoyancy and Ma° is the critical Marangoni number for pure thermocapillarity. According to Parmentier, Regnier and Lebon (1996), when buoyancy is the single responsibility of the convection, only rolls will be observed.

As soon as capillary effects is, however, observed and it appears that a hexagonal structure is preferred at the linear threshold. The more the thermocapillary forces are dominant with respect to the buoyancy forces the larger the size of the region where hexagons are stable. It is shown that the direction of the motion inside the hexagons is directly linked to the value of the Prandtl number and for Pr > 0.23, the fluid moves upward at the center of the hexagons, in accord with experiments. A subcritical region where hexagons are stable has also been displayed by these above authors - the region is the largest when buoyancy does not act and in this case, the value found for the subcritical parameter is in excellent agrement with direct numerical simulation performed by Thess and Orszag (1995).

But, all these above results correspond to the case when the upper free (!) surface is flat. A detailed analysis of the above system (11.1) can be found in Cross (1980, 1982) and in Cross and Hohenberg (1993). Recently, instability of a liquid hanging below a solid ceiling (the so-called Rayleigh-Taylor (R-T) instability) has been considered by Limat (1993) in a short Note, according to lubrication equation derived in Kopbosynov and Pukhnachev (1986), - this author discuss the influence of the initial thickness on the R-T instability and the results are summarized by a diagram giving the different possible regimes. This diagram allows one to predict two different thickness dependence that are selected by the physical properties of the liquid.

For the vertical film, the recent review paper by Chang (1994) gives an excellent survey concerning mostly the various transition regimes on a free-falling vertical film and for an extension of this review, see Chang and Demekhin (1996) survey - but in these both review papers the discussion concerning the Marangoni effect is absent.

The experimental investigation of three-dimensional instabilities of film flows is presented in the paper by Liu, Schneider and Gollub (1995) and several distinct transverse instabilities are found to deform the travelling waves : a synchronous mode (in which the deformations of adjacent wave front are in phase) and a subharmonic mode (in which the modulations of adjacent wave front are out of phase - in this case the herringbone patterns result).

The 3D subharmonic weakly nonlinear instability is due to the resonant excitation of a triad of waves consisting of the fundamental two-dimensional wave and two oblique waves. The evolution of wavy films after the onset of either of these 3D instabilities is complex - however, sufficiently far downstream, large-amplitude solitary waves absorb the smaller waves and become dominant. In Liu, Schneider and Gollub (1995) a detailed study of these instabilities is then presented, along with a qualitative treatment of the further evolution toward an asymptotic "turbulent" regime. In recent review paper by Oron, Davis and Bankoff (1997), the long-scale evolution of thin (macroscopic) liquid films is considered.

By means of long-scale evolution equations, many interesting cases are discussed, giving the reader both an overview and representative behaviors typical of thin films. The first topic is 'bounded films', which have one free surface and one interface with a solid phase. The second topic concerns 'bounded films with slowly varying spatial nonuniformities at the boundaries'. Next, the related problem of the spreading of liquid drops on substrates is considered; the case in which the substrate is inclined to the horizontal and gravity drives a mean flow is also discussed. Finally, an overview is give - the results discussed in this paper give motivation for the development of careful experiments which can be used to test the theories and exhibit new phenomena.

Nonlinear dynamics and breakup of free-surface flows is rewieved in the recent paper by Eggers (1997). The thin film rupture is considered also by Ida and Miksis (1996) and these authors [Ida and Miksis (1995)] are considered also the dynamics of a lamella in a capillary tube. A wavy free-surface flow of a viscous film down a cylinder is considered by A.L. Frenkel (1993).
Concerning the flow and stability of thin films on a rotating disk we mention the papers by : Charwat, Kelly and Gazley (1972), and Higgins (1986). This last author derive an asymptotic solution that describes the thinning of a fluid layer on a rotating disk when the Reynolds number for the flow is small - the solution results from matching a long-time scale expansion that ignores the initial acceleration of the fluid layer with a short-time-scale expansion that accounts for fluid inertia. As more recent papers, concerning'the thin films on a rotating disk, we mention the papers by Needham and Merkin (1987), Sisoev and Shkadov (1987, 1988, 1990) and the Doctoral Thesis by Christel Bailly (1995).
Nonlinear evolution of waves on a vertically falling film (but unfortunately, whitout the Marangoni effect!) is considered by Chang, Demekhin and Kopelevich (1993). The flow of a liquid in a plane channel on the bottom of which a specified temperature distribution is maintained while the free-surface is thermally isolated is considered by Bobkov and Gupalo (1996).In the above cited paper by Thess and Orszag (1995), devoted to the limit of infinite Prandtl number, the case of high Marangoni number is also considered - these authors note that 'the kinematically possible velocity fields can have remarkable complexity when $Ma >> 1$', and it is a challenging problem for future studies to understand Bénard-Marangoni convection in the limit $Ma \to \infty$.
Viscous thermocapillary convection at high Marangoni number is also considered by Cowley and Davis (1983), where a boundary-layer analysis is performed that is valid for large Ma and Pr. In the paper by Nepomnyaschy and Velarde (1994) a dissipation-modified Boussinesq like system of equations governing 3D long wavelength Marangoni - Bénard oscillatory convection in a shallow layer heated from the air side is presented. In this paper, solitary waves and their oblique and head-on interaction are also considered. Marangoni covection and instabilities in liquid mixtures with Soret effects (the inclusion of the so-called 'Soret effect' means that the mass flux is the sum of temperature and concentration gradients - but, usually, the so-called Dufour effect, by which the concentration gradient would contribute to the heat flux is ignored) is considered by Joo (1995) and more recently by Bergeron et al. (1998). In Oron and Rosenau (1994), the authors show that the quadratic Marangoni instability enables an existence of new stable steady states which in variance with the conventional Marangoni induced patterns, are continuous and do not rupture. It is interesting to note that, the inherently unstable viscous liquid film flow down a vertical plate can be stabilized by oscillating the plate at appropriate amplitudes and frequencies although the stabilization is obtainable only over a relatively small range of Re - see, for instance the recent paper by Lin and Chen (1998). The onset of steady Marangoni convection, in a spherical shell of fluid with an outer free surface surrounding a rigid sphere, is analyzed by Wilson (1994) using a combination of analytical and numerical techniques.

In Doctoral thesis by Jean-Marc Vince (1994), the propagative waves in convection systems subject to the surface tension effects are studied very accuratly, a dynamical systems approach is also performed, in particularly via amplitudes equations 'à la Ginsburg-Landau' [see, equations (11.1)]. In Sisoev and Shkadov (1997a,b), recently, dominant waves in a viscous liquid flowing in a thin sheet are analyzed; the principle of selection of the periodic solutions realized experimentally as regular waves is justified and it is shown that, if several periodic solutions exist at a given wave number, a regime characterized by the maximum values of both amplitude and phase velocity is realized - variation of the wave number causes a jump-like transition of the attractor to another regime, which gives rise to two-periodic solution in the vicinity of the bifurcation points. The unsteady spreading of a thin layer of an incompressible viscous liquid over an impermeable curved surface, which occurs under the action of the gravity force, is considered by Grigorian and Khairetdinov (1998) - the solution of the boundary-value problem which arises reduces to solving a Cauchy problem for equations with a small number of independent variables [see also: Grigorian and Khairetdinov (1996)].
An interesting and well documented overview concerning : drops, liquid layers and the Marangoni effect, is the recent paper by Velarde (1998). Spatio-temporal instability in free ultra-thin films is considered, recently, by Shugai and Yakubenko (1998) - the analysis shows that even in the linear approximation, the long-range intermolecular force strongly affect the evolution of initially localized disturbances, but linear theory always overestimates the film life time, due to the explosive nonlinear growth of disturbances at later stages of evolution.
Concerning the more recent papers (published in 1999) we mention the following papers by : Betelù and Diez, Or, Kelly, Cortelezzi and Speyer, and Boos and Thess. An interesting IUTAM Symposium Proceedings (held in Haifa, Israel, 17-21 March 1997) and published in 1999 (edited by D. Durban and J.R.A. Pearson) concerns the 'Non-linear singularities in deformation and flow' - various papers are related with the interfacial effects in fluids and also with the capillary breakup and instabilities. Finally, we mention some books related with the thin films theory : V.Ya. Shkadov (1973), C. Isenberg (1978), T.S. Sorensen (1978, Ed), J. Zierep and H. Oertel (1982, Eds.), R.E. Meyer (1983, Ed.), J.K. Platten and J.C. Legros (1984), R. Finn (1986), M.G. Velarde (1987 and 1988, Ed.), I.B. Ivanov (1988, Ed.), S. Van Vaerenberg, P. Colinet, and J.C. Legros (1990), H.J. Rath (1992, Ed), R.P. Chhabra and D. de Kee (1992, Eds.), S.V. Alekseenko, V.E. Nakoryakov and B.T. Pokusaev (1992), B. Straughan (1992), D.D. Joseph and Yu.R. Renardy (1993, Part I), J.B. Simanovskii and A.A. Nepomnyaschy (1993), R.F. Probstein (1994).

12 References

Alekseenko S.V. (1985). *AIChE J.*, 31, 1446.

Alekseenko S.V. Nakoryakov V.E. and Pokusaev B.T. (1992). *Wave Flow of Liquid Films.* Nauka, Novosibirsk, original Russian ed.

Bénard H. (1900). *Rev. Gen. Sci. Pures Appl.*, 11, 1261.

Bailly Ch. (1995). *Modelisation asymptotique et mumérique de l'écoulement dû à des disques en rotation.* Thése presentée à l'Université des Sciences et Technologies de Lille, Villeneuve d'Ascq Cedex. Soutenue le 18 avril 1995, N° d'ordre 1512, 160 pages.

Benjamin T.B. (1957). Wave formation in laminar flow down an inclined plane. *J. Fluid Mech.* 2, 554-574 .

Benney D.J. (1966). *Journal. Math. and Phys.*, 45, 150.

Bergeron A. et al. (1998). Marangoni convection in binary mixtures with Soret effect. *J. Fluid Mech.*, 375, 143-177 .

Betelù S.I. and Diez J.A. (1999). A two-dimensional similarity solution for capillary driven flows. *Physica D*, 126, 136-140.

Binnie A.M. (1957). Experiments on the onset of wave formation on a film of water flowing down an inclined plane. *J. Fluid Mech.*, 2, 551-553.

Block M.J. (1956). Surface tension as the cause of Bénard cells and surface deformation in a liquid film. *Nature,* 178, 650-651.

Bobkov N.N. and Gupalo Yu.P. (1996). The flow pattern in a liquid layer and the spectrum of the boundary-value problem when the surface tension depends non-linearly of the temperature. *Prikl. Mat. Mekh.*, 60, 6, 1021-1028, Russian original ed.

Boos W. and Thess A. (1999). Cascade of structures in long-wavelengt Marangoni instability. *Phys. Fluids,* 11 (6) 1484-1494.

Bragard J. and Velarde M.G. (1997). Bénard Convection Flows. *J. Non-Equilib. Thermodyn.*, 22, 1-19.

Chang H.-Ch., Demekhin E.A. and Kopelevich D.I. (1993). Nonlinear evolution of waves on a vertically falling film. *J. Fluid Mech.*, 250, 433-480.

Chang H.-Ch. (1994). Wave evolution on a falling film. *Ann. Rev. Fluid Mech.*, 26, 103-136.

Chang H.-Ch. and Demekhin E.A. (1996). Solitary Wave Formation and Dynamics on Falling Films. *Advances in Appl. Mech.* 32, 1-58.

Charwat A.F., Kelly R.E. and Gazley C. (1972). The flow and stability of thin liquid films on a rotating disk. *J. Fluid Mech.*, 53, part2, 227-255.

Chhabra R.P. and de Kee D., Eds. (1992). *Transport processes in bubles, drops and particles.* New York, Hemisphere.

Cowley S.J. and Davis S.H. (1983). Viscous thermocapillary convection at high Marangoni number. *J. Fluid Mech.*, 135, 175-188.

Cross M.C.,(1980). *Phys. Fluids* 23, 1727.

Cross M.C. (1982). *Phys. Rev.* A 25, 1065.

Cross M.C. and Hohenberg P.C. (1993). Pattern formation outside of equilibrium. *Rev. Mod. Phys.*, 65, 851.

Dauby P.C. and Lebon G. (1996). Bénard-Marangoni instability in rigid rectangular containers. *Journal of Fluid Mechanics,* 329, 25-64.
Davis S.H. (1987). Thermocapillary instabilities. *Ann. Rev. Fluid Mech.*, 19, 403-435.
Durban D. and Pearson J.R.A., Eds. (1999). *Non-linear Singularities in deformation and Flow.* Kluwer Academic Publishers, Dordrecht- The Netherlands.
Drazin P.G. and Reid W.H. (1981). *Hydrodynamic Stability.* Cambridge University Press, Cambridge.
Eggers J. (1997). Nonlinear dynamics and breakup of fre-surface flows. *Reviews of Modern Physics,* 69 (3), 835-929 .
Finn R. (1986). *Equilibrium Capillary Surfaces*. Springer-Verlag, New York.
Frenkel A.L. (1993). On evolution equations for thin films flowing down solid surfaces. *Phys. Fluids*, A 5, 2342-2347 .
Garazo A.N. and Velarde M.G. (1991). Dissipative Korteweg- de Vries description of solitary waves and their interaction in Marangoni-Bénard layers. *Phys. Fluids*, A 3, 2295-2300.
Golovin A.A., Nepomnyaschy A.A. and Pismen L.M. (1994). Interaction between short-scale Marangoni convection and long-scale deformational instability. *Phys. Fluids*, 6 (1), 35 - 48.
Grigorian S.S. and Khairetdinov E.F. (1996). Solution of the equations of flow of a thin layer of heavy viscous liquid over a curvilinear surface. *Vestnik MGU, Mat. Mekh.*, 6, 32-36.
Grigorian S.S. and Khairetdinov E.F. (1998). Unsteady spreading of a thin layer of viscous liquid over a curved surface. *Prikl. Mat. Mekh.*, 62, 1, 151-161, Russian original ed.
Higgins B.G. (1986). Film flow on a rotating disk. *Phys. Fluids*, 29 (11), 3522-3529.
Ida M.P. and Miksis M.J. (1995). Dynamics of a lamella in a capillary tube. *SIAM J. Appl. Math.*, 55 (1), 23-57.
Ida M.P. and Miksis M.J. (1996). Thin film rupture. *Appl. Math. Lett.*, 9 (3), 35-40 .
Ida M.P. and Miksis M.J. (1998a). The Dynamics of thin films I : General theory. *SIAM J. Appl. Math.*, 58 (2), 456-473.
Ida M.P. and Miksis M.J. (1998b). The Dynamics of thin films II: Applications. *SIAM J. Appl. Math.*, 58 (2), 474-500 .
Isenberg C. (1992). *The Science of Soap Films and Soap Bubles.* Tieto, Clevendon, Avon, England (1978) - original edition; see also: Dower, New York (1992).
Ivanov I.B., Ed., (1988). *Thin Liquid Films*. Marcel Dekker, Inc.. New York.
Joo S.W. (1995)., Maraéngoni instabilities in liquid mixtures with Soret effects. *J. Fluid Mech.*, 293, 127-145 .
Joseph D.D. (1976). *Stability of Fluid Motions II*. Springer-Verlag, Berlin.
Joseph D.D. and Renardy Yu.R. (1993). *Fundamentals of Two-Fluid Dynamics. Part I:Mathematical Theory and Applications.* Springer-Verlag, New York, Inc.

Kapitza P.L. and Kapitza S.P. (1949). Wave flow of thin layers of a viscous fluid: II. Experimental study of undulatory flow conditions. *J. Exp. Theor. Phys.* 19, 105-120 (in Russian). See also :*Collected Papers of P.L. Kapitza*, ed. Ter Haar D. Pergamon, London, Vol. 2, 690-709, (1965).
Kashdan D. et al. (1995). Nonlinear waves and turbulence in Marangoni convection. *Phys. Fluids,* 7 (11),2679 - 2685.
Kopbosynov B.K. and Pukhnachev V.V. (1986). Thermocapillary flow in thin liquid films. *Fluid Mech. Sov. Res.* 15, 95.
Koschmieder E.L. (1993). *Bénard Cells and Taylor Vortices.* Cambridge Univ. Press, Cambridge, UK.
Landau L.D. (1944). *C.R. Akad. Sci. USSR*, 44, 311.
Landau L.D. (1965). *Collected Papers,* 387-391, Oxford.
Limat L. (1993). Instabilité d'un liquide suspendu sous un surplomb solide: influence de l' épaisseur de la couche. *C.R. Acad. Sci. Paris,* t. 217, Série II, 563-568.
Lin S.P. (1969). Finite amplitude stability of a parallel flow with a free surface. *J. Fluid Mech.*, 36, 113-126.
Lin, S.P. (1970). *J. Fluid Mech.* 40, 307 .
Lin S.P. (1974). Finite amplitude side-band stability of a viscous film. *J. Fluid Mech.*, 63(3), 417-429.
Lin S.P. and Chen J.N. (1998). The mechanism of surface wave suppression in film flow down a vertical plane. *Phys. Fluids,* 10 (8), 1787-1792.
Liu J., Schneider J.B. and Gollub J.P. (1995). Three-dimensional instabilities of film flows. *Phys. Fluids,* 7 (1), 55-67.
Meyer, R.E., Ed. (1983). *Waves on fluid interfaces.* Academic Press, New York.
MyersT.G. (1998). Thin films with high surface tension. *SIAM Rev.*, 40 (3), 441-462.
NeedhamD.J. and Merkin J.H. (1987). The development of nonlinear waves on the surface of a horizontal rotating thin liquid film. *J. Fluid Mech.*, 184, 357-379.
Nepomnyaschy A.A. and Velarde M.G. (1994). A Three-dimensional description of solitary waves and their interaction in Managoni-Bénard layers. *Phys. Fluids,* 6 (1), 187-198.
Newell A.C. (1974). *Envelope equations*, in Lectures in Applied Math., 15, ed. A.C. Newell, 157.
Normand Ch., Pomeau Y. and Velarde M.G. (1977). Convective instability: A Physicist's approach. *Rev. Mod. Phys.*, 49 (3), 581-624.
Or A.C., Kelly R.E., Cortelezzi L. and Speyer J.L. (1999). Control of long-wavelength Marangoni- Bénard convection. *J. Fluid Mech.*, 387, 321-341.
Oron A. and Rosenau Ph. (1992). *Journal Physique* II France, 2, 131.
Oron A. and Rosenau Ph. (1994). On a nonlinear thermocapillary effect in thin liquid layers. *J. Fluid Mech.*, 273, 361-374 .
Oron A., Davis S.H. and Bankoff, S.G. (1997). Long-scale evolution of thin liquid films. *Reviews of Modern Physics,* 69 (3), 931-960 .

Parmentier P.M., Regnier V.C. and Lebon G. (1996). Nonlinear analysis of coupled gravitational and capillary thermoconvection in thin fluid layers. *Physical Review E*, 54 (1), 411-423.
Pavithran S.and Redeekopp L.G. (1994). *Studies in Appl. Maths.*, 93, 209.
Pearson J.R.A. (1958). On convection cells induced by surface tension. *J. Fluid Mech.*, 4 , 489-500.
Pérez- Garcia C. and Carneiro G. (1991). Linear stability analysis of Bénard-Marangoni convection in fluids with a deformable free surface. *Phys. Fluids,* A3, 292 .
Platten J.K. and Legros J.C. (1984). *Convection in Liquids.* 1st ed., Springer-Verlag, Berlin.
Probstein R.F. (1994)., *Physicochemical Hydrodynamics: An Introduction.* Wiley - Interscience Publ., second ed., New York.
Rath H.J., Ed. (1992). *Microgravity Fluid Mechanics.* Springer -Verlag.
Regnier V.C. and Lebon G. (1995). Time-growth and correlation length of fluctuations in thermocapillary convection with surface deformation. *Q. Journal. Mech. Appl. Math.* 48, Pt.1, 57-75 .
Shkadov V.Ya. (1967). *Izvestiya Akad. SSSR, Mech. Zhidkosti i Gaza,* **1**, 43-50.
Shkadov V.Ya. (1973). *Some Methods and Problems of the Theory of Hydrodynamic Stability.* Moscow, Izd. MGU (in Russian).
Scriven L.E. and Sterling C.V. (1964). *J. Fluid Mech.*, 19, 321.
Shugai G.A. and Yakubenko P.A. (1998). Spatio-temporal instability in free ultra-thin films. *Eur. J. Mech. B/Fluids,* 17, n° 3, 371-384.
Simanovskii J.B. and Nepomnyaschy A.A. (1993). *Convective Instabilities in Systems with Interface.* Gordon and Breach.
Sisoev G.M. and Shkadov V.Ya. (1987). Flow stability of a film of viscous liquid on a rotating disk. *J. Eng. Physics* (English ed. , translated from Russian), 52, 671-674.
Sisoev G.M. and Shkadov V.Ya. (1988). *J. Eng. Physics*, 55, N°3, 419-423.
Sisoev G.M. and Shkadov V.Ya. (1990). *J. Eng. Physics*, 58, N°4, 573-577.
Sisoev G.M. and Shkadov V.Ya. (1997). Dominant waves in a viscous liquid flowing in a thin sheet. *Physics-Doklady,* 42 (12), 683-686.
Sisoev G.M. and Shkadov V.Ya. (1997). Development of dominating waves from small disturbances in falling viscous-liquid films. *Fluid Dynamics*, 32, (6), 784-792.
Sorensen T.S. (1978). *Dynamics and Instability of Fluid Interfaces.* Springer-Verlag, Berlin.
Stewartson K. and Stuart J.T. (1971). *J. Fluid Mech.*, 48, 529.
Straughan B. (1992). *The Energy Method, Stability, and Nonlinear Convection.* Springer - Verlag, New York, Inc.
Stuart J.T. (1960). *J. Fluid Mech.*, 9, 353.
Takashima M. (1981a). *J. of the Physical Soc.of Japan,* Vol. 50, n° 8, 2745-2750.
Takashima M. (1981b). *idem,* 2751-2756.

Thess A. and Orszag S.A. (1995). Surface-tension-driven Bénard concetion at infinite Prandtl number. *J. Fluid Mech.*, 283, 201-230.
Trifonov Yu.Ya. and Tsvelodub O.Yu. (1991). *J. Fluid Mech.*, **229**, 531.
VanHook S.J. et al. (1995). Long-wavelength instability in surface-tension -driven Bénard convection. *Phys. Rev. Lett.*, 75, 4397 .
Van Vaerenberg S., Colinet P. and Legros J.C. (1990). *The Role of the Soret Effect on Marangoni-Bénard Stability.* Springer.
Velarde M.G., Ed. (1987). *Physicochemical Hydrodynamics ; Interfacial Phenomena.* Plenum Press, N. Y.
Velarde M.G., Ed. (1988). *Physicochemical Hydrodynamics.* NATO ASI Series, Vol. 174, PLenum Press.
Velarde M.G. and Rednikov A. Ye. (1998). Time-dependent Bénard-Marangoni instability. In: *Time-Dependent Nonlinear Convection*, Chapter 6, 177-218, Tyvand P.A. Editor. Adv. in Fluid Mech., 19, Computational Mech. Publ., Southampton, UK.
Velarde M.G. (1998). Drops, liquid layers and the Marangoni effect. *Phil. Trans. R. Soc.,* London A 356, 829-844.
Vince Jean-Marc. (1994). *Ondes propagatives dans des systèmes convectifs soumis à des effets de tension superficielle.* Doctoral thesis, University Paris 7, N° 345, 185 pages.
Wilson S.K. (1994). The onset of steady Marangoni convection in a spherical geometry. J. of Engineering Math., 28, 427-445.
Wilson S.K. and Thess A. (1997). On the linear growth rates of the long-wave modes in Bénard-Marangoni convection. *Phys. Fluids* 9(8), 2455-2457.
Yih C.-S. (1963). Stability of liquid flow down an inclined plane. *Phys. Fluids* 6(3), 321-330.
Zeytounian R.Kh. (1989). *Intern. Journal Engng. Sciences*, 27 (11), 1361.
Zeytounian R.Kh. (1994). *Modelisation asymptotique en mécanique des fluides newtoniens.* SMAI - Mathématiques et Applications, Vol. 15. Springer-Verlag, Berlin.
Zeytounian R.Kh. (1995). Long-Waves on Thin Viscous Liquid Film : Derivation of Model Equations.In: *Asymptotic Modelling in Fluid Mechanics,* Lecture Notes in Physics (LNP 442),, Springer.
Zeytounian R.Kh. (1997). The Bénard - Marangoni thermocapillary instability problem : on the rôle of the buoyancy. *Int. J. of Enginering Sciences.* 35(5), 455-466.
Zeytounian R.Kh. (1998). The Bénard-Marangoni thermocapillary instability problem. *Uspekhi Fizicheskikh Nauk,* 168 (3), 259 - 286, Russian original ed.
Zierep J. and Oertel H., Eds. (1982). *Convective transport and instability phenomena.* Braun - Verlag, Karlsruhe.

HYDRODYNAMICS OF SLOPPED FALLING FILMS

V.Ya. Shkadov
Lomonosov Moscow State University, Moscow, Russia

Abstract. The stability analysis of the viscous liquid film flows leads to mathematical models which are derived from complete boundary problem for Navier–Stokes equations and include dispersion and dissipation. The principal property of similar flows is there are mechanisms of instability and itself development which form progressive or steady wave structures. The spectral methods have been developed which give possibility to construct the steady wave solutions, namely the nonlinear wave bifurcations, and nonstationary solutions developed from small initial disturbances that is attractors. Using two methods it is succeeded to define nonlinear dominating waves from the complicated picture of wave motions for films of finite thickness and to compare it to experiments. The instability of a falling liquid film along a vertical slop with surfactant adsorption–desorption originating surface stresses (Marangoni effect) is investigated.

1 Introduction

The investigations of waves in falling films have been stimulated by well–known experiments of P.Kapitsa and S.Kapitsa (1949). The capillary film flows demonstrate many types of instabilities and formed wave structures. The interest of investigators is attracted by possibilities of visualization of waves and measurements of their parameters that may be unrealizable in other flows.

There are observed the periodic waves of two types, namely, the slow waves with unique maximum of film thickness and the fast ones with steep oscillating forward front. The shapes of slow waves are similar to harmonic function and fast waves may be compared with solitary signals.

Theoretical investigations of nonlinear waves in downflowing films take their origin from the same time as experimental ones. The small parameter methods were exploited in the great number of papers up to now. But that approach does not lead to quantitative explanations of experimental observations. Another approach based on the Galerkin method was applied by Shkadov (1967). This paper will be refered to as paper Sh in what follows. The basic model system of that paper for local film thickness and flow rate is turned to be an effective theoretical tool to correlate experimental data on nonlinear waves in liquid films. As a first success in that direction an explanation of the nonlinear wave characteristics of Kapitsa's experiments is given. Then in a long series of publications the results of more systematic experiments of paper Alekseenko et al. (1979) and others are constructed theoretically.

We discuss here the space–periodic solutions of the basic model system only leaving aside the small parameter approaches discussed elsewhere (Chang, 1994). Many mathematical questions arise on the path to full explanation of experimental data. The main of these questions are considered below. Generalization of the method by inclusion the Marangoni effect is an essential part of consideration also.

2 Mathematical formulation

2.1 Basic equations

The full mathematical formulation of the problem in dimensionless form has been justified in paper Sh. Navier–Stokes equations in dimensionless form are written as

$$\begin{aligned} &\frac{\partial u}{\partial x}+\frac{\partial v}{\partial y}=0, \\ &\frac{\partial u}{\partial t}+u\frac{\partial u}{\partial x}+v\frac{\partial u}{\partial y}=-\frac{\partial p}{\partial x}+\frac{1}{\kappa \mathrm{Re}}\left(\frac{\partial^2 u}{\partial y^2}+\kappa^2\frac{\partial^2 u}{\partial x^2}\right)+\frac{1}{\kappa \mathrm{Fr}^2}, \\ &\kappa^2\left(\frac{\partial v}{\partial t}+u\frac{\partial v}{\partial x}+v\frac{\partial v}{\partial y}\right)=-\frac{\partial p}{\partial y}+\frac{\kappa^2}{\kappa \mathrm{Re}}\left(\frac{\partial^2 v}{\partial y^2}+\kappa^2\frac{\partial^2 v}{\partial x^2}\right). \end{aligned} \tag{1}$$

Relations of dimensional values denoted by upper wave and dimensionless ones are determined according to

$$\tilde{t}=\frac{H_*}{\kappa U_*}t, \quad \tilde{x}=\frac{H_*}{\kappa}x, \quad \tilde{h}=H_* h, \quad \tilde{u}=U_* u, \quad \tilde{v}=\kappa U_* v. \tag{2}$$

Boundary conditions at the hard wall and free surface are

$$\begin{aligned} y=0: &\quad u=0, v=0, \\ y=h(x,t): &\quad \frac{\partial h}{\partial t}+u\frac{\partial h}{\partial x}=v, \\ &\quad \frac{\partial u}{\partial y}+\kappa^2\left(\frac{\partial v}{\partial x}+\frac{4}{1-b^2}\frac{\partial h}{\partial x}\frac{\partial v}{\partial y}\right)=0, \ b=\kappa\frac{\partial h}{\partial x}, \\ &\quad p-\frac{2\kappa^2}{\kappa \mathrm{Re}}\frac{1+b^2}{1-b^2}\frac{\partial v}{\partial y}=-\frac{\kappa^2}{\mathrm{We}}\frac{1}{(1+b^2)^{\frac{3}{2}}}\frac{\partial^2 h}{\partial x^2} \end{aligned} \tag{3}$$

where $\mathrm{Re}=U_* H_*/\nu, \mathrm{We}=\rho U_*^2 H_*/\sigma, \mathrm{Fr}^2=U_*^2/gH_*$ are Reynolds, Weber and Froude numbers accordingly. Here ρ and ν are density and viscosity of liquid, σ is surface tension, g is the gravity and H_* is mean film thickness. To select two free parameters U_* and κ the equalities following from the fact that capillary, gravity and viscous forces are of the same order in thin film flows are used

$$\frac{\kappa^2}{\mathrm{We}}=\frac{3}{\kappa \mathrm{Re}}=\frac{1}{\kappa \mathrm{Fr}^2}=\frac{1}{5\delta}. \tag{4}$$

It leads to derivation of the formulas for scale U_* and the parameter κ

$$U_*=\frac{gH_*^2}{3\nu}, \quad \kappa=\left(\frac{2025\delta^2}{\gamma^3}\right)^{1/11}, \tag{5}$$

where

$$\gamma = \frac{\sigma}{\rho(\nu^4 g)^{1/3}}, \qquad \delta = \frac{1}{45\nu^2}\left(\frac{\rho H_*^{11} g^4}{\sigma}\right)^{1/3}. \tag{6}$$

Then our problem contains only two free parameters δ and γ.

Following to paper Sh the film flows under condition $\kappa^2 \ll 1$ are studied. In this case relations (1,3) lead to an estimate

$$p = -\frac{\kappa^2}{\mathrm{We}}\frac{\partial^2 h}{\partial x^2} + O\left(\kappa^2\right),$$

— in this approximation the variation of pressure is determined by action of the surface tension. After cutting off members of order $O\left(\kappa^2\right)$ in (1,3) one can derive the boundary layer equations taking into account the self–induced pressure

$$\begin{aligned}
&\frac{\partial u}{\partial x} + \frac{\partial v}{\partial y} = 0,\\
&\frac{\partial u}{\partial t} + u\frac{\partial u}{\partial x} + v\frac{\partial u}{\partial y} = \frac{1}{5\delta}\left(\frac{\partial^3 h}{\partial x^3} + \frac{1}{3}\frac{\partial^2 u}{\partial y^2} + 1\right),\\
y = 0:\ &u = 0, \quad v = 0,\\
y = h(x,t):\ &\frac{\partial h}{\partial t} + u\frac{\partial h}{\partial x} = v, \quad \frac{\partial u}{\partial y} = 0.
\end{aligned} \tag{7}$$

To solve (7) by Galerkin method the velocity u is presented as

$$u = \sum_{i=1}^{N} b_i(x,t)w_i(y)$$

where the basic functions w_i form the full set at $y \in (0, h)$. Further the velocity v is defined by the continuity equation. After substitution u, v in the momentum equation from (7) and its projection on w_i one can derive the system of equations on $b_i(x,t)$, $h(x,t)$. These equations and the kinematic condition at free surface form closed system.

In the case $N = 1$ the accuracy of model may be examined by comparison with experiments only. Presented model is based on the velocity approximation

$$u(x,y,t) = \frac{3q}{h}\left(\frac{y}{h} - \frac{y^2}{2h^2}\right) \tag{8}$$

where q is the flow rate

$$q(x,t) = \int_0^h u\,dy.$$

After substitution (8) in (7) and integration on the film thickness the evolutionary system may be derived (Shkadov, 1967, 1977).

$$\frac{\partial h}{\partial t}+\frac{\partial q}{\partial x}=0,$$
$$\frac{\partial q}{\partial t}+\frac{6}{5}\frac{\partial}{\partial x}\left(\frac{q^2}{h}\right)=\frac{1}{5\delta}\left(h\frac{\partial^3 h}{\partial x^3}+h-\frac{q}{h^2}\right). \tag{9}$$

To find the periodic solutions of (9) the condition to be used

$$\frac{1}{L}\int_{x_0}^{x_0+L} h dx = 1 \tag{10}$$

where L is the wavelength. It means that the scale H_* is the mean film thickness on the wavelength.

The basic model system (9) contains the only external parameter δ. Liquid properties and film thickness H_* effect on wave development through parameter δ simultaneously. Scales of time and space depend on $\kappa(\delta,\gamma)$ where γ is determined in (6) by liquid properties only. If spatial–periodic solutions are sought the system contains also inner parameter that is a wave number.

2.2 Linear stability

As follows from (9) there is solution $h=1, q=1$ corresponding to a waveless flow. To study linear stability the perturbed solution in form $h=1+\hat{h},\ q=1+\hat{q}$ is considered where

$$\left(\hat{h},\hat{q}\right)=\left(\check{h},\check{q}\right)\exp\left(i\alpha\left(x-\omega t\right)\right),$$

$\alpha=2\pi/L$ is given wavenumber, $\omega=\omega_r+i\omega_i$ is unknown complex eigennumber (here and below down indexes r and i denote real and imaginary components of value). After linearization (9) and substitution $\hat{h}$, $\hat{q}$ one can derive the dispersion relation

$$\omega^2+\left(\frac{i}{5\alpha\delta}-\frac{12}{5}\right)\omega+\frac{6}{5}-\frac{\alpha^2}{5\delta}-\frac{3i}{5\alpha\delta}=0. \tag{11}$$

For neutral perturbations (11) leads to

$$\omega_r=3,\quad \omega_i=0,\quad \alpha_n=\sqrt{15\delta},$$

where α_n is neutral wavenumber. The waveless flow is unstable at $\alpha\in(0,\alpha_n)$. It is convenient to use normalized wavenumber $s=\alpha/\alpha_n$ together with α.

Nonlinear solutions of (9) are presented as finite Fourier series

$$h=\sum_{k=-N}^{N} h_k(t)\exp i\alpha k x,\ q=\sum_{k=-N}^{N} q_k(t)\exp i\alpha k x, \tag{12}$$
$$h_k=h^*_{-k},\ q_k=q^*_{-k},\ h_0=1.$$

Equations for unknown coefficients $h_k(t)$, $q_k(t)$ with help of Galerkin procedure are formulated.

In the case of the nonstationary periodic waves computations the initial conditions $h_k(0)$, $q_k(0)$ have to be added to equations for Fourier coefficients (12).

2.3 Examination of the basic model system

As it was shown in many papers the system (9) is turned to be the main tool to obtain numerous solutions for real nonlinear waves observed in experiments. The reason for this is that all simplifications during system derivation are well grounded on *the principle of uniform approximations.* This system will be regular if coefficient $\delta \leq 1$.

The main question arises if there are experiments in which two proposed conditions $\kappa^2 \ll 1$ and $\delta \leq 1$ are realized simultaneously. In Figure 1 experimental conditions quoted

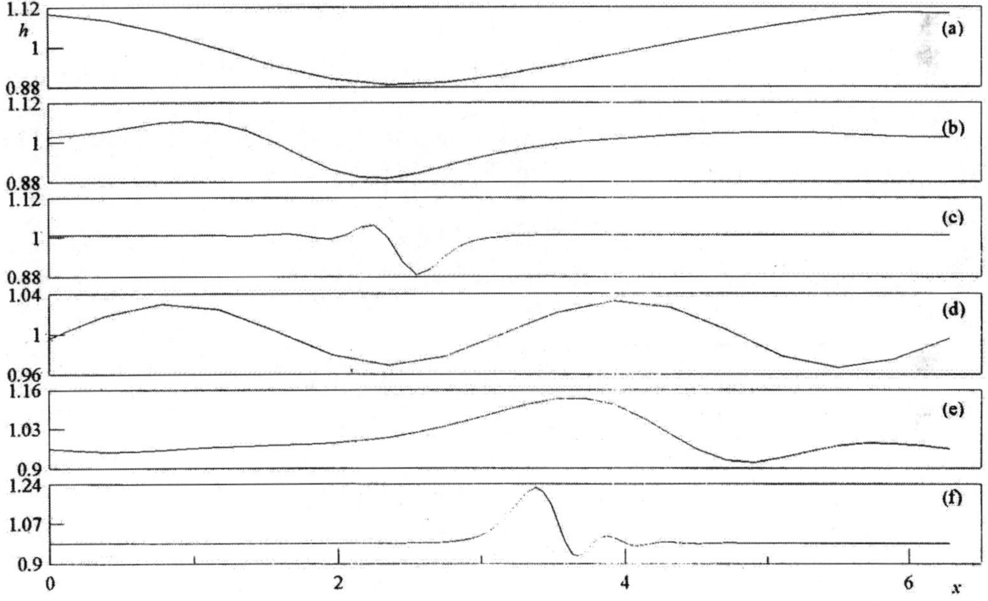

Figure 1. (a) Functions $\mathrm{Re}(\gamma)$ at some δ values and intervals of Re_A in experiments: $\gamma = 3906$ (1), $\gamma = 3298$ (2), $\gamma = 1129$ (3), $\gamma = 1098$ (4), $\gamma = 491$ (5), $\gamma = 465$ (6), $\gamma = 195$ (7), $\gamma = 105$ (8), $\gamma = 722$ (9), $\gamma = 1147$ (10), $\gamma = 1116$ (11), $\gamma = 2850$ (12). (b) Functions $\kappa^2(\gamma)$.

on the plane (γ, δ) are represented (Alekseenko et al., 1979). Two Reynolds numbers are used in Figure 1a , namely, $\mathrm{Re} = \left(5^9 3^7 \gamma^3 \delta^9\right)^{1/11}$ together with Re_A from experiments and these values are related by $\mathrm{Re}_A = \mathrm{Re} q_0$. Since for nonlinear waves $q_0 \in (1, 1.4)$ it is seen from Figure 1 that two main conditions for the most experiments with regular waves are satisfied except region of small γ. Therefore our system is justified and the problem is to investigate its manifold of solutions. Except Kapitsa's experiments (curve 12 in Figure 1a) in what $\delta \in (0.04, 0.25)$ most experimental conditions are exposed inside the domain $\delta \in (0.1, 0.5)$.

It is relevant to mention about theoretical approaches to justification of the basic model system (9) the full bounary layer equations (7) are solved for the first family waves, so as the solutions of that equations are obtained for various wave regimes by Shkadov et al. (1970, 1973a, 1973b), Demekhin et al. (1983, 1987a, 1989). The correlations of results with according solutions of the basic model system (9) are very good in all the cases investigated, for the space–periodic and solitary waves. The solutions of linearized system (9) were compared successfully with that ones of the Orr–Sommerfeld equation, following from the full Navier–Stokes formulation for small disturbances.

2.4 Weakly nonlinear asymptotics

In connection with the fact that $\delta > 0.04$ for experimentally observed regular waves (P.Kapitsa and S.Kapitsa, 1949; Alekseenko et al., 1979), some remarks on the case of small δ must be done. Let us introduce new variables $t_1 = \alpha_n t, x_1 = \alpha_n x, \alpha_n \equiv \sqrt{15\delta}$. After substitution in the second equation (9) we derive for $\delta \ll 1$

$$q = h^3 + o\left(\alpha_n^3\right).$$

Then the equations (9) lead to

$$\frac{\partial q}{\partial t_1} = 3h^2 \frac{\partial h}{\partial t_1} + o\left(\alpha_n^3\right) = -3h^2 \frac{\partial q}{\partial x_1} + o\left(\alpha_n^3\right) = -9h^4 \frac{\partial h}{\partial x_1} + o\left(\alpha_n^3\right).$$

After substitution last relation in the second equation (9) one can receive

$$q = h^3 + \alpha_n^3 \left(h^3 \frac{\partial^3 h}{\partial x_1^3} + h^6 \frac{\partial h}{\partial x_1} \right) + o\left(\alpha_n^6\right).$$

and the first equation (9) leads to within $o\left(\alpha_n^6\right)$ to Benney equation

$$\frac{\partial h}{\partial t_1} + 3h^2 \frac{\partial h}{\partial x_1} + \alpha_n^3 \frac{\partial}{\partial x_1} \left(h^3 \frac{\partial^3 h}{\partial x_1^3} + h^6 \frac{\partial h}{\partial x_1} \right) = 0. \tag{13}$$

With the help of variables replacement

$$x_2 = x_1 - \left(3 + \alpha_n^3 C\right) t_1, \; t_2 = \alpha_n^3 t_1, \; h = 1 + \alpha_n^3 H,$$

we obtain from (13) to within $o\left(\alpha_n^3\right)$ the equation

$$\frac{\partial H}{\partial t_2} + \frac{\partial}{\partial x_2} \left(3H^2 - CH + \frac{\partial^3 H}{\partial x_2^3} + \frac{\partial H}{\partial x_2} \right) = 0. \tag{14}$$

Nonlinear wave solutions of (14) were constructed by Shkadov (1973a) in the first time; full theory by Demekhin et al. (1991) was developed. Popular weakly nonlinear equations (13) and (14) which follow from basic system (9) as $\delta \to 0$ don't contain physical parameters and their solutions describe mathematical waves only.

It is necessary to note the essential physical difference between the system (9) and its asymptotic approximation at $\delta \to 0$ that is the equation (14). The system (9) at finite values δ describes the physical waves and is suitable to comparison with experiments. The asymptotic equation (14) simulates the mathematical waves of unbounded length and infinitesimal amplitude and is not included parameters connecting with experimental conditions. This circumstance defines the preference of (9) before (14). It is important to note that mathematical model for nonlinear waves is reduced to single equation in the limiting case $\delta \to 0$ only. That model includes a system of two equations for finite δ values.

2.5 Some generalizations

The basic model system (9) for vertically falling film are applied. It is easy to generalize this equations for slopped falling films with tangential force τ acting along the free surface. For the velocity profile $u(x,y,t)$ and self–induced pressure $p(x,y,t)$ we have

$$u(x,y,t) = \frac{3q}{h}\left(\eta - \frac{\eta^2}{2}\right) - \frac{\tau h}{2}\left(\eta - \frac{3\eta^2}{2}\right), \quad \eta = \frac{y}{h},$$

$$p = -\frac{\kappa^2}{\mathrm{We}}\frac{\partial^2 h}{\partial x^2} + \frac{\cos\theta}{\kappa^2 \mathrm{Fr}^2}(h - y) \tag{15}$$

where θ is the inclination angle of a rigid wall. By integration the equations on the interval (o,h) we obtain the modified basic model system

$$\frac{\partial h}{\partial t} + \frac{\partial q}{\partial x} = 0, \tag{16}$$

$$\frac{\partial q}{\partial t} + \frac{\partial}{\partial x}\left(\frac{6q^2}{5h} + \frac{h\tau q}{20} + \frac{h^2\tau^2}{120}\right) = \frac{1}{5\delta}\left(h\frac{\partial^3 h}{\partial x^3} + h - \frac{q}{h^2} + \frac{\tau}{2}\right) - \chi h\frac{\partial h}{\partial x}.$$

One need to put $g \to g\sin\theta$ in the relations (5,6) and $\chi = 3\cot\theta/\mathrm{Re}$.

The system (16) was derived by Esmail and Shkadov (1971) in the first time. Numerical solutions of (16) for waves in films on incline and in tube with counter flowing gas were obtained (Demekhin et al., 1989).

The extention of the basic model system (9) for three-dimensional flow along incline was derived. Let us introduce parabolic flow along z direction and according flow rate r additionally; then by the method of paper Sh described above we obtain

$$\frac{\partial h}{\partial t} + \nabla\cdot\omega = 0, \tag{17}$$

$$\frac{\partial \omega}{\partial t} + \frac{6}{5}\left[\omega\cdot\left(\nabla\frac{\omega}{h}\right) + \left(\frac{\omega}{h}\nabla\right)\cdot\omega\right] = \frac{1}{5\delta}\left[\mathbf{f}h + h\nabla(\Delta h) - \frac{\omega}{h^2}\right] - \chi h\nabla h.$$

Here $\boldsymbol{\omega} = (q, r)$ is a flow rate vector, $\mathbf{f} = \mathbf{g}/|\mathbf{g}|$ is a unit gravity vector. Investigations of generelized basic system properties as well as of the structure of 3D waves were carried out (Demekhin and Shkadov, 1984, 1988).

Generalization of the basic model system (9) by taking into account electrical forces is given (Shkadov and Sisoev, 2000a). Film flows with Marangoni effect are considered in section 6.

3 Regular periodic and solitary waves

Experimental observations of regular waves and instability of the main flow lead to search nonlinear steady regimes. The problem of bifurcations of nonlinear steady periodic solutions of (9) was formulated in paper Sh, where the first family of waves was found. This family softly bifurcates from the waveless flow on neutral curve and exists at $s \in (0, 1)$. Because nonlinear waves of first family move with phase velocity $c < 3$ they were named as *slow waves.* The second family of *fast waves* together with some other bifurcating solutions are obtained by Shkadov et al. (1981), see also Bunov et al. (1984). The full study of intermediate bifurcations is fulfilled as well as full two–parametric manifold of bifurcations is constructed up to now (Sisoev and Shkadov, 1997a, 1999) .

3.1 Families of periodic solutions

The system (9) possesses a sequence of steady–state periodic solutions. To found ones the substitution $h(\xi)$, $q(\xi)$, $\xi = \alpha(x - ct)$ is used and nonlinear eigenvalue problem is formulated from (9)

$$\begin{gathered} 3s^3h^3\frac{d^3h}{d\xi^3} + \frac{s}{5}\left[6\left(q_0 - c\right)^2 - c^2h^2\right]\frac{dh}{d\xi} + \frac{1}{5\delta\alpha_n}\left[h^3 - q_0 - c\left(h - 1\right)\right] = 0, \\ h(\xi) = h(\xi + 2\pi), \quad h'(\xi) = h'(\xi + 2\pi), \quad h''(\xi) = h''(\xi + 2\pi), \\ \frac{1}{2\pi}\int\limits_{\xi}^{\xi+2\pi} h d\eta = 1. \end{gathered} \tag{18}$$

where c and q_0 are unknown values of velocity and dimensionless mean flow rate respectively and the normalized wavenumber $s \equiv \alpha/\alpha_n$ is used instead of α. The last condition in (18) expresses the constancy of mean film thickness. Because relations (18) are invariant to transformation $\xi \to \xi + \triangle\xi$ the problem is stated correctly.

The search of solutions of (18) includes two stages, namely the determination of solution at some values δ, s and continuation of that according to varying parameters. To realize this procedure it is supposed to introduce $s(\lambda)$, $\delta(\lambda)$, $h(\xi, \lambda)$, $c(\lambda)$, $q_0(\lambda)$ where λ is a variable characterizing the curve in the plane s, δ. By differentiating the equation (18) one can derive

$$b_1\frac{\partial}{\partial\lambda}\left(\frac{\partial^3 h}{\partial\xi^3}\right) + b_2\frac{\partial}{\partial\lambda}\left(\frac{\partial h}{\partial\xi}\right) + b_3\frac{\partial h}{\partial\lambda} + b_4\frac{dc}{d\lambda} + b_5\frac{dq_0}{d\lambda} + b_6\frac{ds}{d\lambda} + b_7\frac{d\delta}{d\lambda} = 0. \tag{19}$$

where functions $b_j,\ j = 1, \ldots, 7$ depending on

$$h, \quad \frac{\partial h}{\partial \xi}, \quad \frac{\partial^2 h}{\partial \xi^2}, \quad \frac{\partial^3 h}{\partial \xi^3}, \quad c, \quad q_0, \quad s, \quad \delta$$

are known. The solution of (19) is presented as

$$h(\xi, \lambda) = \sum_{k=-M}^{M} h_k(\lambda) \exp ik\xi, \quad h_k = h^*_{-k}, \quad h_0 = 1. \tag{20}$$

After substitution (20) in (19) it is possible to derive the system of ordinary differential equations for Fourier coefficients $h_k(\lambda)$ and functions $c(\lambda)$, $q_0(\lambda)$, $s(\lambda)$, $\delta(\lambda)$. At any value of λ the solution can be corrected by Newton's method applied to nonlinear algebraic system following from (19).

The boundary eigenvalue problem (18) includes two governing parameters, namely, $\delta \in (0, \infty)$ and the wavenumber $s \in [0, 1]$. Described above numerical method allows to continue solution along any curve $\delta(\lambda)$, $s(\lambda)$ if it has been known at some λ value. There are infinite set of solutions $\{h(\xi), c, q_0\}$ corresponding to different pairs of parameters (s, δ) and a few solutions may be exist at given pair.

The set of solutions (18) which may be parametrized by scalar variable λ for $\delta = const$ is named *family of solutions*. If some property of solutions described by scalar parameter is considered then any family may be identified by function $f = f(\lambda)$. It is accepted to use wavenumber s as the bifurcation parameter of family which corresponds to curve $s(\lambda)$. By eliminating λ it is possible to find the trace curve of family in plane (s, f) where this curve is originated at some bifurcation point s_b. Phase velocity c, wave amplitude a or maximal wave height h_{max} may be applied as function f and λ is natural parameter of curve $s(\lambda)$ calculated from bifurcation point s_b. Relation $f = c$ is applied below to presentation the solutions in the plane (s, c) as it done in Fihure 2 for small δ values.

The most difficult problem is connected with seeking of any initial solution of family. This problem has been solved by three methods, namely: (a) analysis of attractors of (9) at numerical integration from small initial perturbations of trivial solution, (b) using the solitary waves as trace of periodic solutions at $s = 0$, (c) investigation of branching points. Numerous systematic computations have led to very complicated structure of periodic solutions presented below.

Any family of solutions originates at bifurcation point $s = s_b$ and continues to $s \to 0$ in plane (s, c). The bifurcation points are in vicinities of $s_n \approx 1/n$, $n = 1, 2, \ldots$. Any family arises at bifurcation point as $2\pi/n$-harmonic wave with its amplitude are finite at $n \geq 2$ (hard bifurcation). At $s \to 0$ every family transforms to solitary wave. Examples of the wave shapes for the first and second families are presented in Figure 3 The limiting periodic wave profiles as $s \to 0$ are the fast or solutary waves with one or two main ma--s.

3.2 Solitary waves

The equations (9) possesses the solutions in the form of steady solitary waves too. Existence of solitary waves as the solutions of (9) by Shkadov (1977) is discovered. It is shown

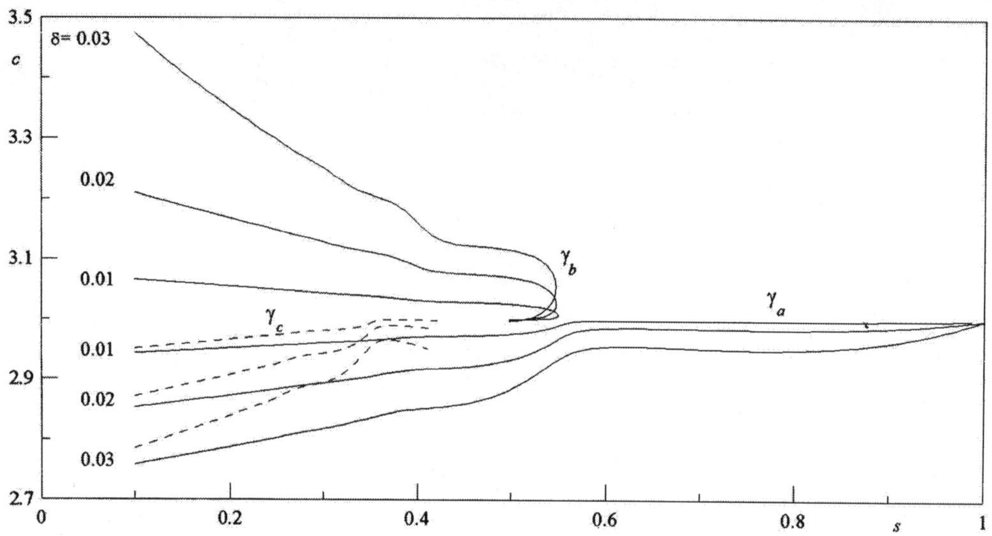

Figure 2. Phase velocity of steady waves: first family γ_a, second family γ_b, additional slow family γ_c.

infinite set of solitary waves to exist on a countable set of the phase velocity intervals. Detailed investigation of solitary waves are conducted up to now. The correspondence between thc periodic and solitary waves as wavenumber is diminishing is established, while the full spectrum of solitary waves to the two–parametric manifold of wave solutions is included (Shkadov et al., 1981; Bunov et al., 1984, 1986; Demekhin and Shkadov, 1983, 1986, 1991; Sisoev and Shkadov, 1999).

To calculate solitary waves the transformation of (9) to the coordinate system $\eta = x - ct$ moving with velocity c is done and steady solutions $h(\eta)$, $q(\eta)$ are sought. After some transformations the equation similar to (18) may be derived

$$h^3 \frac{d^3 h}{d\eta^3} + \delta \left[6\,(c-1)^2 - c^2 h^2 \right] \frac{dh}{d\eta} + h^3 - 1 - c\,(h-1) = 0. \tag{21}$$

This equation must be solved at boundary conditions

$$\eta \to \pm\infty : \quad h \to 1. \tag{22}$$

To solve (21,22) the numerical shooting Shkadov's (1977) method is applied. To realize this method it is necessary to consider the asymptotic form $h(\eta)$ at $\eta \to \pm\infty$. Because at this regions $h = 1 + \hat{h}$, $|\hat{h}| \ll 1$ it is possible to linearize (21)

$$\frac{d^3 \hat{h}}{d\eta^3} + \delta \left(6 - 12c + 5c^2\right) \frac{dh}{d\hat{\eta}} + (3 - c)\hat{h} = 0. \tag{23}$$

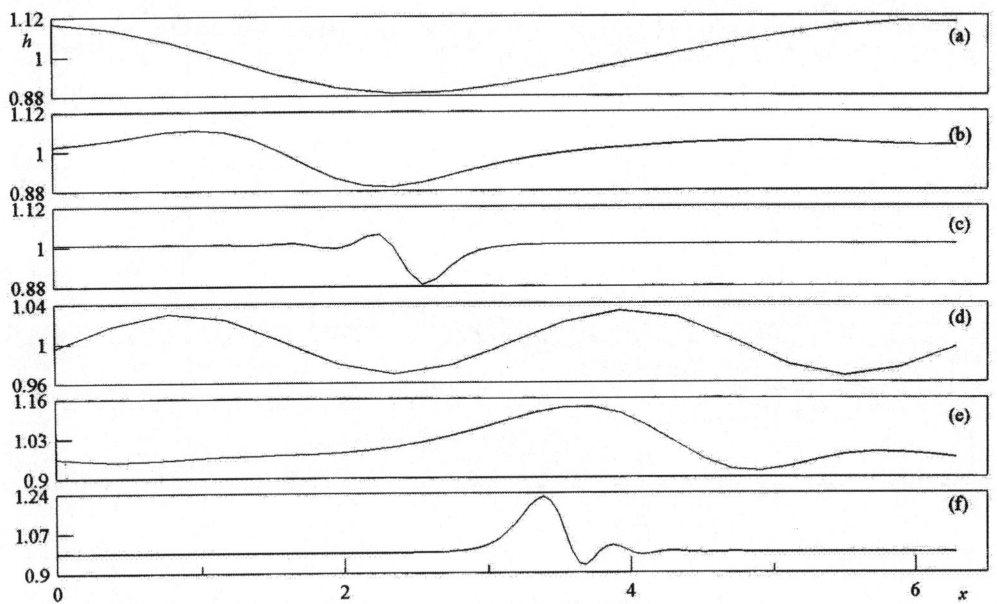

Figure 3. $\delta = 0.03$. The first family waves: (a) — $s = 0.9$, (b) — $s = 0.4$, (c) — $s = 0.1$. The second family waves: (d) — $s = 0.497$, (e) — $s = 0.4$, (f) — $s = 0.1$

Solutions of (23) define the asymptotic behavior $h(\eta)$

$$\begin{aligned} c < 3:\ & h \approx 1 + A_- \cos(\varsigma\eta - \phi) \quad \text{at} \quad \eta \to -\infty, \\ & h \approx 1 + A_+ \exp(-2\sigma) \quad \text{at} \quad \eta \to +\infty, \\ c > 3:\ & h \approx 1 + A_- \exp(-2\sigma) \quad \text{at} \quad \eta \to -\infty, \\ & h \approx 1 + A_+ \cos(\varsigma\eta - \phi) \quad \text{at} \quad \eta \to +\infty. \end{aligned} \tag{24}$$

Here σ and $\varsigma > 0$ form the complex root $\chi = \sigma + i\varsigma$ of characteristic equation and ϕ is the phase shift.

To find solitary wave in the case $c < 3$ the small value A_+ is taken and the equation (21) is integrated to negative direction of η until some value at which h, $dh/d\eta$, $d^2h/d\eta^2$ have to coincide with asymptotic values. These conditions are determined the values c, ϕ, A_-. In the case $c > 3$ the procedure is similar but the value A_- is taken and values c, ϕ, A_+ are found after integration to the positive direction of η.

In Figure 4 the eigenvalues c for homogeneous problem (21,22) describing solitary waves are shown as functions of parameter δ. For every δ fixed there are several eigen solutions which could be arranged in two groups of the fast ($c > 3$) and slow ($c < 3$) waves. The tedious numeric analysis proved that for every δ solitary waves exist in a countable set of segments $[c_1, c_1']$, $[c_2, c_2']$ where $c_n, c_n' \to 3$ as $n \to \infty$. The examples of wave profiles are shown in Figure 5. are that ones at the segments $n = 1, 2$. Each segment

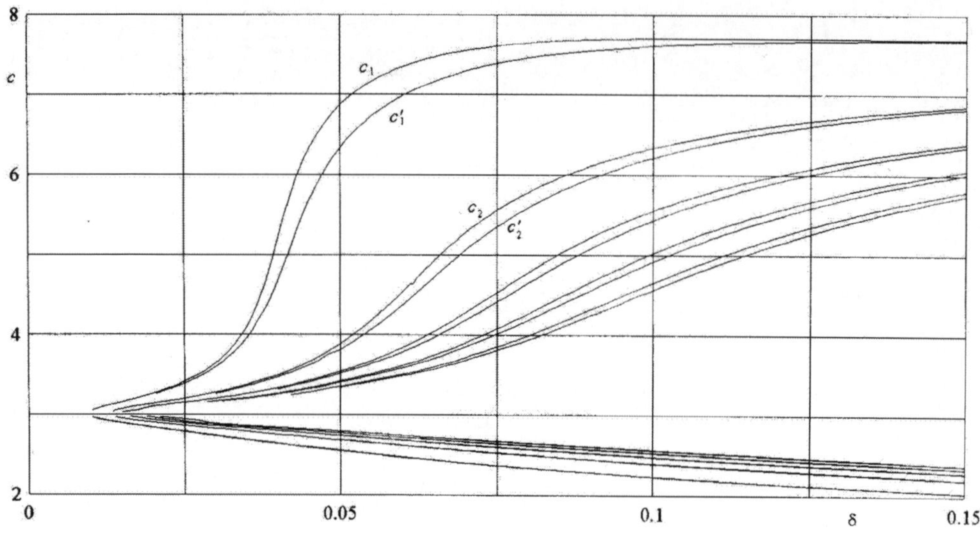

Figure 4. Phase velocity of steady solitary waves

has a complicated structure, there are multihumped solutions within n–segment. These solutions are combined waves of complicated structure consisting of solitary waves and periodic ones. The limiting periodic wave profiles as $s \to 0$ are the fast or slow solitary waves with one or two main humps (see Bunov et al., 1986).

The basic model system (9) has also solutions of compression level jump type. In this case solitary solution for (23) follows to boundary conditions

$$\eta \to \mp\infty : \quad h \to 1 \pm A, \quad , A > 0. \tag{25}$$

It is very convenient to represent the eigen solutions of (23) as trajectories in the phase space. Then solitary and level jump waves are represented by homoclinic and heteroclinic trajectories accordingly. Periodic and biperiodic solutions in phase space by circles and invariant torii are represented.

3.3 Full picture of bifurcations

The first family of 2π–periodic solutions of (18) bifurcates softly from trivial solution $h = 1$, $q = 1$ at neutral value $s = 1$ and continues into a domain of linear instability $s \in [0, 1]$. Let us consider the periodic soluiton including the first two harmonics only

$$h = 1 + \rho \sin \xi + \rho^2 \left(\phi_{20} \sin 2\xi + \phi_{21} \cos 2\xi\right). \tag{26}$$

Then nonlinear algebraic system for Fourier coefficients in paper Sh had been derived. That nonlinear system in the followin form could be represented

$$(3 - c)\,\phi_{21} + 2\alpha \left(k - 4\alpha^2\right) \phi_{20} = \frac{3}{2},$$

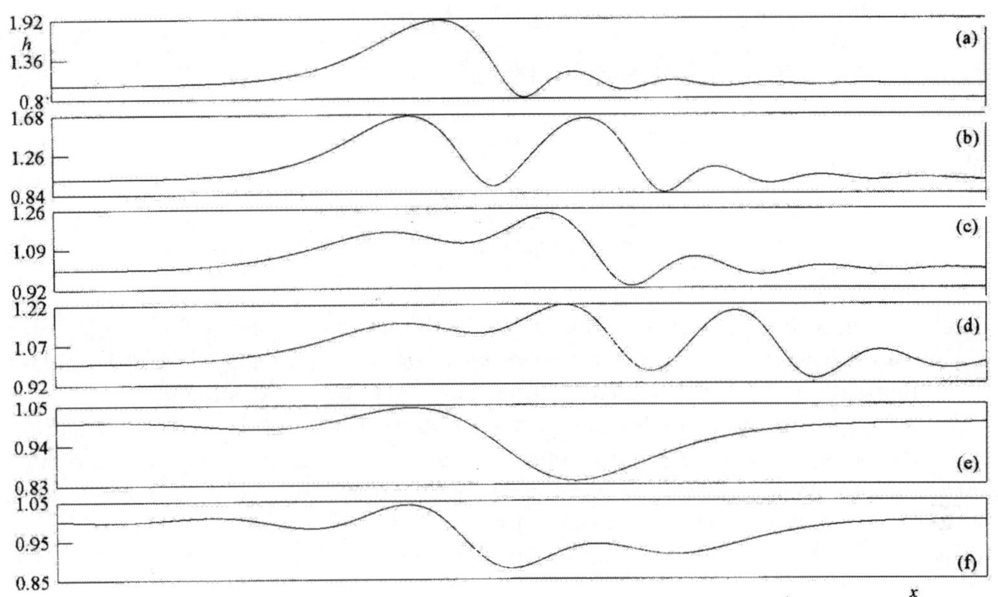

Figure 5. $\delta = 0.04$. Fast solitary waves: (a) — $c_1 = 5.194$, (b) — $c_1' = 4.667$, (c) — $c_2 = 3.496$, (d) — $c_2 = 3.469$. Slow solitary waves: (e) — $c_{-1} = 2.633$, (f) — $c_{-2} = 2.749$.

$$
\begin{aligned}
&-2\alpha\left(k-4\alpha^2\right)\phi_{21}+(3-c)\,\phi_{20}=\frac{3}{2}\alpha^3+\delta c^2\alpha,\\
&c=3+\rho^2\left[\frac{3}{4}-3\phi_{21}+\left(\frac{21}{2}\alpha^3+\delta c^2\alpha\right)\phi_{20}\right],\\
&\rho^2=\alpha\left(k-\alpha^2\right)\left[\left(\frac{3}{4}-\frac{21}{2}\phi_{21}\right)\alpha^3+\delta c^2\alpha\left(\frac{1}{4}-\phi_{21}\right)-3\phi_{20}\right]^{-1},\\
&k=\delta\left[6\left(q_0-c\right)^2-c^2\right],\quad q_0=1+\frac{3}{2}\rho^2.
\end{aligned}
\tag{27}
$$

The full parametric investigation (27) by numerical method was conducted. Good results were obtained for $s \in (0.5, 1)$, while for $s < 0.5$ we need to construct solution by direct numerical investigation of (18). This numerical solution exists at any value δ and coincide with one of solutions of solitary wave problem (one–humped slow solitary wave) at $s \to 0$. For points (α, δ) disposed on the neutral curve $s = 1$ $(\alpha = \alpha_n)$ we obtain from (27)

$$
\alpha_n = \sqrt{15\delta}, \quad k_n = \alpha_n^2, \quad c_n = 3, \quad q_n = 1, \quad , \rho_n = 0.
$$

In the vicinity of $s = 1$ wave solution of the first family (27) may be written analytically. For $s = 1 - \varepsilon^2$ the approximated representation of solution are

$$
h = 1 + 2\varepsilon\beta \sin\xi + (\varepsilon\beta)^2\left(-\frac{1}{\alpha_n^3}\sin 2\xi + \frac{7}{5}\cos 2\xi\right),
$$

$$c = 3 - 12.3(\varepsilon\beta)^2, \quad q_0 = 1 + 6(\varepsilon\beta)^2, \qquad (28)$$
$$\beta^2 = \frac{2\alpha_n^6}{3}\left(1 + 4.14\alpha_n^6\right)^{-1}, \rho^2 = 4\left(\varepsilon\beta\right)^2.$$

To find the solutions of (21) in vicinity of $s \approx 1/n$ it is conveniently to rewrite the solution (28) in form

$$h(\xi) = \sum_{k=-\infty}^{\infty} h_k(\lambda) \exp iks\xi, \quad h_k = h^*_{-k}, \quad h_0 = 1 \qquad (29)$$

where $s = 1/n$. The computation solution (29) with additional nonzero members means determination new $2\pi n$–periodic solution bifurcated off 2π–periodic one. After replace $\xi \rightarrow n\xi$ bifurcation of 2π–periodic solution off $2\pi/n$–periodic wave is found.

Accurate systematic computations confirm these concepts about the new solution bifurcations at $s_n \approx 1/n$. The analysis of bifurcation conditions and construction of new solution by analytical method at $n = 2$ is given. Very tedious computations of bifurcating solutions may be carried out numerically only (Shkadov et al., 1981).

The traces of the first fifteen bifurcated families in plane (s, c) at $\delta = 0.15$, are presented in Figure 6.

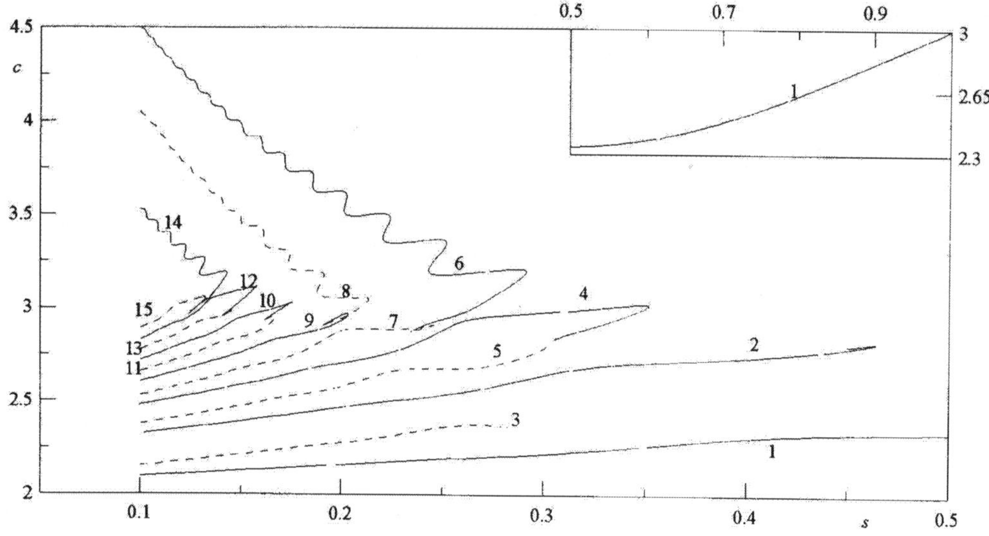

Figure 6. Phase velocity of steady periodic waves at $\delta = 0.15$: the first family γ_a (curve 1), the second family γ_b (6), the intermediate families $\gamma_{c,1}$ (2) and $\gamma_{c,2}$ (4). Solid and dashed curves denote families of one– and two–humped limiting waves correspondingly.

The birth of new families takes place by couples in vicinity s_n where phase velocity of waves of one family increases with receding from bifurcation point and it of other family decreases. The families of first and second types may be named *fast* amd *slow families* and denoted γ_+^n and γ_-^n correspondingly. These bifurcations belong to two–sided type. The functions $s(\lambda)$ are not monotone in small vicinity $\lambda = 0$, i.e. there are two solutions of one family, as a rule, and maximal value of wavenumber s_{max} that solution exist at $s \le s_{max}$ only.

It is necessary to note that values s_b for couple of fast and slow families of one order n are disposed very close. The unique exception is the waves of second bifurcation $n = 2$. Points of bifurcation of two families are closed at small values δ, but as δ growth the bifurcation point of slow family moves to small wavenumbers. It may be noted here that fast wave of the second bifurcation γ_b plays important role in classification of solutions and their comparison with experiments. It is shown below that waves of this family have maximal phase velocity and maximal amplitude among all solutions at given δ, s. This family has been firstly and named *the second family* as additionally to the first family γ_a (see Bunov et al., 1984).

3.4 Dominating waves

The first family γ_a go on from $s = 1$ and there are not solutions with smaller velocity c at given δ. The second family γ_b which go on from $s = s_b$ consists of waves with maximal velocities which belong to the small wavenumber domain $s_b < 0.5$. New additional families γ_c appear at intermediate bifurcations between $s = 1$ and $s = s_b$. The number of these families depends on δ value and grows together with δ. There are two intermediate slow families $\gamma_{c,1}, \gamma_{c,2}$ which waves are fastest in two intervals of s in Figure 6.

Now from all solutions in plane (δ, s) we'll consider only those ones which have extremely properties, namely, their phase velocity c, amplitude a and mean flow rate q_0 are maximal at every fixed wave number s. We refer to this solutions as dominating waves. ¿From Figure 6 it is clear that at $\delta = 0.15$ the set of dominating waves includes 4 pieces belonging to described families $\gamma_a, \gamma_b, \gamma_{c,1}, \gamma_{c,2}$. There are jumps of the dominating waves parameters at values of s dividing one piece from the other.

As δ grows the point of bifurcation of the second base family γ_b moves to $s \to 0$ and the quantity of intermediate families increases According to this movement a number of the pieces constituting set of dominating waves is grown. For example, this number is equal to 2,3,4,5,6,7 at $\delta = 0.04, 0.1, 0.15, 0.2, 0.225, 0.247$ and so on (Sisoev and Shkadov 1997a, 1997b).

4 Unsteady waves and global attractors

Nonuniqueness of solutions of (18) at given governing parameters δ and s leads to necessity to select such ones that are realized in experiments. Initial study has been carried out by Shkadov (1968) where linear stability analysis of the first family waves has been done and evolution of unsteady solutions from arbitrary small initial conditions into regular steady waves is discovered. With discovery of new families a selection problem was complicated.

Now we proceed to prove that from all regular wave solutions existing at given governing parameters (δ, s) the dominating waves are suitable only to compare with experimental observations. Validity of this statement is based on fact that the dominating waves are stable and have capacity to attract nonstationary solutions of (9) from small vicinity of the main flow and another regular solution. Systematic computations of unsteady solutions from space–periodic and localized initial disturbances are realized (Demekhin et al., 1981, 1987b,c, 1991). The global attractor for (9) was investigated as well the method of nonstationary developing solutions was applied to investigate instability of regular waves. To clear it numerous integrations of Coshy problem for equations (9) have been carried out with two types of initial data.

Firstly, in case of initial data as small harmonic or stochastic perturbations of the main flow $h = 1, q = 1$ it has been shown that limiting in time solutions belong to set of dominating waves and also this result doesn't depend on a form of initial data. In addition the oscillating regimes have been discovered in vicinity of values s and δ closed to jump points between different pieces of dominating waves set. Some examples of evolution in time are shown in Figure 7a where h_m is maximal wave height.

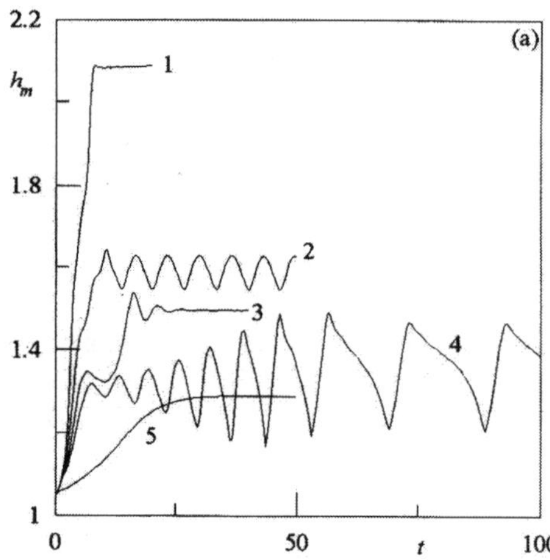

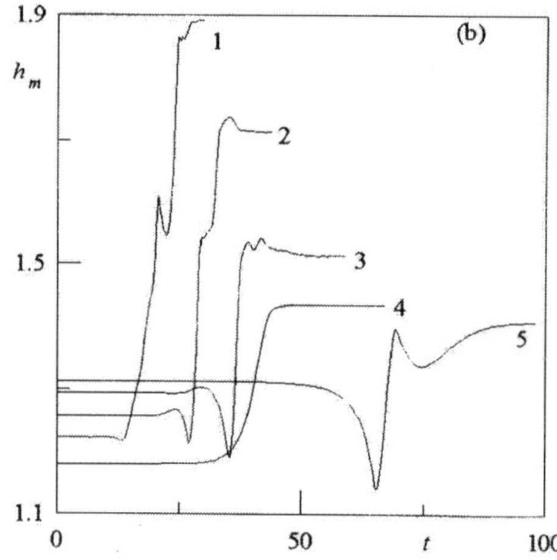

Figure 7. Formation of limiting regimes from small disturbances: (a) main flow at $\delta = 0.15$: 1 — $s = 0.15$, 2 — $s = 0.3$, 3 — $s = 0.4$, 4 — $s = 0.466$, 5 — $s = 0.8$; (b) steady nonlinear waves of family $\gamma_{c,1}$ at $\delta = 0.1$: 1 — $s = 0.2$, 2 — $s = 0.25$, 3 — $s = 0.318$, 4 — $s = 0.477$, 5 — $s = 0.5$.

Secondly, instability of solutions (18) that don't belong to the set of dominating waves has been studied too by their using as initial data for Coshy problem for equations (9). Because accuracy of computation of steady waves is controlled then these initial data correspond to small random perturbations of solutions (18). Results of numerical experiments confirm that limiting regimes belong to set of dominating waves. Examples of development in time are presented in Figure 7b. It may be noted that initial and dominating waves at $s = 0.477$ and $s = 0.5$ belong to the same family $\gamma_{c,1}$ (Sisoev and Shkadov, 1997a, 1997b, 2000).

Thus it was shown by direct numerical experiments that main flow and all the wave regimes except the dominating ones are unstable. They are transformed spontaneously to the dominating wave for chosen wave number s and similarity parameter δ. Hence *dominating waves are global attractors* of initial value problem for (9) and have to be used for comparison with experiments.

Attractive properties of dominating waves are weakened for very long waves ($s \ll 1$) and high δ values ($\delta > 0.4$). In the first case the waves are solitonlike and their shapes have lengthy unperturbed part that is the cause of instability similar to one of main flow. In particular an alternation of coherent solitonlike structures and nonstationary periodic waves has been revealed by Sisoev and Shkadov (1998).

In the case of large δ a weakness of attractors appears as a rule in the form of nonstationary waves with parameters that are varied in narrow intervals closed to values of the dominating waves. At any case we'll use a name *the dominating wave* in accordance with definition introduced above but its global attractive capacity must be studied in numerical experiments in frame of (9). If the limiting wave is formed then it will be referred as *the tested dominating wave.*

Now we have tables of dominating waves for several δ values to compare with experimental observations. It allows to formulate numerical experiments instead of physical ones.

5 Nonlinear waves: theoretical and experimental data correlations

The first success in comparing the solutions of system (9) with Kapitsa's experimental data on nonlinear waves was achieved in paper Sh. Papers by Demekhin et al. (1985, 1987b, 1987c) contain numerous examples of solutions to explain various experimental observations. The full numerical procedure which provides an equivalence of theoretical and experimental data on regular waves is developed by Shkadov and Sisoev (2000b).

In Figure 8a the sets of dominating waves for several values of δ is demonstrated in plane (a, c). There are three main groups of regular waves. First group is composed by the waves which belong to the first main family at different value of δ and fall to one separate correlation curve I. Pieces of intermediate slow families are placed separately and form second group II. Third group III of fast waves is presented by solutions of the second main family. It is seen that at fixed δ value the waves belonging to the intermediate families and to the second family dispose in narrow band which boundaries are nearly straight lines closing with growth of wavelength. It corresponds qualitatively and numerically to experimental functions $c(a)$, which linear correlation with some dispersion is clearly demonstrated.

In Figure 8b the comparison of theory and experiments are shown in plane (Q, S), $Q = (45\delta)^{9/11} q_0/3$, $S = 15\delta q_0^{4/3}\alpha$. These parameters for representation of experimental observations were used by Alekseenko et al. (1979). There are depicted the domains restricted by curves l_1, l_2, l_3, l_4 in which waves are observed in experiments. The accordance

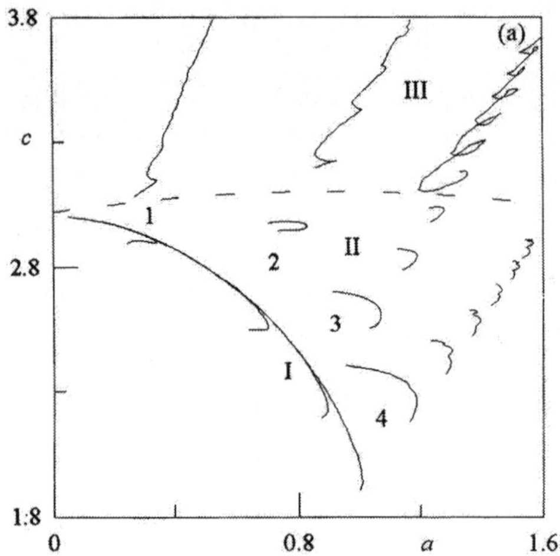

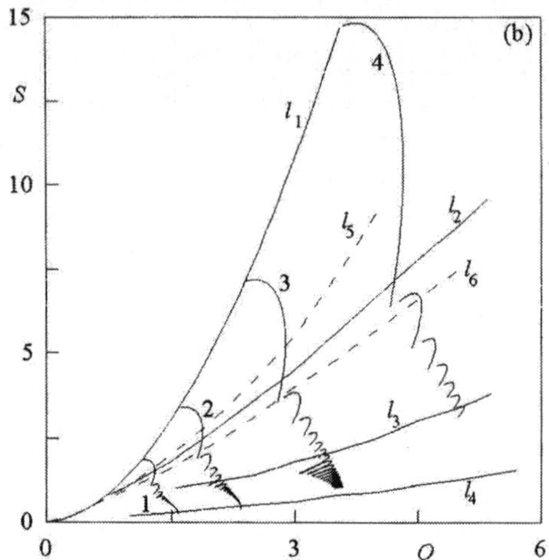

Figure 8. Dominating waves: (a) 1 — $\delta = 0.04$, 2 — $\delta = 0.1$, 3 — $\delta = 0.2$, 4 — $\delta = 0.4$; (b) 1 — $\delta = 0.1$, 2 — $\delta = 0.15$, 3 — $\delta = 0.247$, 4 — $\delta = 0.4$.

of experimentally established dividing curves with theoretical boundaries of mentioned above groups of regular waves is very accurate. The curve l_1 coincides with neutral curve of equations (9) $S = 3^{1/3}Q^{11/6}$. The waves of the first main family belong to the domain between curves l_1 and l_2. Only in Kapitsa's experiments by special selection of the disturbance frequency were genereted nonlinear waves of the first family γ_a from interval $\delta \in (0.04, 0.2)$. All experimental points fall into to one correlation curve of optimal regimes l_5 revealed in paper Sh. Also regular waves have been observed in two domains between curves l_2, l_3 and l_3, l_4. Artificial exiting waves may be generated in both domains but natural developing ones are formed in domain l_2, l_3 only. It is very interesting to see that slow dominating waves of intermediate families strictly belong to domain between curves l_2, l_3 and fast dominating waves of the second base family γ_b belong to domain between l_3, l_4. Also there is depicted curve l_6 corresponding to waves with maximal amplification factors in frame of linear stability analysis of main flow for equations (9).

We proceed now to discuss *numerical experiments* to model the regular waves by comparing concrete quantitative data from computations and observations. In Figure 9a results for slow waves in different liquids in plane (Q, V), $V = c/q_0^{2/3}$ are shown. The lower boundary of experimental data in Figure 9a corresponds to nearly harmonic Kapitsa's (1949) waves and coincides with the optimal regimes of the paper Sh. Also for few values of δ we marked curves of the dominating waves. As there are an accordance of theoretical and experimental points in principal we see also that the unique correlation curve does not exist for these waves. There is dispersion of points as inherent property of the regular waves set. Accounting $Q = \mathrm{Re}_A/\gamma^{3/11}$ it is surprising example when two wave velocities c are registered for one Reynolds number Re_A, i.e. for equal experimental conditions.

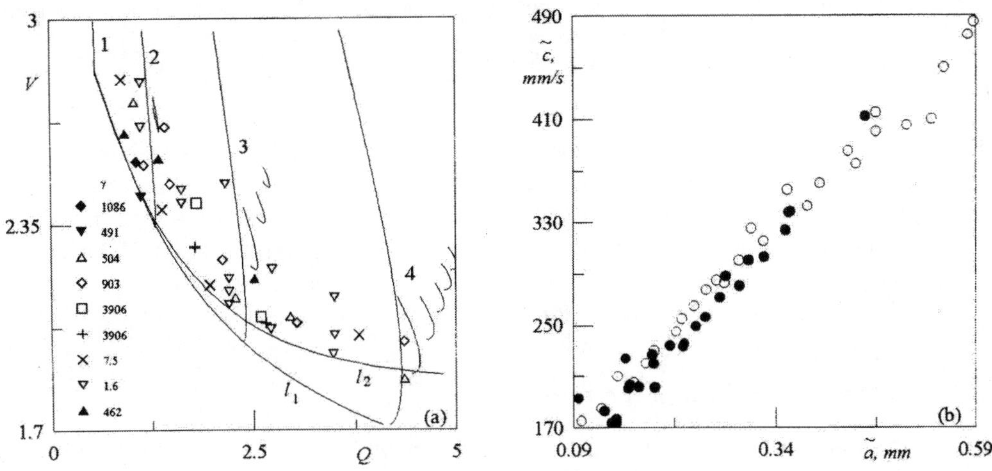

Figure 9. (a) Regular waves in experiments (denoted by symbols) and the dominating waves in computations: 1 — $\delta = 0.04$, 2 — $\delta = 0.1$, 3 — $\delta = 0.2$, 4 — $\delta = 0.4$. Curves of linear waves with maximal amplification factors (l_1) and optimal regimes (l_2) are depicted too. (b) Comparison of experimental (○) and numerical (•) results for $\gamma = 3274$.

Moreover the phase velocity of developed nonlinear regime is close to one of linear disturbance that is typical property of the dominating waves belonging to the intermediate families.

Let us proceed now to discuss fast waves. Contrary to slow waves there are more clear correlation curves in plane (a, c) for the fast dominating waves set, see domain III in Figure 8a. The dispersion of points is decreasing in plane (a, c) as we move from branch to branch towards the second main family γ_b or in other words from slow to fast waves. As an example dimensional values of amplitude $\tilde{a}$ and phase velocity $\tilde{c}$ in experiments with liquid having $\nu = 0.0103$ cm^2/s, $\sigma/\rho = 72.9$ cm^3/s^2 are indicated by empty circles in Figure 9b where the tested dominating waves are denoted by black points for several δ values.

Now we can reproduce numerically points which have been observed for artificially disturbed waves. Let us assume that liquid properties and flow rates are known. Based on this data the similarity parameter δ could be calculated. Then from the list of the dominating waves for this δ value we pick up the dimensional parameters of waves in acceptable physical plane. Thus numerical calculations could be exploited to obtain base points on the properties of regular waves instead of physical experiments.

6 Falling films instability and the Marangoni effect

Surface tension gradients due to mass or heat transfer along or across an interface (Marangoni effect) either create flows or alter existing ones or even trigger instabilities eventually leading to flow motions, steady or otherwise. For film flow the influence of

Marangoni effect appears only as tangential stress in boundary condition on the surface

$$\mu\frac{\partial u}{\partial y} = \frac{\partial \sigma}{\partial x}. \tag{30}$$

This relation for tangential forces with the correctness accepted for the boundary layer approximation (7) is represented. As boundary condition (30) represents a partial case of surface force τ, discussed above in formulas (15–16), it is possible to use the basic model equations (16) for hydrodynamic part of the problem under consideration. We need now to discuss boundary value problem to obtain model equation for σ. We follow here the recent investigations(Velarde et al., 2000).

6.1 Mathematical model for masstransfer

Let us confine ourselves by the solutal surface tension effect consideration following to the paper Hennenberg et al. (1992). We need to examine three values for surfactant mass transfer in this case: c — the bulk concentration, $\overline{c}$ — the bulk concentration on the free surface, Γ — the concentration in the absorbed layer. For the bulk concentration of surfactant we use Fick's diffusion equation

$$\frac{dc}{dt} = D\triangle c \tag{31}$$

and boundary conditions on the rigid wall

$$y = 0: \quad \frac{\partial c}{\partial y} = 0 \tag{32}$$

On the surface of the liquid film equation for adsorption–desorption kinetics, the equations of mass conservation of surfactant and Fick's law are employed

$$y = h(x,t): \; j = -D\frac{\partial c}{\partial y}, \quad k_a\overline{c} - k_d\Gamma = j,$$

$$j = \frac{k_g}{m}\overline{c} + \frac{\partial \Gamma}{\partial t} + \frac{\partial u\Gamma}{\partial x} - D_s\frac{\partial^2 \Gamma}{\partial x^2} \tag{33}$$

(with differentiation along the surface). In the initial section uniform distribution of surfactant is specified

$$x = x_0: \quad c = c_0 = const \tag{34}$$

The first term on the right part of last equation (33) is the desorption or evaporation of the surfactant from the liquid phase into the gas phase, k_g — gas phase masstransfer coefficient of surface active solute, m — the ratio of the concentration in the liquid phase to the concentration in the gas phase at equilibrium. These equations have been employed by Palmer and Berg (1972), Hennenberg et al. (1992) for the stability of a horizontal liquid layer with solutal Marangoni effect. In Ji and Setterwall (1994) the instability of falling film for the partial case of $\Gamma = const$ is investigated. The coefficient of surface tension must be specified to obtain closed mathematical formulation. Two functional dependencies will be used here

$$\sigma = \sigma(\overline{c}), \quad \sigma = \sigma(\Gamma).$$

To pose the problem in dimensionless form we introduce suitable scales

$$\begin{aligned} &x, y, h, t \to \frac{1}{\kappa} H_* x, \quad H_* y, \quad H_* h, \quad \frac{H_*}{\kappa U_*} t, \\ &u, v, c, \Gamma \to U_* u, \kappa U_* v, c_* (1+c), \Gamma_* (1+\Gamma). \end{aligned} \tag{35}$$

The scales H_*, U_* are defined by (5), c_* and Γ_* are yet to be prescribed. After introducing (35) to full equations and boundary conditions and omitting all terms which have an order κ^2, the boundary layer approximation for diffusion part of the problem is deduced

$$\frac{\partial c}{\partial t} + u\frac{\partial c}{\partial x} + v\frac{\partial c}{\partial y} = \frac{1}{\kappa \mathrm{Pe}}\frac{\partial^2 c}{\partial y^2}, \tag{36}$$

$$y = 0: \quad \frac{\partial c}{\partial y} = 0,$$

$$y = h: \quad \frac{\partial c}{\partial y} + \mathrm{Bi}\,(1+\overline{c}) + \kappa \mathrm{G}\left(\frac{\partial \Gamma}{\partial t} + \frac{\partial u(1+\Gamma)}{\partial x}\right) - \kappa^2 \mathrm{Di}\frac{\partial^2 \Gamma}{\partial x^2} = 0,$$

$$-\frac{\partial c}{\partial y} = \pi_1 (1+\overline{c}) - \pi_2 (1+\Gamma).$$

The boundary layer approximation with self–induced pressure to full Navier–Stokes formulation includes hydrodynamic and diffusion parts which are connected by Marangoni stress in boundary condition

$$\tau = -\kappa \mathrm{Ma}\frac{\partial \overline{c}}{\partial x}. \tag{37}$$

The following ten dimensionless parameters have been introduced

$$\begin{aligned} &\mathrm{Re} = \frac{U_* H_*}{\nu}, \quad \mathrm{Pe} = \frac{U_* H_*}{D}, \quad \mathrm{We} = \frac{\rho H_* U_*^2}{\sigma}, \quad \mathrm{Fr} = \frac{U_*^2}{g H_*}, \\ &\mathrm{Ma} = -\frac{d\sigma}{d\overline{c}}\frac{c_*}{\mu U_*}, \quad \mathrm{G} = \frac{\Gamma_* U_*}{c_* D}, \quad \mathrm{Bi} = \frac{k_G H_*}{m D}, \\ &\mathrm{Di} = \frac{D_s \Gamma_*}{D H_* c_*}, \quad \pi_1 = \frac{k_a H_*}{D}, \quad \pi_2 = \frac{k_d \Gamma_* H_*}{c_* D}. \end{aligned} \tag{38}$$

Here Re, Pe, We, Fr, and Bi are numbers of Reynolds, Peclet, Weber, Froude and Biot, Di refers to diffusion. In (38) the relation $\sigma(\overline{c})$ is assumed to be known. If rather the relation $\sigma(\Gamma)$ is considered then in (37) the term $\mathrm{Ma}_1 \Gamma_x$ appears instead of $Ma\overline{c}_x$ with

$$\mathrm{Ma}_1 = -\frac{d\sigma}{d\Gamma}\frac{\Gamma_*}{\mu U_*}.$$

The parameter κ is the ratio of derivatives of the unknown functions with respect to x to those with respect to y.

By omitting all the terms of order κ^2 the boundary value problem for the masstransfer of surfactant in the solute film is derived.

The small parameter $\varepsilon = (\kappa\mathrm{Pe})^{-1/2}$ in equation (36) gives an indication on diffusive boundary layer to exist near the film surface. Let us introduce the stretched coordinate near the surface $y = h(x,t)$

$$y = h - \varepsilon\zeta.$$

For concentration field $c(x,y,t)$ inside diffusive boundary layer the following boundary conditions could be formulated

$$\zeta = 0: \; c = \overline{c}(x,t),$$
$$\zeta = \Delta(x,t): \; c = 0, \; \frac{\partial c}{\partial \zeta} = 0.$$

Here $\Delta(x,t)$ is usual normalized thickness of diffusive boundary. The simplest polynomial representation of $c(x,y,t)$ which satisfies boundary conditions is

$$c = \overline{c}\left(1 - \frac{\zeta}{\Delta}\right)^2. \tag{39}$$

By integrating equation (36) from $\zeta = 0$ to $\zeta = \Delta$ we obtain

$$\frac{\partial}{\partial t}\int_0^{\Delta} c d\zeta + \frac{\partial}{\partial x}\int_0^{\Delta} uc d\zeta = \left.\frac{\partial c}{\partial \zeta}\right|_0^{\Delta}. \tag{40}$$

If we put (39) into (40) then from (36,40) the system of equations for $\overline{c}$, Γ, φ, Δ as functions of x and t is derived

$$\begin{gathered}\frac{\partial \varphi}{\partial t} + \frac{\partial}{\partial x}\left[(A\overline{u} + BM_1)\varphi\right] = 2\frac{\overline{c}}{\Delta}, \\ \kappa\mathrm{G}\left(\frac{\partial \Gamma}{\partial t} + \frac{\partial}{\partial x}(\overline{u}\Gamma) + \frac{\partial \overline{u}}{\partial x}\right) - \kappa^2\mathrm{Di}\frac{\partial^2 \Gamma}{\partial x^2} + \mathrm{Bi}\,(1+\overline{c}) = -2\frac{\overline{c}}{\varepsilon\Delta}, \\ \pi_1(1+\overline{c}) - \pi_2(1+\Gamma) = -2\frac{\overline{c}}{\varepsilon\Delta}, \quad \varphi = \frac{1}{3}\overline{c}\Delta.\end{gathered} \tag{41}$$

where

$$\varphi = \int_0^{\Delta} c d\zeta, \quad M_1 = Mh\overline{c}_x, \quad M = \kappa\mathrm{Ma},$$
$$A = 1 - \frac{1}{10}\left(\frac{h_1}{h}\right)^2, \quad B = \frac{1}{4}\frac{h_1}{h} - \frac{1}{10}\left(\frac{h_1}{h}\right)^2, \quad h_1 = \varepsilon\Delta.$$

Now we have closed basic model system for h, $\overline{u}$, $\overline{c}$, Γ, Δ which includes hydrodynamic part (16) and diffusion part (41). These two parts by Marangoni number Ma in (37) are connected.

For stationary film flow we have $h = 1$, $\overline{u} = 3/2$. Using the stretched variables $x_1 = \varepsilon^2 x$, $\varphi_1 = \varepsilon\varphi$ the system (41) to the form is transformed

$$\begin{aligned} &\overline{u}\frac{\partial \varphi_1}{\partial x_1} - \frac{1}{2}\varepsilon^2 M \frac{\partial}{\partial x_1}\left(\frac{\overline{c}\varphi_1}{s}\overline{c}_x\right) = -s, \\ &s = \mathrm{Bi}\,(1+\overline{c}) + \varepsilon^2 \kappa \mathrm{G}\overline{u}\frac{\partial \Gamma}{\partial x_1} - \varepsilon^4 \kappa^2 \mathrm{Di}\frac{\partial^2 \Gamma}{\partial x_1^2}, \\ &s = \pi_1 (1+\overline{c}) - \pi_2 (1+\Gamma), \\ &\varphi_1 = -\frac{2}{3}\frac{\overline{c}^2}{s}. \end{aligned} \tag{42}$$

Neglecting the terms of order ε^2, it follows from (42)

$$\begin{aligned} &\frac{d}{dx_1}\left(\frac{\overline{c}^2}{1+\overline{c}}\right) = \mathrm{Bi}^2 (1+\overline{c}), \\ &\pi_1 (1+\overline{c}) - \pi_2 (1+\Gamma) = \mathrm{Bi}\,(1+\overline{c}), \\ &s = \mathrm{Bi}\,(1+\overline{c}). \end{aligned} \tag{43}$$

Differential equation (42) with initial condition (34) has solution in the closed form

$$\ln(1+\overline{c}) + \frac{1}{2}\left(\frac{1}{1+\overline{c}}\right)^2 = \frac{1}{2} + \mathrm{Bi}^2 x_1. \tag{44}$$

From the second equation (43) we obtain

$$\pi_1 - \pi_2 = \mathrm{Bi}, \quad \Gamma = \overline{c}. \tag{45}$$

Now the concentrations $\overline{c}$, Γ for the stationary diffusion conditions could be calculated from (44) for every x_1 value, if parameter Bi is specified. From (44) it follows that the evaporation of surfactant to the gas phase is the only reason for adsorption–desorption to deviate from equilibrium state which takes place for Bi=0. Thus the posed problem deals with the development of the surfactant masstransfer from the equilibrium conditions at the initial section $x_1 = 0$. To estimate the accuracy of the approximate solution (44) the comparison of (44) with the exact solution of stationary diffusion problem given in Ji and Setterwall (1994) can be done. For the dimensionless parameters values $\mathrm{Pe} = 10^6$, $\mathrm{Bi} = 10$, $\overline{c}_0 = -0.25$ exact solution gives $x = 1000l$, $h_1 = 0.09$ (h_1 is obtained only approximately from graphical dependence $c(y)$). The appropriate values from (44) are $x = 1012l$, $h_1 = 0,0666$.

6.2 Diffusion instability

Let us now consider the behavior of the small amplitude disturbances when the diffusion part of the general problem considered. The base state corresponds to the diffusion boundary layer near the surface of the falling film. The linear or nonlinear behavior of the solute film system with evaporating surfactant is described by equations (41). For the main conditions we take

$$\overline{u}_0 = \frac{3}{2}, \quad h_0 = 1, \quad \overline{c}_0 = \overline{c}\left(\varepsilon^2 x_0\right), \quad \Gamma_0 = \overline{c}\left(\varepsilon^2 x_0\right),$$
$$s_0 = \mathrm{Bi}\left(1 + \overline{c}_0\right), \quad \varphi_0 = -\frac{2}{3}\frac{\overline{c}_0^2}{s_0}. \tag{46}$$

For the hydrodynamic instability the main state (46) is assumed to have no x–dependence as $\overline{c}\left(\varepsilon^2 x_0\right)$ is a slowly varying function. Because of x enters in (46) as a parameter, the stability analysis must be fulfilled for various sections $x = x_0$.

Small disturbances of h, φ, $\overline{u}$, Γ, $\overline{c}$ as functions of x, t are introduced

$$(h, \varphi, \Gamma, \overline{c}, \overline{u}) = (h_0, \varphi_0, \Gamma_0, \overline{c}_0, \overline{u}_0) + (h', \varphi', \Gamma', \overline{c}', \overline{u}').$$

After linearization (41) the equations for disturbances are

$$\frac{\partial \varphi'}{\partial t} + \overline{u}_0 \frac{\partial \varphi'}{\partial x} + \varphi_0 \frac{\partial u'}{\partial x} - \frac{1}{2} M \frac{\overline{c}_0 \varphi_0}{s_0} \frac{\partial^2 \overline{c}'}{\partial x^2} = -s',$$
$$\frac{1}{\varepsilon} s' = \mathrm{Bi}\overline{c}' + \kappa \mathrm{G}\left(\frac{\partial \Gamma'}{\partial t} + (1 + \Gamma_0)\frac{\partial u'}{\partial x} + \overline{u}_0 \frac{\partial \Gamma'}{\partial x}\right) - \kappa^2 \mathrm{Di}\frac{\partial^2 \Gamma'}{\partial x^2}, \tag{47}$$
$$\frac{1}{\varepsilon} s' = \pi_1 \overline{c}' - \pi_2 \Gamma'.$$

We seek the solutions of (47) as normal modes

$$(h', \varphi', \Gamma', \overline{c}', \overline{u}') = \left(\tilde{h}, \tilde{\varphi}, \tilde{\Gamma}, \tilde{c}, \tilde{u}\right) \exp i\alpha\,(x - \omega t). \tag{48}$$

Then, the amplitudes obey the following algebraic system

$$\beta\left(\overline{u}_0 - \omega\right)\tilde{\varphi} + \left(B_1 - M_2 \varphi_0 \alpha^2\right)\tilde{c} + G_1\left[\beta\left(\overline{u}_0 - \omega\right) + D_1 \alpha^2\right]\tilde{\Gamma} =$$
$$-\beta\left[\varphi_0 + G_1\left(1 + \Gamma_0\right)\right]\tilde{u}, \tag{49}$$
$$\tilde{\varphi} - \gamma_1 \tilde{c} - \frac{\gamma_2}{B_1}\left[G\beta\left(\overline{u}_0 - \omega\right) + D_1 \alpha^2\right]\tilde{\Gamma} = \gamma_2 \frac{G_1}{B_1}\left(1 + \Gamma_0\right)\tilde{u},$$
$$\tilde{c} - \left[1 + \mathrm{T}\beta\left(\overline{u}_0 - \omega\right) + \mathrm{T}D_1 \alpha^2\right]\tilde{\Gamma} = \mathrm{T}\beta\left(1 + \overline{c}_0\right)\tilde{u}.$$

The linearized version of the hydrodynamic part (16) is

$$\frac{\partial h'}{\partial t} + \frac{\partial q'}{\partial x} = 0,$$
$$5\delta\left(\frac{\partial q'}{\partial t} + \frac{\partial Q'}{\partial x}\right) = \frac{\partial^3 h'}{\partial x^3} + 2h' - \frac{2}{3}u' - \frac{2}{3}M\overline{c}'_x, \tag{50}$$
$$q' = h' + \frac{2}{3}u' + \frac{1}{6}M\overline{c}'_x, \quad Q' = \frac{6}{5}h' + \frac{8}{5}\overline{u}' + \frac{7}{20}M\overline{c}'_x.$$

Using (48) we get from (50) the following equations for unknown amplitudes $\tilde{u}$, $\tilde{h}$, $\tilde{c}$

$$\frac{2}{3}\tilde{u} + (1-\omega)\,\tilde{h} + \frac{1}{6}z\tilde{u} = 0,$$

$$\left(\frac{2}{3}\theta - \frac{2}{3}\omega\beta + \frac{8}{5}\beta\right)\tilde{u} + \left[-2\theta + \beta\alpha^2\theta + \beta\left(\frac{6}{5} - \omega\right)\right]\tilde{h} + \tag{51}$$

$$z\left(\frac{2}{3}\theta - \frac{1}{6}\omega\beta + \frac{7}{20}\beta\right)\tilde{c} = 0.$$

The following notations are used in (49, 51)

$$B_1 = \varepsilon V \mathrm{Bi}, \quad G_1 = \varepsilon\kappa V \mathrm{G}, \quad D_1 = \varepsilon\kappa^2 \mathrm{Di}, \quad M_2 = -\frac{\kappa\bar{c}_0 \mathrm{Ma}}{2\mathrm{Bi}\,(1+\bar{c}_0)},$$

$$\theta = \frac{1}{5\delta}, \quad \beta = i\alpha, \quad z = M\beta,$$

$$V = \sqrt{\frac{3}{2}\frac{1+\bar{c}_0}{|\bar{c}_0|}}, \quad \gamma_1 = \frac{2+\bar{c}_0}{\bar{c}_0}, \quad \gamma_2 = -\frac{2\bar{c}_0^2}{3\mathrm{Bi}\,(1+\bar{c}_0)^2}.$$

Now we have closed homogeneous system (49, 51) for $\tilde{u}$, $\tilde{h}$, $\tilde{c}$, $\tilde{\Gamma}$, $\tilde{\varphi}$. By assuming the determinant of this system to be zero, the dispersion equation could be obtained. This dispersion equation determines four eigenvalues ω_k depending on the wavenumber α and on the free dimensionless parameters Bi, G, T, Di, δ, κ, $\bar{c}_0$, Ma, ε.

Some classification of parameters in their connection with physical or mechanical processes is to be done. The main parameter connecting hydrodynamic and diffusion parts of the film flow problem with surfactant is Marangoni number Ma. The both variants of positive (Ma > 0) and negative (Ma < 0) solutal systems are considered. The main hydrodynamic parameters are Re, γ or equivalently δ, γ. This two values determine the mean film thickness H_*, mean velocity U_* and flow rate H_*U_* as well as parameter κ. The diffusion parameters Pe, $\bar{c}_0$ determine the local thickness of diffusion boundary layer h_1 and smallness parameter ε. Two values T, D_1 characterize the masstransfer of surfactant by the adsorption–desorption and the intensity of dissipation by the surface diffusion. Besides the limiting case of fast desorption (T $= 0$) the more general case (T ~ 1) are considered. Intensity of the surfactant evaporation by parameter Bi is determined. The remaining parameter G gives an indication to the typical value of surface excess concentration Γ_* in comparison with c_*.

A few remarks about parameter T are of interest. The disturbed equation of the adsorption–desorption kinetics (49) contains T, D_1, $\bar{c}_0$ together with the wavenumber α, where

$$\mathrm{T} = \kappa\frac{U_*}{k_d H_*}, \quad D_1 = \kappa^2\frac{\mathrm{Di}}{k_d H_*^2}. \tag{52}$$

As it is clear from (52) parameter T does characterize the ratio of rates of the surface excess concentration Γ by two transfer processes: one is convective flow along the film surface, other is desorption inside bulk liquid. For T $\to 0$ the case of diffusion controlled adsorption–desorption kinetics is obtained

$$\Gamma = \bar{c}.$$

This equation corresponds to a fast desorption process Brian (1971) leading to local kinetic equilibrium for $j \neq 0$. In the opposite limiting case, $\mathrm{T} \to \infty$, it follows from (49)

$$\tilde{\Gamma} = \frac{1 + \bar{c}_0}{\bar{u}_0 - \omega - \beta D_1} \tilde{u}. \tag{53}$$

This equation corresponds to the kinetically frozen desorption. Only for $1+\bar{c}_0$ it is possible to consider the surface excess concentration of surfactant Γ as constant and hence $\tilde{\Gamma} = 0$ for $j \neq 0$.

6.3 Numerical experiments: eigenvalues

For every prescribed set of free physical parameters the eigenvalues $\omega(\alpha)$ could be computed for arbitrary high $\alpha > 0$ values. But we must keep in mind that assumptions of the long wave approximation introduce some limitations on α. For the problem under consideration there exist indeed two lengths h and h_1. By virtue of inequality $h_1 \ll h$ short waves in h scale could be considered as long wave in h_1 scale. Some arbitrary upper boundary $\alpha < 10$ in the following numerical experiments is used. Let us consider the case of constant surface excess concentration Γ.

Positive Marangoni numbers. First of the all we try to cross–check our method for eigenvalues given by dispersion equation following from (49, 51) by comparing with exact numerical solutions of the Orr–Sommerfeld formulation presented in Ji and Setterwall (1994) for $\Gamma = const$. Note that our problem is a more general formulation of instability problem as we include nonequilibrium surfactant adsorption–desorption kinetics of surfactant. Besides, their case is somehow artificial as it does not follow from the general formulation (36) as $\mathrm{T} \to \infty$. Furthermore, when $\mathrm{T} \to \infty$ the term $1 + \bar{c}_0$ in (49) must be put to equal zero. For $\Gamma_0 = 0$, $\tilde{\Gamma} = 0$ the dispersion equation is the third–order algebraic equation for this case. For the $\mathrm{Ma} > 0$ two of three roots yield unstable modes. For illustration and comparison we choose the following numerical values of the main parameters

$$\mathrm{Re} = \frac{40}{3}, \quad \mathrm{Pe} = \frac{2}{3}10^6, \quad \mathrm{Bi} = 10, \quad \bar{c}_0 = -0.25.$$

This set of parameter values fits well a liquid metal with $\gamma = 29.2$, $\mathrm{Ca} = 0.2$. The calculations of eigenvalues are completed for various values Ma, T, α. In Figure 10 typical curves for $c_r = c_r(\alpha)$ and $\alpha c_i = \alpha c_i(\alpha)$ are plotted. As amplification factor $\alpha c_i(\alpha)$ of various instability modes could differ to several orders, a normalized amplification factor $f = m\alpha c_i$ where m is appropriate scale is used in figures. One of these growing modes is easily identified as the hydrodynamic mode of the falling film with small wavenumber and is indeed the same when $\mathrm{Ma} = 0$. The phase velocity c_r of this wave mode diminishes from $c_r = 3$ as α grows form $\alpha = 0$, takes a minimum value, and then increases. Amplification factor αc_i is positive in the interval $0 < \alpha < \alpha_*$ and has maximum value $(\alpha c_i)_m$ inside of this interval. Other growing modes, which are referred to as diffusion modes appear only at $\mathrm{Ma} \neq 0$. The term "diffusion" could be applied for any mode which disappears at

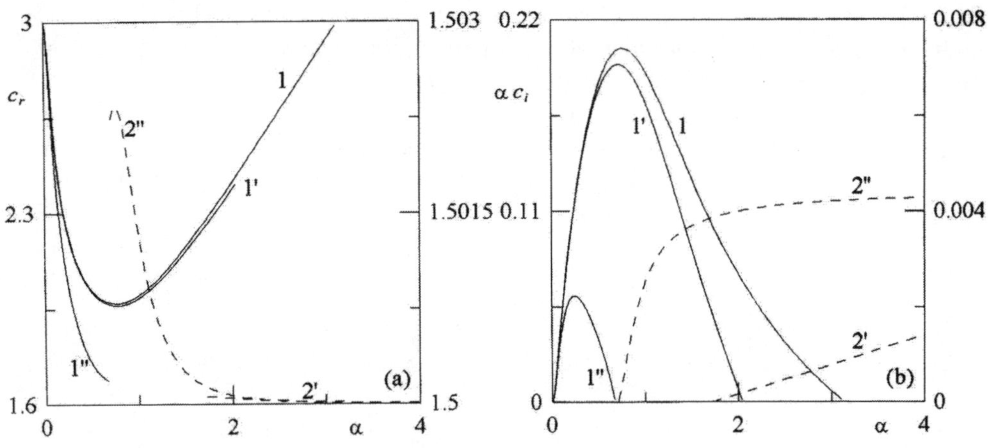

Figure 10. Phase velocities (a) and amplification factors (b) for hydrodynamic (solid curves 1, 1' and 1" connected with left axes) and diffusive modes (dashed curves 2'and 2" connected with right axes) at $\gamma = 29.4$: 1 — Ma = 0; 1',2' — Ma = 0.015; 1",2" — Ma = 1.5.

Ma $\to 0$. Diffusion modes in the case under consideration exist as solutions of dispersion equation for high enough wavenumbers $\alpha > \alpha_{**}$, where α_{**} is to be determined by computations. The wave velocity of a diffusion mode is equal to 3/2 with great accuracy. Thus this wave moves with the velocity of liquid on the film surface. This mode which can be identified as a monotone instability mode on the liquid interface leads to patterned interface. The amplification factor αc_i of this diffusive mode is two to three orders lower that of hydrodynamic mode and tends to its maximum value as α grows.

For the chosen main variant with $\alpha = 1.355$ the eigenvalues are

$$\omega_1 = 1.313 - 0.106i, \quad \omega_2 = 1.253 - 0.1125i,$$
$$\omega_1 = 0.999 + 0.00045i, \quad \omega_2 = 1.000 + 0.00167i.$$

The solutions of dispersion equation discussed here are on the right side while the solutions of Ji and Setterwall (1994) of the full Orr–Sommerfeld formulation are on the left side. The reasonable accordance of the two approaches to the eigenvalue problem could be observed.

The calculations of the eigenvalues for various α, Ma, T values are completed. The values $D_1 = 0.01; 0.0001$ have been tested. The structure of spectrum $\omega(\alpha)$, which in Figure 10 is demonstrated remains for the all values Ma > 0. Existence of hydrodynamic and diffusion instability modes is the main feature of that spectrum. The type of the surfactant influence over the hydrodynamic instability mode on such figures could be seen. The amplification factor of the most unstable mode is diminishing with Ma increasing. If δ is small and Ma is sufficiently high diffusion mode grows more fast than hydrodynamic mode grows. The critical value of the wavenumber α_* moves to the long waves and as a result the region of instability cancels ($\alpha_* < \alpha_0$). Phase velocity to some degree diminishes. The salient features of surfactant influence on the hydrodynamic instability mode which follow from solutions of dispersion equation agree with the results of Lin (1970).

Negative Marangoni numbers. Very interesting results for the solute system model $\Gamma = const$, Ma < 0 were obtained.As it could be seen from Figure 11 there are three

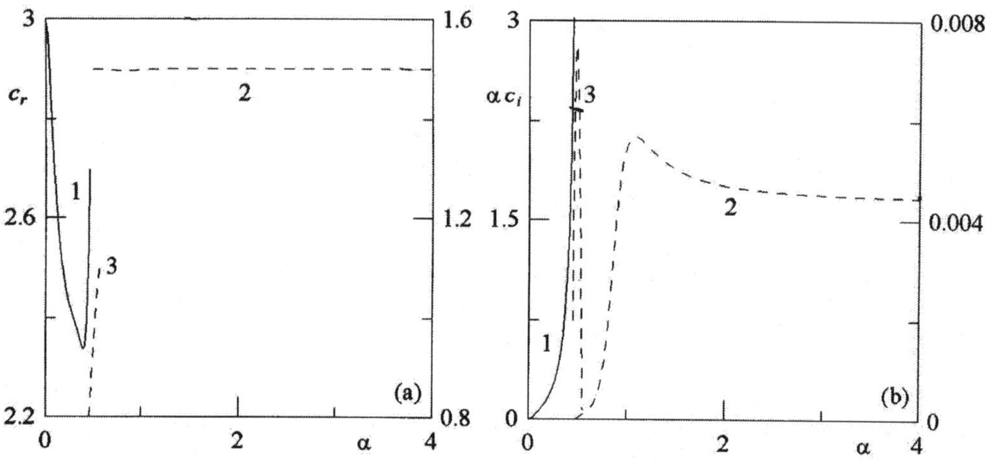

Figure 11. Phase velocities (a) and amplification factors (b) for hydrodynamic (solid curves 1 connected with left axes) and diffusive modes (dashed curves 2 and 3 connected with right axes) at $\gamma = 29.4$, Ma $= -1.5$.

growing modes according to α. That modes by the phase velocity values could be identified. Amplification factor of diffusive mode with phase velocity $c_r = 1.5$ has inessential changes with transition from Ma $= 1.5$ to Ma $= -1.5$. There is second particular diffusive mode on finite interval wavenumbers with phase velocity of that mode varies linearly in the vicinity of point $c_r = 1$, $\alpha = 0.5$. The third branch of solutions begins at $\alpha = 0$ as hydrodynamic instability mode with phase velocity $c_r = 3$, but then converts to the mode of explosive growing as α tends to critical value α_k. To say more precisely $\alpha c_i \to \infty$ as $\alpha \to \alpha_k$. Let us show that appearance of the explosive growing mode is inherent property of the film flows for the solute system model $\Gamma = const$, Ma < 0. By equating to zero coefficient before highest member in dispersive equation we obtain

$$\alpha_k^2 = 4\frac{2+\bar{c}_0}{\bar{c}_0}\frac{\mathrm{Bi}}{\kappa^2 \mathrm{GMa}}. \tag{54}$$

For the values given to the dimensionless parameters in Figure 11 the equation (54) becomes

$$\alpha_k^2 = -\frac{0.385}{\mathrm{Ma}}$$

and gives $\alpha_k \approx 0.507$. Let us introduce the smallness parameter ε_1, so that $\omega = \Omega/\varepsilon_1$, $\varepsilon_1 \ll 1$, $|\Omega| = o(1)$. Then from equations (49, 51) it follows

$$\tilde{h} = \varepsilon_1^2 \check{h}, \quad \tilde{\Gamma} = \varepsilon_1 \check{\Gamma}, \quad \tilde{\varphi} = \varepsilon_1 \check{\varphi}, \quad \tilde{c} = \check{c}, \quad \tilde{u} = \check{u}, \tag{55}$$

where all values $\check{h}$, $\check{\Gamma}$, $\check{\varphi}$, $\check{c}$, $\check{u}$ on the right side of (55) are of the same order. Neglecting terms $o\left(\varepsilon_1^2\right)$, from the first equation (49) we obtain

$$\check{u} + i\mathrm{Ma}\alpha\check{c} = 0. \tag{56}$$

The unstable mode actually represents a longitudinal wave since surface deformations are small as it is clear from (55). The disturbances $\overline{u}'$, $\overline{c}'$ are phase shifted by $\varphi = \pi/2$ as it follows from (56). They can be expressed in the following way

$$\overline{c}' = |c| \exp ix_1, \quad \overline{u}' = \frac{\alpha\,|\mathrm{Ma}c|}{4} \exp i\left(x_1 - \frac{\pi}{2}\right), \quad x_1 = \alpha\,(x - \omega t)\,. \tag{57}$$

As (57) shows the longitudinal oscillations $\overline{u}'$ carry surface active agent out of the points where $\overline{c}'$ is minimal and bring it to the points where $\overline{c}'$ is maximal. Some sort of resonance takes place. The unbounded solutions appear at the critical wavenumber α_k, defined by (54). When $\mathrm{Ma} < 0$ and $|\mathrm{Ma}|$ is small the critical wavenumber α_k of the explosive models outside of the hydrodynamic mode instability interval exposed ($\alpha_k > \alpha_n$), with $|\mathrm{Ma}|$ growing α_k moves to α_n.

For $\alpha \in (0, \alpha_k)$ a special combined mode with singularity at point $\alpha = \alpha_k$ occurs as a result of hydrodynamic and explosive mode interaction. The main feature of combined mode is abnormally high values of amplification factor αc_i near the boundary point $\alpha = \alpha_k$ of the interval of instability $(0, \alpha_k)$ where that mode exists. Phase velocity c_r grows with α approaching to α_k. This is an example of the wave transformations under resonant conditions. From the physical arguments the occurrence of singularity behavior for the growing coefficient could give an indication to some shortcomings of the model $\Gamma = const$ applied. It is necessary to examine the model of adsorbed layer with $\Gamma \neq const$ for $\mathrm{Ma} < 0$.

Veriable surface excess concentration. In general case $\Gamma \neq const$ of a falling film of weak volatile surfactant solution in which the surfactant mass transfer is governed by the diffusion, evaporation and adsorption–desorption processes in the near-surface layer the development of instability depends on nine dimensionless external parameters. We can take γ, δ, c_0, Pe, G, Bi, T, Di, and Ma as these independent parameters. If the parameters are given, the problem reduces to the numerical solution of the dispersion relation for various of the wave number α and the spectral analysis of $\omega = \omega(\alpha)$.

For a finite adsorption–desorption rate ($\mathrm{T} \neq 0$) the dispersion relation determines four eigenvalues. The number of possible particular solutions with increasing amplitude and their properties substantially depend on Ma and T, but the clear separation of the solutions into hydrodynamic and diffusion instability modes is conserved.

Transition to the general model with $\Gamma = \Gamma(x,t)$ and $\mathrm{T} \neq 0$ leads to regularization of the solution. This means that there is no singular explosive growth mode. Let as take an example of the calculation of the eigenvalues for $\mathrm{Ma} = 1$ and $\mathrm{T} = 1$. Together with the hydrdynamic mode there are two diffusion instability modes for which the values of $\omega_r(\alpha)$ and $\omega_i(\alpha)$ differ only slightly. The phase velocity of these waves almost coincides with the velocity of the liquid on the free surface and the amplification coefficient increases with α. With variation of the parameters Ma and T for fairly large values of α we also obtain growing solutions in the form of waves propagating with respect to the liquid.

The Marangoni effect driven by the surfactant mass transfer is important for the onset and development of instability in a falling film of weak solution. The calculations carried out show that the derived model system of evolutionary equations (16),(41) is a qualitatively correct reflections of the instability properties of a falling film with a surfactant and its solutions agree with the individual results obtained by means of another method. As distinct from the cumbersome approach based on the direct solution of the system of Navier–Stokes equations it gives opportunities for studying multiparameter problem, calculating numerous variants, and draving generalizing conclusions with allowance for the kinetics of the adsorption–desorption processes.

New diffusion instability modes whose development is completely determined by the Marangoni effect are revealed. Depending on the phase velocity, these modes can be combined into two groups. Perturbations propagating with the velocity of the liquid on the film surface form structures stationary with respect to that liquid. The other perturbations are traveling waves. As distinct from the transverse waves of the ordinary hydrodynamic mode, the longitudinal waves have the greatest phase velocities and amplification coefficient. The influence of the Marangoni effect leads to weakening of the hydrodynamic forces and surfactant mass transfer are also formed.

For a given thickness of the falling film the number and type of instability modes in it substantially depend on the surfactant transfer model adopted. For the systems described by the insoluble surfactant model solutions of the explosive–growth wave type exist. In the systems in which the complete model of the adsorption–desorption processes is taken into account the instability is described only by solutions with finite amplification coefficients.

The wavenumber interval on which new diffusion instability modes exist is bounded from below and not bounded from above. Consequently, these perturbations are mainly shortwave. At the same time, the formation of regular nonlinear wave structures in falling film of a pure liquid is associated with the fact that only perturbations on the bounded interval $0 < \alpha < \alpha_k$ are unstable.

Although in the problem considered there is a second linear scale (the diffusion boundary layer thickness) which corresponds to short waves, the presence of the short waves does not completely agree with the assumptions concerning longwave perturbations adopted in deriving the basic system of model equations. Therefore, it is necessary to compare the investigation of the instability spectra of a falling film containing a surfactant with the solutions of the complete linearized Navier–Stockes equations.

7 Conclusions

To give theoretical explanation of regular real nonlinear waves in films we apply adequate mathematical model (9). In frame of that we have found manifold of regular wave solutions and investigated its properties (bifurcations, attractors). Then we have revealed the dominating waves that compose the subset of this manifold and possess extremely values of main wave parameters. Tables of the tested dominating waves have been computed. On the basis of these tables it is possible to reproduce experimental data of physical waves in falling liquid films.

The basic system (9) enables the investigation to be extended to nonstationary evolution and interactions of the wave structures in films such as deterministic–stochastic transitions (Sisoev and Shkadov, 1998). Note that extension of system (9) to films on inclined planes, axisymmetric bodies, to flows with tangential forces up to now are accomplished. For every such case the weakly nonlinear asymptotics (13,14) from extended system (9) could be deduced.

The method of paper Sh to films under influence of Marangoni effect is applied. Generalization of the basic model system (9) gave an effective tool for linear instability consideration. It is attractive task to apply this multiparametric model as to new instability modes so to nonlinear waves investigations.

There are great deal of works on downflowing films up to now which exploit the basic model system (9) or some generalizations of that model. The term *Shkadov's model* is used in these papers to characterize the method of investigation. Below a short list of such publications accomplished by observations are given.

Webb (1972) is the first one to carry out an adequate analysis of the basic model system (9) in English: "the Shkadov method is the only consistent solution to falling film flow using Kapitsa approach". After considering some current publications on the film flows Koulago and Parséghian (1995) summarize: " it seems that this real pioneer work of Shkadov (and other of his important works) has not been taken into account in the West side". As for weakly nonlinear approximation (14) to the full system (9) the authors give the conclusion:" One of the form of the equation of Shkadov (9) has been found by Kuramoto on one hand and by Sivashinsky on the other hand in works that have been published between 1974 and 1976".

The authors of monograph Alekseenko et al. (1992) the great number of known numerical solutions of the basic model system (9) have gathered to compare with their experimental data. They give the conclusion concerning system (9): "for moderate Reynolds numbers this approach gives splendid results". That conclusion by numerous figures is accompanied.

Papers by Trifonov and Tzvelodub (1985, 1991, 1992), Chang (1993, 1995a) contain numerous value of numerical solutions of system (9) for nonlinear waves under various partial combinations of parameters δ, s. For some cases two or more wave solutions are constructed. These works include also many examples of successful correlation of theoretical and experimental values.

Some articles to examination of the properties of nonlinear periodic and solitary waves in dissipative medium as nonstationary solutions of basic system (9) are devoted (Chang et al., 1995b, 1996, Cheng and Chang, 1995).

The papers by Yu et al. (1995), Ruyer-Quil and Manneville (1998), Zeytounian (1998) are important in the sense they are devoted to some successful generalizations of the basic model (9) for the film flows.

References

Alekseenko, S. V., Nakoryakov, V. E., and Pokusaev, B. T. (1979). Waves on the surface of vertically flowing liquid film. Preprint 36–79, IT SO AN USSR, Novosibirsk.

Alekseenko, S. V., Nakoryakov, V. E., and Pokusaev, B. T. (1992). *Wave flow of films.* Novosibirsk: Nauka.

Brian, P. L. T. (1971). Effect of Gibbs adsorption on Marangoni instability. *AIChE J.* 17:765–772.

Bunov, A. V., Demekhin, E. A., and Shkadov, V. Y. (1984). On the nonuniqueness of nonlinear wave–solutions in a viscous layer. *Prikl. Mat. Mekh.* 48(4):691–696.

Bunov, A. V., Demekhin, E. A., and Shkadov, V. Y. (1986). Solitary wave bifurcation in a falling liquid layer. *Vestn. MGU, Mat. Mekh.* 2:73–78.

Chang, H.-C., and Demekhin, E. A. (1996). Solitary wave formatiom and dynamics on falling film. *Adv. Appl. Mech* 32:1–58.

Chang, H.-C., Demekhin, E. A., and Kopelevich, D. I. (1993). Nonlinear evolution of waves on a vertically falling film. *J. Fluid Mech.* 250:433–480.

Chang, H.-C., Demekhin, E. A., and Kalaidin, E. (1995a). Interaction dynamics of solitary waves on a falling film. *J. Fluid Mech.* 294:123–154.

Chang, H.-C., Demekhin, E. A., and Kopelevich, D. L. (1995b). Stability of a solitary pulse against wave packet disturbances in an active medium. *Phys. Rev. Lett.* 75(9):1747–1750.

Chang, H.-C. (1994). Wave evolution on a falling film. *Annu. Rev. Fluid Mech.* 26:103–136.

Cheng, M., and Chang, H.-C. (1995). Competition between subharmonic and sideband secondary instabilities on a falling film. *Phys. Fluids* 7(1):34–54.

Demekhin, E. A., and Shkadov, V. Y. (1981). On unsteady waves in a viscous liquid layer. *Izv. Akad. Nauk SSSR, Mekh. Zhidk. i Gasa* 3:151–154.

Demekhin, E. A., and Shkadov, V. Y. (1984). On three–dimensional nonstationary waves in downflowing liquid film. *Izv. Akad. Nauk SSSR, Mekh. Zhidk. i Gasa* 5:21–27.

Demekhin, E. A., and Shkadov, V. Y. (1985). Two–dimensional wave regimes of a thin film of viscous liquid. *Izv. Akad. Nauk SSSR, Mekh. Zhidk. i Gasa* 3:63–67.

Demekhin, E. A., and Shkadov, V. Y. (1986). Theory of solitons in systems with dissipation. *Izv. Akad. Nauk SSSR, Mekh. Zhidk. i Gasa* 3:91–97.

Demekhin, E. A., Demekhin, I. A., and Shkadov, V. Y. (1983). Solitons in falling viscous films. *Izv. Akad. Nauk SSSR, Mekh. Zhidk. i Gasa* 4:9–16.

Demekhin, E. A., Kaplan, M. A., and Shkadov, V. Y. (1987a). Mathematical models of the theory of viscous liquid films. *Izv. Akad. Nauk SSSR, Mekh. Zhidk. i Gasa* 6:73–81.

Demekhin, E. A., Tokarev, G. Y., and Shkadov, V. Y. (1987b). On the existence of critical reynolds number for liquid film flowing under the action of gravity. *Teor. Osnovy Khim. Teknol.* 21(3):555–589.

Demekhin, E. A., Tokarev, G. Y., and Shkadov, V. Y. (1987c). Two–dimensional unsteady waves on a vertical liquid film. *Teor. Osnovy Khim. Teknol.* 21(2):177–183.

Demekhin, E. A., Tokarev, G. Y., and Shkadov, V. Y. (1988). Numerical investigation of the three-dimensional wave evolution in downflowing liquid film. *Vestn. MGU, Mat. Mekh.* 2:50–54.

Demekhin, E. A., Tokarev, G. Y., and Shkadov, V. Y. (1989). Instability and nonlinear waves on a vertical liquid film flowing countor to a turbulent gas flow. *Teor. Osnovy Khim. Teknol.* 23(1):64–70.

Demekhin, E. A., Tokarev, G. Y., and Shkadov, V. Y. (1991). Hierarchy of bifurcations of space–periodic structures in a nonlinear model of active dissipative media. *Physica D* 52:338–361.

Esmail, N. M., and Shkadov, V. Y. (1971). To nonlinear theory of waves in viscous liquid film. *Izv. Akad. Nauk SSSR, Mekh. Zhidk. i Gasa* 4:54–59.

Hennenberg, M., Chu, X.-L., Velarde, M. G., and Sanfeld, A. (1992). Transverse and longitudinal waves at the air liquid interface in the presence of an adsorption barrier. *J. Colloid and Interface. Sci.* 150:7–21.

Ji, W., and Setterwall, F. (1994). On the instabilities of vertical falling liquid films in the presence of surface–active solute. *J. Fluid Mech.* 278:297–323.

Kapitsa, P. L., and Kapitsa, S. P. (1949). Wave flow of thin viscous liquid films. *Zh. Eksp. Teor. Fiz.* 19:105.

Koulago, A. E., and Parséghian, D. (1995). A propos d'une equation de ls dynamique ondulatoire dans les films liquids. *Journal de Phisique III France* 5:309–312.

Lin, S. P. (1970). Stabiliting effects of surface–active agents on a film flow. *AIChE J.* 16:375.

Palmer, H. J., and Berg, J. C. (1972). Hydrodynamic stability of surfactant solutions heated from below. *J. Fluid Mech.* 51(2):385–402.

Ruyer-Quil, C., and Manneville, P. (1998). Modeling film flows down inclined planes. *Eur. Phys. J. B* 6:277–292.

Shkadov, V. Y., and Sisoev, G. M. (2000a). Influence of electric field to nonlinear waves on downflowing liquid films. In Atten, P., and Denat, A., eds., 2nd *International Workshop. Electrical Conduction, Convection and Breakdown in Fluids*, 146–149. Grenoble, France: CNRS and Université Joseph Fourier.

Shkadov, V. Y., and Sisoev, G. M. (2000b). Wavy falling liquid films: theory and computation instead of physical experiment. In Chang, H.-C., ed., *IUTAM Symposium on Nonlinear Waves in Multi–Phase Flow*, volume 57 of *Fluid Mechanics and Its Applications*, 1–10. Notre Dame, USA: Notre Dame University.

Shkadov, V. Y., Kholpanov, L. P., Malyusov, V. A., and Zhavoronkov, N. M. (1970). Nonlinear theory of liquid film wave flows. *Teor. Osnovy Khim. Teknol.* 4(6):859–867.

Shkadov, V. Y., Epikhin, V. E., Demekhin, E. A., Bunov, A. V., and Filyand, I. V. (1981). Stability of the flows with contact surfaces (liquid layers, capillary jets). Report 2564, Institute of Mechanics of Lomonosov Moscow State University, Moscow.

Shkadov, V. Y. (1967). Wave flow regimes of a thin layer of viscous fluid subject to gravity. *Fluid Dynamics* 1:43–51.

Shkadov, V. Y. (1968). Towards a theory of wave flows of thin viscous–liquid layer. *Izv. Akad. Nauk SSSR, Mekh. Zhidk. i Gasa* 2:20–25.

Shkadov, V. Y. (1973a). *Problems of nonlinear hydrodynamic stability of viscous liquid layers, capillary jets, and internal flows.* Diss. D. Sci., Dept. of Mechanics and Mathematics, Lomonosov Moscow State University, Moscow.

Shkadov, V. Y. (1973b). Some methods and problems of the theory of hydrodynamic stability. Scientific Proceedings 25, Institute of Mechanics of Lomonosov Moscow State University, Moscow.

Shkadov, V. Y. (1977). Solitary waves in a viscous liquid layer. *Izv. Akad. Nauk SSSR, Mekh. Zhidk. i Gasa* 1:63–66.

Sisoev, G. M., and Shkadov, V. Y. (1997a). Development of dominating waves from small disturbances in falling viscous–liquid films. *Fluid Dynamics* 32(6):784–792.

Sisoev, G. M., and Shkadov, V. Y. (1997b). Dominant waves in a viscous liquid flowing in a thin sheet. *Physics–Doklady* 42(12):683–686.

Sisoev, G. M., and Shkadov, V. Y. (1998). Instability and coherence of nonstationary solitary waves. *Physics–Doklady* 43(4):489–493.

Sisoev, G. M., and Shkadov, V. Y. (1999). On two-parametric manifold of the waves solutions of equation for falling film of viscous liquid. *Physics–Doklady* 44(7):454–459.

Sisoev, G. M., and Shkadov, V. Y. (2000). Instabilities and reorganizations of regular waves for falling films of viscous liquids. *Moscow University Mechanics Bulletin* 55(4):44–48.

Trifonov, Y. Y., and Tsvelodub, O. Y. (1991). Nonlinear waves on the surface of a falling liquid film. Part 1. Waves of the first family and their stability. *J. Fluid Mech.* 229:531–554.

Trifonov, Y. Y., and Tsvelodub, O. Y. (1992). Nonlinear waves on the surface of a falling liquid film. Part 2. Bifurcations of the first-family waves and other types of nonlinear waves. *J. Fluid Mech.* 244:149–169.

Trifonov, Y., and Tzvelodub, O. Y. (1985). Nonlinear waves on the surface of a liquid film falling down a vertical plane. *Zh. Prikl. Mekh. Tekh. Fiz.* 5:15–19.

Velarde, M. G., Shkadov, V. Y., and Shkadova, V. P. (2000). The influence of surfactants on the instability of downflowing liquid film. *Izv. Akad. Nauk SSSR, Mekh. Zhidk. i Gasa* 4:56–67.

Webb, D. R. (1972). A note on periodic solutions to flow in a liquid film. *AIChE J.* 18(5):1068–1069.

Yu, L.-Q., Wasden, F. K., Duckler, A. E., and Balakotaiah, V. (1995). Nonlinear evolution of waves on falling film at high Reynolds numbers. *Phys. Fluids* 7(8):1886–1902.

Zeytounian, R. K. (1998). The Bénard–Marangoni thermocapillary instability problem. *Uspekhi Fizicheskikh Nauk* 168:259–286.

E EFFECT OF INTERFACIAL PHENOMENA ON MATERIALS PROCESSING

K.C. Mills
Imperial College, London. UK

1 INTRODUCTION

Interfacial phenomena play an important role in materials processing. However, until recently, their importance has not been widely recognised. To date few processes have been designed to use interfacial phenomena to improve process control or product quality and knowledge of interfacial phenomena has been largely limited to the explanation of occurrences, such as variable weld penetration in TIG/GTA welding. However, as this review shows, interfacial phenomena affect a wide spectrum of materials processes.

2 FUNDAMENTALS

2.1 SURFACE ACTIVITY

Certain elements such as sulphur and oxygen are very surface active [1] in liquid metals. These elements prefer interfacial sites and are present in large concentrations at interfaces (for instance, oxygen concentrations at the interface of liquid iron ore can be 100 times that in the bulk). They cause the following effects in Fe and other metals such as Ni, Cu etc.

(i) a dramatic decrease in surface tension (γ) as can be seen from Figure 1a where 50 ppm O (or S) results in a 30% decrease in γ for Fe [2];

(ii) a change in the temperature coefficient ($d\gamma/dT$) from negative to positive which can result in a change in the magnitude and direction of the fluid flow of the metal in the vessel [2].

Surface activity can be ranked in the hierarchy:

(i) Group VI elements > Group V > Group IV

(ii) Within any group the heavier elements are more surface active than lighter elements e.g. Te > Se > S $\simeq$ O.

It has been suggested that since the outer electron rings are fully filled, the surface sites tend to be filled with molecules which have a large number of electrons in the outer ring, e.g. Group VI elements.

It should also be noted that it is the soluble O and not the combined O (e.g. oxides) which affects surface tension. Certain elements such as Ca, Al, Mg react strongly with O and reduce the soluble O (denoted $\underline{O}$) to very low levels e.g. (a few ppm) and form stable metallic oxides (Figure) [3]. The total O is not a measure of the soluble O ($\underline{O}$%) in such cases. Thus very low concentrations of these reactive elements (e.g. Ca) can have a marked effect on the process because of their effect on the surface active elements present. Thus for the Fe-Al-O system in Figure 2a, 10^{-1}% of Al will lower the soluble O to < 5 ppm but 10^{-3}% Ca (10 ppm) (Fe-Ca-O system) will reduce

$\underline{O}$ to < 1 ppm.

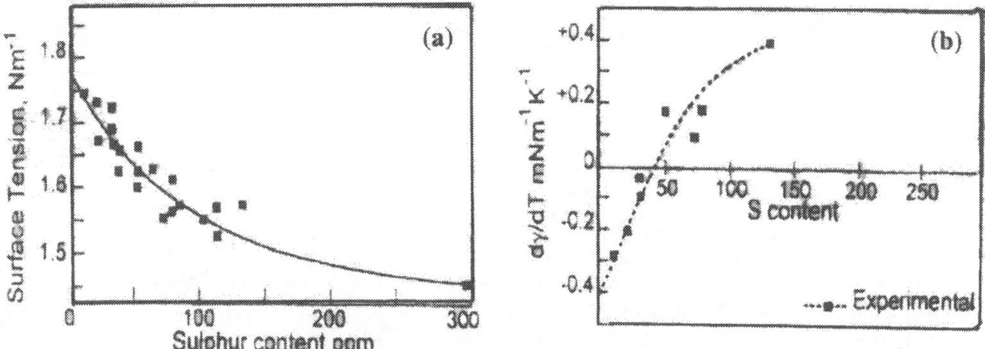

Figure 1 Effect of S content on (a) surface tension (γ) of Fe and (b) temperature dependence of surface tension (dγ/dT) of austenitic stainless steels [2].

2.2 SURFACE AND INTERFACIAL TENSION

The interfacial tension is usually referred to as the surface tension when one of the two phases involved is a gas. In metallurgical industry it is common to use a liquid slag, this carries out the following functions:

(i) it protects the surface of the metal from oxidation.

(ii) it absorbs inclusions, such as Al_2O_3, from the liquid metal.

(iii) it reacts with the metal to remove harmful (surface active) elements which sit preferentially at the grain boundaries (e.g. S, Bi etc) of the solidified metal.

(iv) in certain cases e.g. continues casting it lubricates the metal.

The slag plays a crucial rôle. Thus for processes involving metal and slag it is the interfacial tension (γ_{ms}) which is involved and which frequently plays an important part in the process.

The interfacial tension between metal and slag (γ_{ms}), shows the same dependence upon the sulphur content [4] as that of the surface tension of the metal, (γ_m), as shown in Figure 1a. This is not surprising since (i) Cramb and co-workers [4,5] have found the Girafalco-Good relation (Equation 1) to be an effective way of determining γ_{ms} and (ii) $\gamma_m \simeq 4\,\gamma_s$ and hence γ_m tends to be the most important factor

$$\gamma_{ms} = \gamma_m + \gamma_s - \phi(\gamma_m \gamma_s)^{0.5} \tag{1}$$

where γ_s = surface tension of slag and ϕ = interaction coefficient which is given by the ratio $\left(W_A^{ms} / W_C^m W_C^s\right)$ where W_A and W_C are the work of adhesion and cohesion, respectively (see Section 2.4.1).

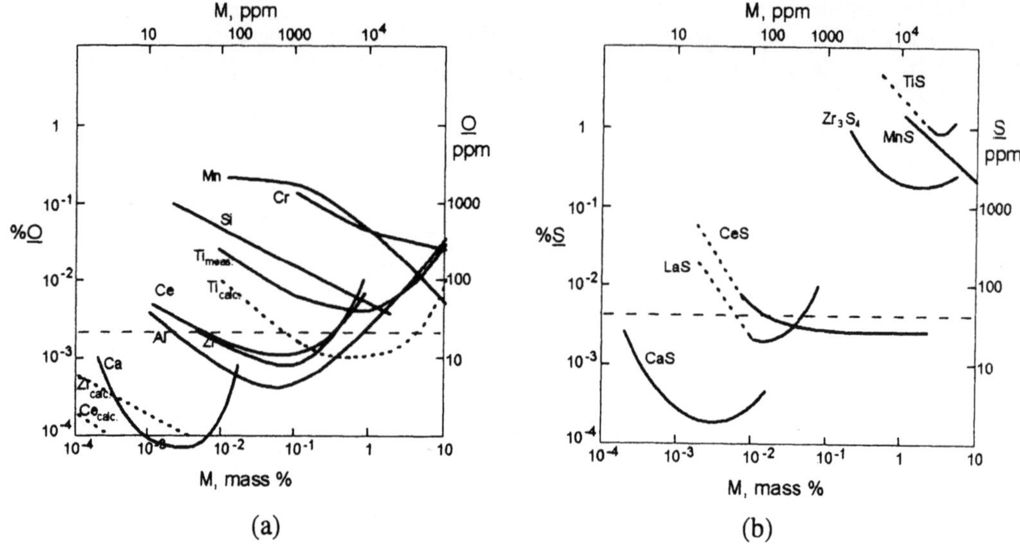

Figure 2 The effect of different alloying additions (M) on (a) the soluble O (O%) content in the Fe-M-O system and (b) the soluble S (S%) content in the Fe-M-S system [4].

The interfacial phenomena affecting metallurgical processes can be divided into the factors affecting the following:

(i) Marangoni flows

(ii) Wettability and contact angle

(iii) Formation of emulsions

(iv) Formation of foams

(v) The formation of jets and surface waves.

2.3 EFFECT OF MARANGONI FORCES

When there is a surface tension gradient along a surface there is a fluid flow along the surface from a region of low surface tension to high surface tension (Figure 3). This is frequently referred to as Marangoni flow. Surface tension gradients can arise from:

(i) temperature differences along the surface which cause *thermocapillary* flow;

(ii) composition differences along the surface which cause *diffusocapillary* flows;

(iii) electric potential differences along the surface which cause *electrocapillary* flow.

Figure 3 Schematic diagram showing Marangoni flows

2.3.1 Thermocapillary forces

Most substances have negative (dγ/dT) temperature coefficients. However, when concentrations of surface active elements such as S or O in liquid metals exceed a certain critical concentration these systems exhibit positive (dγ/dT) values. This is shown in Figure 1b In some cases there is a limit to O solubility (e.g. Si) (i.e. saturation) and thus (dγ/dT) remains negative (Figure 4a) and γ has a fixed value for the O-saturated condition.

Most inorganic and organic liquids have negative temperature coefficients (dγ/dT) but organic liquids such as nonanol and decanol have both negative and positive coefficients (e.g.Figure 4b).

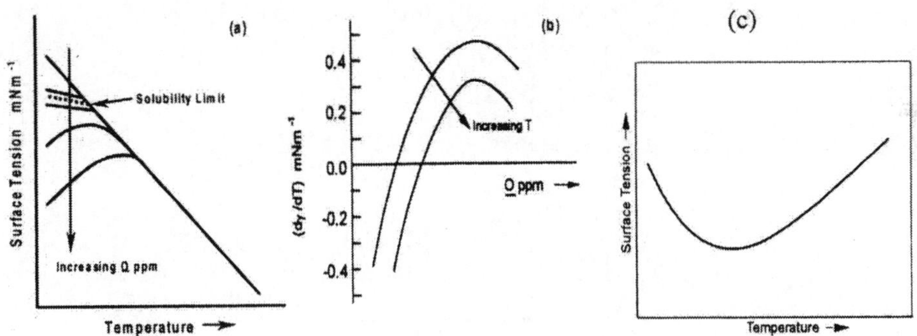

Figure 4 Schematic diagrams showing the effect of surfactants (a) on surface tension (γ)-temperature relations eg Oxygen on molten Si,- - -,represents saturation limit of O in metal; (b) on dγ/ dT as a function of soluble O in metal and (c) on surface tension of decanol as a function of temperature

Since thermocapillary flow occurs from regions of low to high surface tension, the direction of flows will occur from high temperature regions to low temperature when (dγ/dT) is positive and from high to low temperature when (dγ/dT) is negative. Thus the direction of flow is dependent upon the sign of (dγ/dT).

2.3.2 Diffusocapillary

Surface active elements (e.g. S in Fe) reduce the surface tension, consequently, Marangoni flow occurs from regions of high to low surfactant concentrations. Local changes in surface tension may occur through evaporation or dissolution of a specific species or compound.

2.3.3 Electrocapillarity effects

When an electrical potential is applied to a two liquid system, such as mercury and aqueous NaCl solution, changes in surface tension occur. These changes result in a change in the curvature of the meniscus.

Even when there is no applied electrical potential there is a "double layer" of ionic charges formed at the meniscus. There is (i) a net positive charge on the metal (mercury) side and (ii) a net negative charge on the aqueous side since H_2O molecules orientate themselves in such a way that their negative ends face the positive charges of the metal. The introduction of ions, such as Na^+ Cl^-, to the solution enhances these effects. The Cl^- ions are absorbed on the surface of the metal and Na^+ ions occur preferentially in the outer layer. This arrangement of ions (Figure 5) is known as a double layer or the Helmholtz zone.

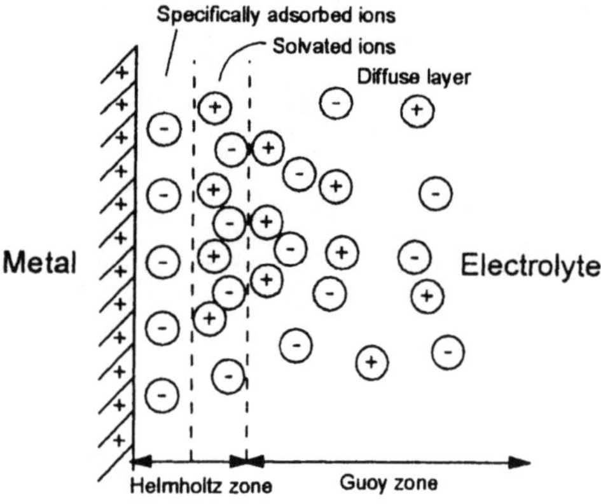

Figure 5 Schematic representation of distribution of ions and charges in the interfacial region between a metal and an aqueous electrolyte.

When we change the electrical potential by δE, work is done in changing the potential of the charges at the interface and altering the charge itself. This results in a change in surface tension ($\partial\gamma$) and is given by the Lippman equation.

$$(\partial\gamma / \partial E)_{T,P,\mu} = -q_E \tag{2}$$

where q_E is the surface change density and the subscript T,P,μ denotes the temperature, pressure and chemical potential which is kept constant. If we differentiate with respect to electrical potential

$$(\partial^2\gamma/\partial E^2)_{T,P,\mu} = -(\partial\sigma/\partial E)_{T,P,\mu} = -C \qquad (3)$$

If C is a constant for the double layer when the γ versus E (electrocapillary curve) will have a parabolic form (Figure 6).

As the mercury is made more negative, electrons are forced to the interface where they neutralise the positive charges. The maximum surface tension occurs when q_E is zero.

Electrocapillarity curves have been obtained for interfaces between molten metals, and either molten salts or molten slags, which are ionic in nature [6]. It has been established that the metal side of the interface has a net positive charge and the slag a net negative charge. In this case there are no solvent molecules to separate the ions from one another. It is believed that there is an excess charge at the interface (+ for metal and – for the slag) and this excess falls to zero over 3-4 ionic, or metal, molecule layers.

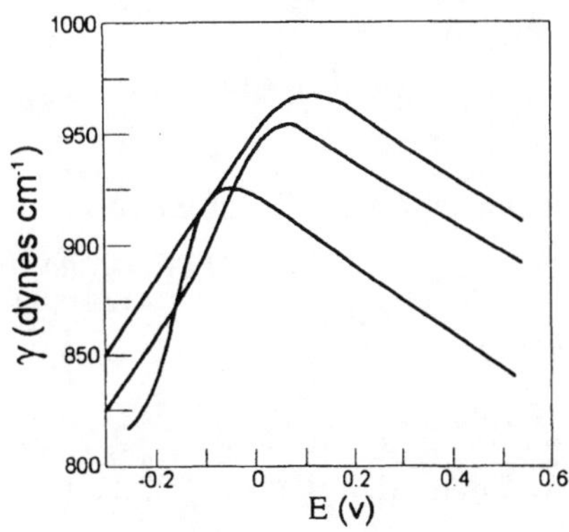

Figure 6 Electrocapillary curves at 1400-1420°C for iron containing 13% carbon, and 3.5% silicon in various SiO2, Al2O3 slag compositions

2.3.4 Marangoni flows perpendicular to the surface

When a thin layer of fluid (Figure7) is heated from below, any instability which occurs will result in the transfer of hot liquid to the surface. For most liquids surface tension decreases with increasing temperature, thus this transport of hot liquid will result in a lower surface tension at the point of emergence. Consequently, there will be a radially outward flow of liquid along the surface. Such behaviour results in the formation of cells (known as Bénard cells) with a hexagonal or polygonal geometry. Obviously the direction of such flows formed by liquid metals could be affected by small changes in the surfactant concentrations. Bénard cells are formed by diffusocapillary flows resulting from differences in chemical composition. The best known example is the hammer (polygonal) finish of metallic paints, resulting from the surface-tension gradients produced by the evaporation of solvents.

When surface active species are present at an interface, they can suppress the flow of fluid perpendicular to the interface. For an interface rich in surfactants (see Figure 8) the flow of liquid will transport the surfactant away from the interface and this decrease in surfactant concentration cannot be instantaneously replaced from the bulk of the liquid. This results in a high surface tension at point of emergence (A) and a diffusocapillary flow will be established in the direction B $\rightarrow$ A i.e. counter to that of the original flow.

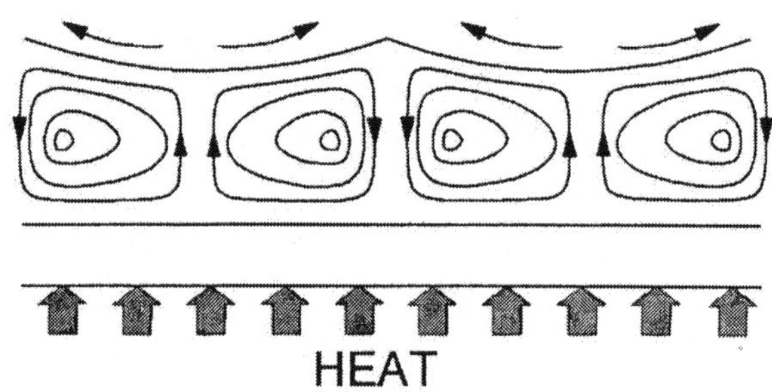

Figure 7 Schematic representation of the formation of Bénard cells

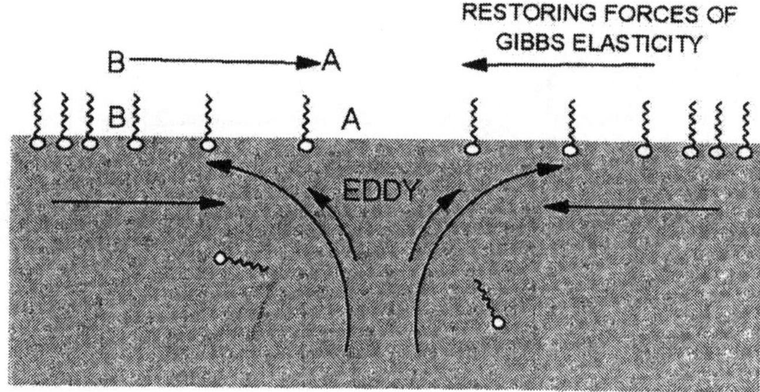

Figure 8 Schematic representation of the creation of a counter-flow by the dispersion of the surfactants in the interface

2.4 WETTABILITY, CONTACT ANGLE (θ)

Non-metallic inclusions, such as Al_2O_3 particles, have a deleterious effect on the mechanical strength of steels and alloys; mechanical strength decreasing with increases in both concentration and magnitude of the inclusions. Gas bubbling is often used to remove inclusions. The situation is shown in Figure 9.

Young's equation can be obtained from a balance of forces where M = metal, I = inclusion and G = gas (bubble).

$$\gamma_{IM} + \gamma_{MG} \cos\theta - \gamma_{IG} = 0 \tag{4}$$

Interfacial tension values pertaining to solid phases are exceedingly difficult to measure and consequently Young's equation is frequently used to obtain these data in terms of measurable properties, namely surface tension (γ_{MG}) and contact angle (θ).

In many cases at high temperatures, wettability is associated with reactivity between the liquid and the solid substrate.

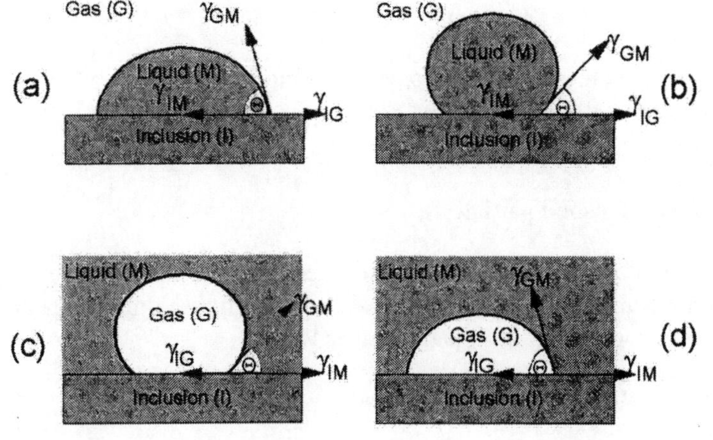

Figure 9. Schematic drawing for (a) and (b) for bubble containing inclusion for wetting and non-wetting condition, respectively, (c) and (d) sessile drop experiments for wetting and non-wetting conditions, respectively.

2.4.1 Work of adhesion

The work of adhesion, W_A, is the work which must be done to separate the metal from the inclusion (and thereby remove the metal/inclusion interface) and create two new surfaces (metal/gas and inclusion/gas).

$$W_A = \gamma_{MG} + \gamma_{IG} - \gamma_{IM} \tag{5}$$

Combination of Equations (4) and (5) yields

$$W_A = \gamma_{MG} (1 + \cos\theta) \tag{6}$$

2.4.2 Flotation coefficient

For good flotation it is necessary that the Flotation coefficient (Δ) (defined in Equation 7) should be both positive and have a high value [7].

$$\Delta = \gamma_{IM} + \gamma_{MG} - \gamma_{IG} \equiv \gamma_{MG} (1 - \cos \theta) \tag{7}$$

2.4.3 Spreading coefficient (S*)

Consider an inclusion at the slag/metal interface. For an inclusion to be removed it is necessary for it to travel through the slag/metal interface and on into the slag phase. The spreading coefficient, S*, is a measure of the ability of a liquid (denoted M but could be metal or slag phase) to spread across the solid and is defined by Equation 8.

$$S^* = \gamma_{IG} - \gamma_{IM} - \gamma_{MG} = \gamma_{MG} (\cos \theta - 1) \tag{8}$$

The spreading will increase as S* becomes more positive, i.e. $0 > S^* > -2\gamma_{MG}$ and this would be favoured by (i) a low γ_{MG} (or high S and O contents) and (ii) a low value of θ ($\theta<90^o$).

2.4.4 Emergence of a solid particle

From a viewpoint of alloy cleanness it is also important that the solid should emerge from the liquid metal and not get entrained in the metal flow. It has been shown [8] that this is favoured when ΔG in Equation 9 is negative.

$$\Delta G = \gamma_{IG} - \gamma_{MG} - \gamma_{MI} < 0 \tag{9}$$

i.e. when (i) γ_{IG} is low, (ii) the surface tension of metal (γ_{MG}) is high (low $\underline{O}$, $\underline{S}$ contents) and (iii) interfacial energy (γ_{MI}) is high (no adsorption layer or chemical reaction at the interface).

2.4.5 Agglomeration of inclusions

The agglomeration of inclusion is also dependent upon interfacial properties. For two inclusions to join it is necessary to replace two γ_{MI} bonds by the two γ_{IG} bonds being formed [8].

$$\Delta G = 2(\gamma_{IG} - \gamma_{IM}) = 2\gamma_{MG} \cos \theta \tag{10}$$

From which it can be shown that $\theta>90^o$ is the criterion for agglomeration and it has been proposed that this mechanism is responsible for the observation that in molten Fe, particles of Al_2O_3 ($\theta = 140^o$) readily agglomerate whereas particles of TiO_2 ((θ =

78°) do not [9].

2.5 FORMATION OF EMULSIONS

Many modern processes involve the formation of emulsions to create a high (surface area/mass) ratio and thereby produce fast kinetics which results in a high production rate.

Kozakevitch [10] followed the desulphurisation of liquid steel by molten slag by using X-rays to observe the change in shape of a sessile drop of the steel in liquid slag (held in crucible). It was noted that there was a marked change in shape of the drop from non-wetting to wetting and then back to non-wetting conditions (Figure 10). These changes are equivalent to the following changes in the metal-slag interfacial tension (γ_{ms}):

(i) a high initial γ_{ms};

(ii) a reduction to a very low γ_{ms} whilst there is a rapid desulphurisation;

(iii) an increase in γ_{ms} to its initial value when desulphurisation is complete.

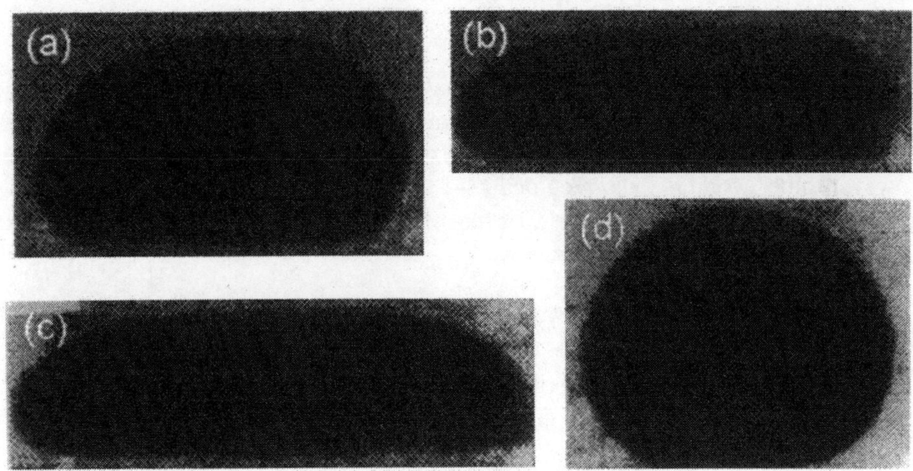

Figure 10 Images showing change in shape of metal drop during the desulphurisation of steel (a) initial; (b), (c) during desulphurisation and (d) when desulphurisation nearly complete (2).

Similar behaviour has been seen for the dephosphorisation of steel [11]. Riboud and Lucas [12] followed on Kozakevitch's work and simultaneously measured mass transfer of Al from the metal to the slag and the slag/metal interfacial tension, γ_{ms}. It can be seen from Figure 1 that (i) there was a massive drop in γ_{ms} (to a very low value) which was associated with high mass transfer rates and (ii) γ_{ms} started to increase when the mass transfer rate slowed, with γ_{ms} eventually reaching its initial value. When γ_{ms} is very low any disturbance or turbulence can cause droplets of one phase to move into the other phase. This is usually referred to as emulsification and the formation of a metal emulsion in the slag phase (or vice versa) leads to very fast kinetics for the refining reactions because of the huge surface area/mass ratio. Thus from a refining process viewpoint a low interfacial tension is very advantageous and modern metal production processes make use of this.

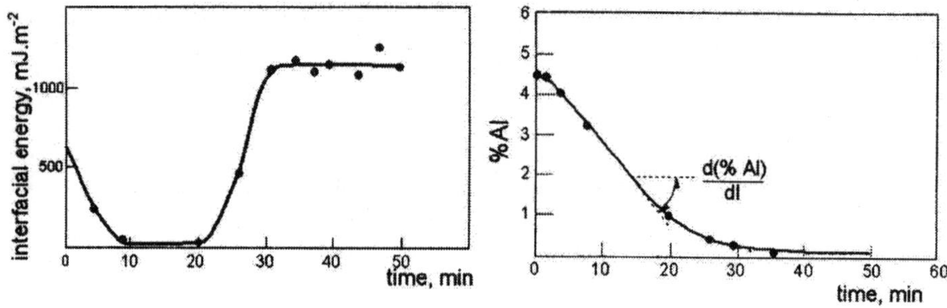

Figure 11 The effects of time (a) on interfacial tension and (b) on the mass transfer of Al from steel (to the slag) [12].

It has been proposed [12,13] that the mass transfer of oxygen from the slag could be responsible when the oxygen stream flux exceeds a value of 0.1 g atom m^{-2} s^{-1}. However, other workers have recorded [14] that there was no dramatic decrease in γ_{ms} with a flux of 0.2 g atom m^{-2} s^{-1}. It has been proposed that the kinetics could be controlled by one of the following steps:

(i) the flux of O in the slag towards the slag

(ii) the flux of reactive element (e.g. Al, Ti) in the metal towards the interface

and

(iii) the flux of the oxidation product (e.g. Al_2O_3) dissolving the slag.

Recent work has indicated that the mass transfer of oxides (SiO_2, FeO, MnO) in the slag is the rate controlling step [4].

Yongsun Chung [4] has suggested that spontaneous emulsification may occur if the relative velocities of the metal and slag phases (Kelvin-Helmholtz instability) is systematically different. This could arise from fluid flow originating from

(i) natural convection, due to local temperature differences caused by exothermic reactions.

(ii) Marangoni flow resulting from local differences in either concentration or temperature (diffuso- and thermocapillary flows, respectively).

Yongsum Chung [4] concluded that Marangoni flow was involved and that emulsification occurred by (a) the upward movement of 'pin-shaped regions' of metal and downward movements of cup-shaped regions of slag (Fpgure 12a) and (ii) 'necking' and emulsification by eddy currents created by Marangoni flow (Figure b).

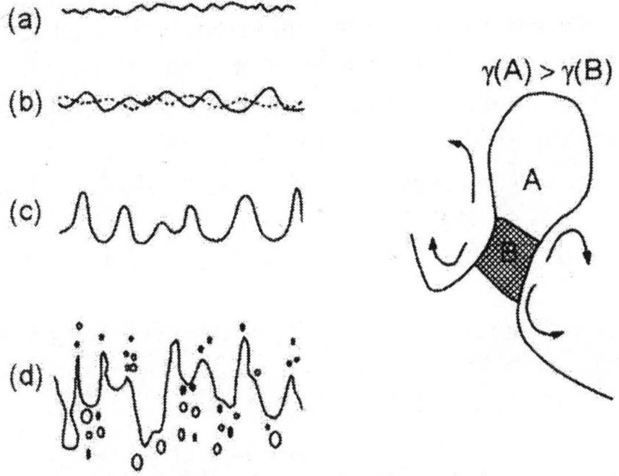

Figure 12 Schematic diagrams (a) showing the growth of perturbation and (b) the Marangoni flows producing 'necking' and emulsification [4].

2.6 FORMATION OF FOAMS

Foams are widely used in iron- and steelmaking process since they provide rapid refining of the metal droplets held in the foam as a result of the enormous (surface area/mass) ratio. Foaming slags are also used on Electric Furnaces (EAF) to stabilise the arc and improve energy efficiency.

There is general agreement that the principal factors promoting foaming are [15-19]:

(i) a low surface tension of slag (γ_s) (P_2O_5, Na_2O, iron oxide additions cause a decrease in γ_s);

(ii) high bulk and surface viscosities which retard draining of the slag film (P_2O_5 and SiO_2 are both surface active and will congregate preferentially at the surface and will tend to increase the surface viscosity);

(iii) the presence of solids in the slag (since this will increase viscosity and tend to lock the bubbles and prevent their escape);

(iv) decreasing temperature since this would increase slag viscosity and encourage the formation of solids (a high solidification temperature of the slag would also be beneficial);

(v) high surface elasticity.

Although there is a general agreement on the factors affecting foam stability there is still some disagreement about the relative importance of the various factors [18,19].

2.7 SURFACE WAVES, FORMATION OF JETS

Wave motion at a surface or interface can result from both gravitational and capillary forces. Lord Rayleigh studied the break-up of a jet of liquid into a gas [20]. Capillary forces make liquid jets unstable when their length, L, exceeds their circumference. Axisymmetric disturbances of the surface grow in amplitude until the jet is pinched off (Figure 13). Rayleigh's theory correctly predicted the drop size and the dependence of L on surface tension. Subsequent work [21] for liquid/gas and liquid/liquid jets has shown that the breakup is dependent upon:

(i) mass transfer through the interface e.g. the transfer of a surfactant from jet to the fluid (jet) causes the length, L, to increase substantially whereas transfer from the fluid to the jet causes a reduction in L.

(ii) the presence of a surface film (with high viscosity) tends to damp out the surface instabilities.

(iii) the presence of surfactants on the jet surface tends to reduce the mass transfer effects, e.g. a high S cast of steel would tend to reduce the effects of oxygen transfer on jet stability [21].

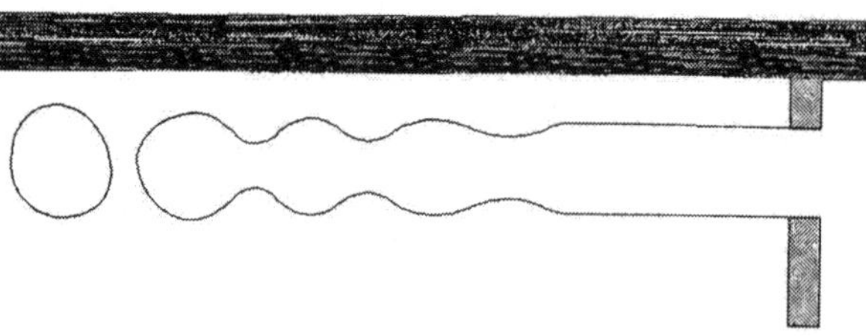

Figure 13 Schematic diagram of the break-up of jets, showing flow within the jet.

3 INTERFACIAL EFFECTS IN WELDING AND JOINING

3.1 VARIABLE WELD PENETRATION IN TIG/GTA WELDING

3.1.1 Introduction

The problem of '*cast-to-cast*' variations in weld penetration produced during autogenous tungsten inert gas (TIG) or GTA welding of stainless and ferritic steels was first noted in the 1960s. The problem is particularly severe in robotic processes requiring thousands of repetitive welds where it is customary to establish the welding parameters which promote deep penetration joints. However, it has been found (as can be seen in Figure 14) that certain batches of steel produced welds with much lower weld penetration than the norm, despite fully meeting the material specifications. Welds can be partial-penetration welds (Figure 14a,b) or full-penetration welds (Figure 14c). It is customary to express penetration in partial welds by the depth (D)/width (W) ratio.

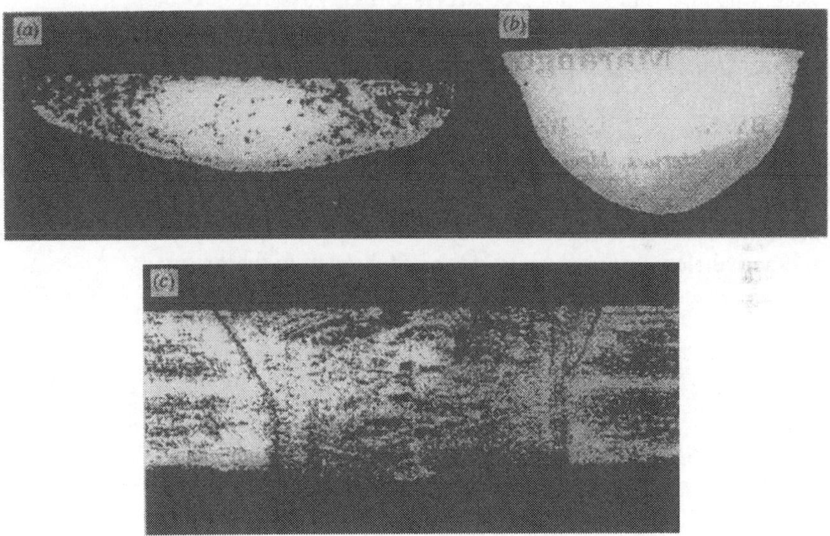

Figure 14 Comparison of the cross section of TIG weld fusion zones in (a) shallow- and (b) deep-penetration welds in stainless steel (x10) and (c) full-penetration weld, in this case penetration defined by the ratio of widths of back and front welds (Wb/Wf).

There had been several attempts to establish a correlation between 'cast-to-cast' variations and systematic variations in the concentrations of specific minor or impurity elements in the metal. However, where such a relationship could be identified it was noted that any such variations in the element concentrations were very small. Thus, any theory proposed to account for variable weld penetration must explain why such small differences in chemical composition can have such a large effect on 'weldability'. Several theories have been proposed in which it was suggested that small differences in minor element concentrations in the steel produced changes in (a) the arc characteristics [22,23] and (b) the surface properties of the weld pool by affecting either the interfacial energies [24], or the fluid flow motion in the weld pool [25]. However, it has also been found that variable weld penetration occurred in non-arc processes such as laser and electron beam welding [26,27] where there are no arc effects. Consequently, cast-to-cast variation could not be explained solely on the basis of changes in the arc characteristics and thus attention has been focused mostly on changes in the surface properties of the melt.

As can be seen from Figure 1a, small differences in the concentrations of surface-active elements, such as sulphur [28] and oxygen [29], cause substantial changes in the surface tension (γ) of iron and other elements. Friedman [30] developed a model of the weld pool in which it was proposed that the surface-tension forces operating in the pool opposed the combined effects of gravity and arc pressure; thus a high surface tension would lead to poor weld penetration. Other theories have focused on the effect of the surface tension on the fluid flow in the weld pool. Ishizaki [31] suggested that the surface-tension gradient ($d\gamma/dT$) across the pool could affect the convective flow in the weld pool. Heiple and Roper [25] developed this theory, and postulated that variable weld penetration is a result of differences in the fluid flow in the weld pool resulting from differences in both the direction and magnitude of thermocapillary forces, and that these were controlled by the concentrations of surface-active elements such as sulphur and oxygen in the metal.

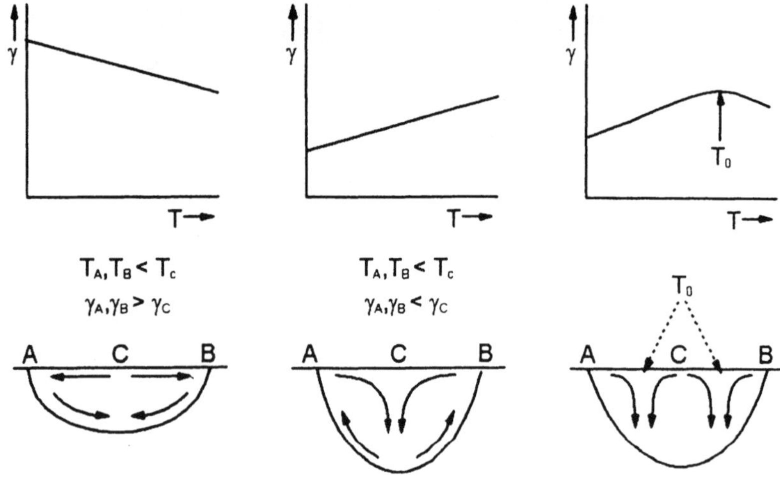

Figure 15 Schematic diagram illustrating the Heiple-Roper theory for variable weld penetration [25].

Heiple and Roper [25] also pointed out that when the sulphur or oxygen concentration exceeded a certain critical value (around 50 ppm), the temperature coefficient of surface tension (dγ/dT) changed from a negative to a positive value (Figure 15). They suggested that since a large temperature gradient exists between the centre and the edges of the weld pool (of the order of 500 K mm^{-1}), a large surface-tension (γ) gradient will be produced across the surface. The resulting Marangoni flow will occur from a region of low γ to a region of high γ. These surface flows subsequently trigger circulation flows in the molten weld pool, as shown in Figure 15. For most pure metals, including iron and steels with low O and S contents, the surface tension decreases with increasing temperature, which results in a negative-surface-tension-temperature coefficient (dγ/dT) (Figure 15a). In this case, the surface tension will be greatest in the cooler regions at the edge of the weld pool and this induces a radially-outward surface flow which carries hot metal to the edge of the pool where the consequent melt-back results in a wide shallow weld. In contrast to this, in Fe-based melts with S (or O) > 60 ppm, (dγ/dT) will be positive (Figure 15b) and thus the surface tension is greatest in the high-temperature region at the centre of the pool and this induces a radially-inward flow. This, in turn, produces a downward flow in the centre of the weld pool (Figure 15b) which transfers hot metal to the bottom of the pool where melt-back of the metal results in a deep and narrow pool.

Keene et al [32] have pointed out that systems which exhibit a positive (dγ/dT) must go through a maximum at some temperature and thus produce a complex flow similar to that shown in Figure c.

3.1.2 Forces affecting the fluid flow in the weld pool

The Heiple-Roper theory makes two assumptions: (i) that the heat transfer in the weld pool is controlled by the fluid flow in the pool and not the heat conduction in the workpeice; and (ii) that the fluid flow is dominated by the thermocapillary forces.

However, there are several other fluid flow mechanisms operating in the weld pool, namely electromagnetic (or Lorentz), aerodynamic drag and buoyancy forces (Figure 16). Under certain welding conditions these forces can have a significant effect on the fluid flow in the weld pool.

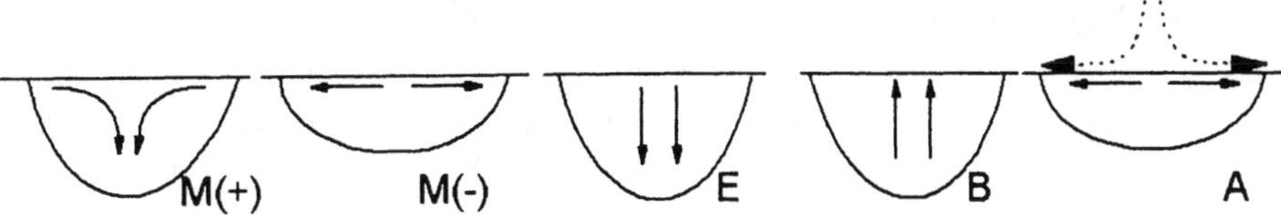

Figure 16 (a) Thermocapillary (Marangoni) forces M(+) or M(-); (b) electromagnetic (Lorentz) forces E, resulting from interaction of current; (c) buoyancy forces B, resulting from density differences caused by temperature gradients; (d) aerodynamic drag forces A

(a) Marangoni forces

These are, for the most part, thermocapillary forces but diffusocapillary forces can arise when welding steels with different sulphur contents (see Section 4.1.7). The direction of the thermocapillary flow is determined by the concentration of O or S in the alloy.

The strength of the thermocapillary flow is determined by the non-dimensional Marangoni number (Ma) defined in Equation 11 where ($d\gamma/dx$) is the temperature gradient, η is the viscosity, a is the thermal diffusivity and L is the characteristic length:

$$Ma = \frac{d\gamma}{dT}\frac{dT}{dx}\frac{L^2}{\eta a} \tag{11}$$

(b) Electromagnetic or Lorentz forces

The Lorentz forces are caused by the interaction of the induced magnetic field and the current carried by a conductor. The welding current induces a magnetic field around the conductor and the Lorentz force acts inwards and downwards in the weld pool (Figure b).

(c) Buoyancy forces

Buoyancy forces are caused by the density differences due to temperature gradients in the weld pool and result in an upward flow (Figure 6c). However, it has been shown that buoyancy forces are generally very small in relation to the other forces in weld pools of less than 10 mm depth.

(d) Aerodynamic drag forces

These forces are produced by the action of the arc plasma flowing over the surface of the weld pool, which induce an outward flow along the surface of the pool (Figure 16d).

However, the fluid flow in the weld pool is exceedingly complex since, thermocapillary, Lorentz, aerodynamic and buoyancy forces can all influence the flow. The situation is further complicated by (i) the front-to-back flow resulting from the relative motion of the workpiece to that of the electrode which is particularly important at high welding speeds and (ii) the 'spin' developed by the liquid metal under conditions of radially inward flow, which tends to reduce the magnitude of the radially inward flow [33].

The mathematical modelling of the relative strengths of the four forces affecting the fluid flow has become a subject of great interest in recent years and about twenty models have been reported [3]. Virtually all these models predict that the Marangoni forces are predominant under normal welding conditions and have a decisive effect on the weld profile [34,35].

3.1.3 Relation between penetration and surface tension

As mentioned previously, penetration is usually expressed by the ratios of the (depth/width) (D/W) and the (back/front) widths (W_b/W_f) for partial- and full-penetration welds, respectively; the latter parameter is subject to welding characteristics and is a less satisfactory measure than (D/W).

The link between weld penetration and surface tension of the alloy has been demonstrated by Mills et al [36] who used the levitated drop method to measure the surface tension of casts with good and bad penetration, i.e. high and low (D/W) ratios, respectively. A typical example is shown in Figure 17a. It was found that:

(i) good weld penetration correlated with low values of surface tension (γ) and positive values of ($d\gamma/dT$) as found in steels with high-sulphur (HS) contents;

(ii) poor weld penetration correlated with high values of γ and negative values of ($d\gamma/dT$) as found in steels with LS contents.

Mills and Keene [3] subsequently correlated ($d\gamma/dT$) with the S contents of the steels and showed that the 'cross-over' point where ($d\gamma/dT$) = 0 occurred around 40 ppm (Figure 17b). Thus good weld penetration is obtained in steels with greater than 60 ppm S and poor penetration in casts with less than 30 ppm S.

Mathematical models have shown that fluid flow in the weld pool is complex, despite the fact that thermocapillary forces tend to be dominant. Nevertheless, on the basis of the Heiple-Roper theory, some correlation between (D/W) and ($d\gamma/dT$) might be expected. The γ-T relationship was determined for three steels with different S contents and the (D/W) ratio was derived for partial welds carried out on 6 mm plates of these steels. It can be seen from Figure 18 that there is a good correlation between (D/W) and ($d\gamma/dT$).

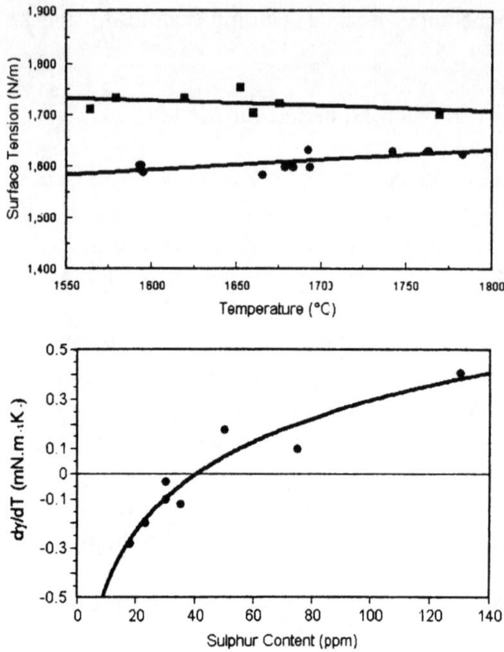

Figure 17 (a) Surface tension of stainless steels with low sulphur (30 ppm) giving poor penetration and high sulphur (80 ppm) and good penetration and (b) dependence of dγ/dT on sulphur content.

3.1.4 Effect of various elements on weld penetration

Elements can be classified into the following three classes:

(i) *Surface-active* elements (e.g. S, O, Se, Te), which affect the magnitude and direction of the fluid flow.

(ii) *Reactive* elements (e.g. Ca, Ce, Al), which react with the surface-active elements and thereby reduce the concentrations of soluble O and S.

(iii) *Neutral* elements which have little effect on the fluid flow in the weld pool.

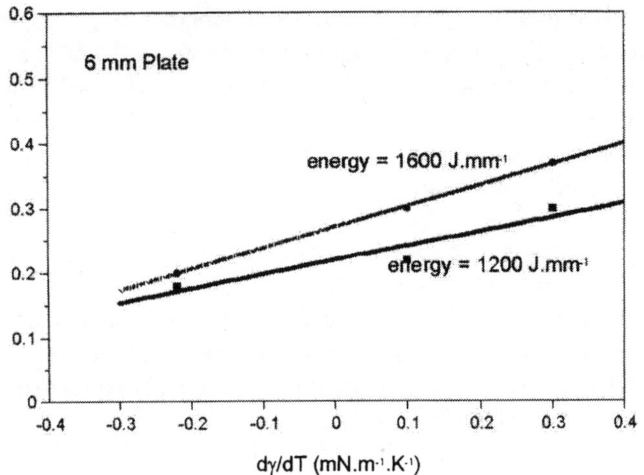

Figure 18 Penetration (D/W) as a function of (dγ/dT) for 6 mm-thick plates

(a) *Surface-active elements*

(i) *Oxygen*

Although oxygen is almost as surface active as sulphur, Robinson and Gould [37] have shown that it does not always have as great an effect on weld penetration as sulphur. It should be noted that it is the concentration of *soluble* O or S, denoted $\underline{O}$, or $\underline{S}$, which affects the surface tension, since the *combined* oxygen (in the form of oxides) has little effect on the surface tension. However, it can be seen from Figure 2a,b that the concentrations of scavenging elements, such as Al, in steel will hold the soluble $\underline{O}$ concentration below 10 ppm, whereas they do not have the same effect on the soluble $\underline{S}$ unless the steel contains large concentrations of Ca or Ce, which is rare. Thus, $\underline{O} \ll O_{total}$ and $\underline{S} \sim S_{total}$ and therefore it is the sulphur and not the oxygen which has the greatest effect on the weld penetration.

(ii) *Sulphur*

Recent work on the effect of sulphur (Shirali and Mills [38]) on weld penetration was carried out on both high-sulphur (HS) and low-sulphur (LS) steels using various doping techniques; the results are shown in Figure . It was found that for all the steels, an increase in S content produced an increase in the depth/width (D/W) ratio. These results are in essential agreement with Heiple-Roper theory and with results obtained by other investigators.

(b) *Reactive elements*

It can be seen from Figure 2a that Al additions in excess of 20 ppm will react with the soluble oxygen present in the steel to form Al_2O_3 and thus reduce the soluble $\underline{O}$

concentration to a very low level. Under these conditions ($d\gamma/dT$) would become negative. Consequently, the thermocapillary forces would be expected to produce a radially outward surface flow resulting in a reduction of the (D/W) ratio of the weld. Calcium and cerium behave in a similar manner. Thus, providing there are sufficient amounts of these elements to react with the oxygen present, the soluble $\underline{O}$ concentration in steels will be below 5 ppm.

The Ca, Ce, La will react with soluble $\underline{S}$ to give their respective sulphides and reduce the $\underline{S}$ to very low levels, but this is not the case for Al (which is sited near to the Zr curve in Figure 2b). Thus, providing the steel does not contain significant levels of Ca, Ce, La, the soluble $\underline{S}$ concentration will be only slightly less than the total S content. This is the reason why penetration can be readily correlated with S content but is less readily correlated with total O content.

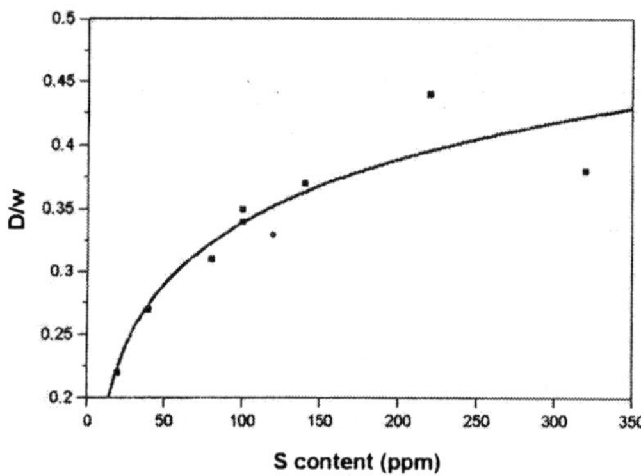

Figure 19 Effect of sulphur content on the (D/W) ratio of TIG welds [38].

3.1.5 Effects of 'slag spots' and oxide films

'Slag spots' are formed by the floating of non-metallic inclusions on the surface of the metal and they are produced by reactions of the metal with O and S. Pollard [39] showed that they attract the arc and reduce the size of the anode root and thus increase the current density. In casts with S > 50 ppm the fluid flow will be radially inward and the slag will be sited in the centre of the pool; thus the resultant high current density will result in increased temperature gradients and, consequently, better penetration. However, in LS casts the flow will be radially outward and the slag spot will be swept to the periphery of the pool. The consequent attraction of the arc will result in deeper penetration at the edge of the pool, which leads to an erratic weld seam (see Section 4.1.8).

When the steel contains significant concentrations of Ca (greater than 20 ppm), the Ca forms an oxide film on the edge of the weld pool. These surface films tend to suppress surface flows and thus produce stagnant regions at the edges of the weld pool.

3.1.6 Effect of welding parameters on weld penetration

Burgardt and Heiple [40] pointed out that since Marangoni forces are usually dominant in the weld pool, the effects of altering welding conditions can be explained in terms of what effects these changes would have on the temperature gradient (and hence the strength of Marangoni forces operating in the weld pool). Thus any change which brings about an increase in temperature gradient would cause increased penetration in high S (HS) casts and reduced penetration in low S (LS) casts. Although this proposition ignores the effect of welding parameters on the other forces operating in the weld pool, these workers did show that it could account for their observations.

Mills and Keene [3] analysed the effect of changes in the welding parameters, such as arc length, welding speed (S_w) and current (I) and voltage (V), etc, on all four forces affecting the fluid flow. The thermocapillary forces are effected by the temperature gradient ($d\gamma/dx$), which is related to the power input (IV). The travel speed (S_w) affects the rate of heat input to the weld and it is this quantity which controls ($d\gamma/dx$). Consequently, the effect of welding parameters on the Marangoni forces can best be studied from measurements of the heat input per unit length of weld. This parameter is referred to as the linear energy and is defined as (IV/S_w).

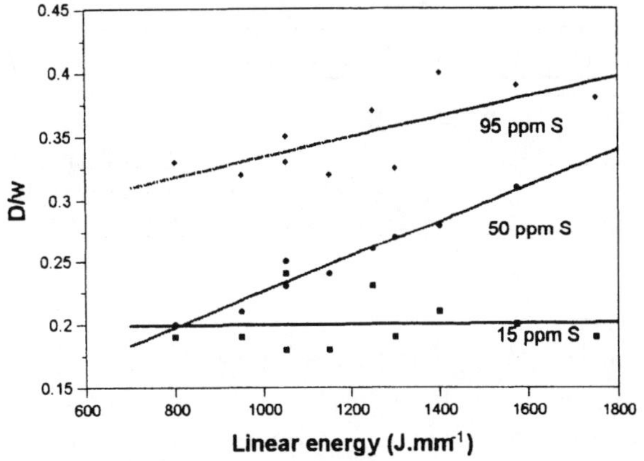

Figure 20 The (Depth/Width) ratio of a weld as a function of the linear energy (heat input per millimetre of weld length).

The linear energy allows the effect of current and welding speed on the Marangoni forces to be taken into account simultaneously but it should be noted that it does not account for either increased Lorentz forces with increasing current or the increased front-to-back motion in the weld pool at high welding speeds. The effect of increased linear energy on high- (HS) medium- (MS) and low-sulphur (LS) steels is shown in Figure 20. It can be seen that increases in the linear energy resulted in increased penetration for HS and MS casts but have little effect on the LS casts. Thus these results are in essential agreement with Burgardt and Heiple's [40] proposition that the effect of welding parameters on penetration can be explained in terms of their effect on the temperature gradients.

3.1.7 Off-centre welding

Tinkler et al [41] showed that when welding a 30 ppm sulphur (LS) plate to a 90 ppm sulphur (HS) plate, the resulting weld was off-centre and displaced towards the LS side. This can be accounted for if it is assumed that Marangoni forces dominate the fluid flow in the weld pool. It can be seen from Figure 21 that the thermocapillary forces in the LS and HS will be from left to right and the diffusocapillary forces will also operate from left to right. Thus these surface flows will cause hot metal to be carried to the LS side and melt back off the steel will result in an asymmetric weld.

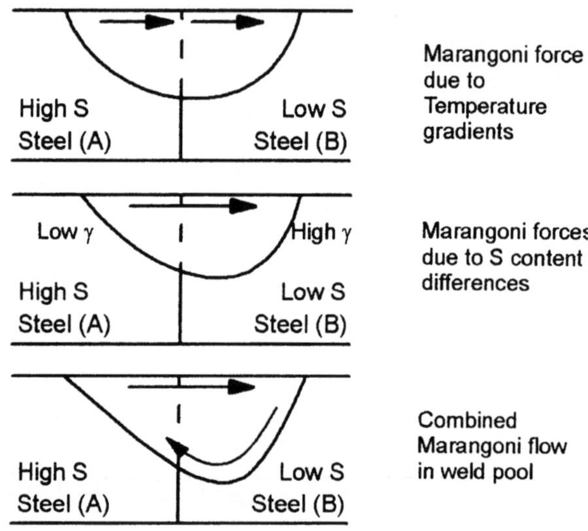

Figure 21 Schematic drawings showing the formation of a non-axisymmetric weld when welding steels have different sulphur contents: (a) Marangoni force due to temperature gradients; (b) Marangoni forces due to S content differences; (c) combined Marangoni flow

3.1.8 Arc wander

It has been mentioned above that certain casts of steel exhibiting poor weld penetration tend to give an erratic weld seam. This is known as 'arc wander' and can be seen inFigure 22. It frequently occurs with steels containing greater than 20 ppm Ca. It is caused by slag spots attracting the arc. For radially inward flows the slag spot will be centred in the centre of the pool and the attraction of the arc will result, sequentially, in a greater current density (because of the smaller anode root) and deeper penetration. With a radially outward flow the slag spot will be swept to the edge of the pool, since it attracts the arc, the hottest part of the pool will become the region close to slag spot and thus the position of the weld seam will change.

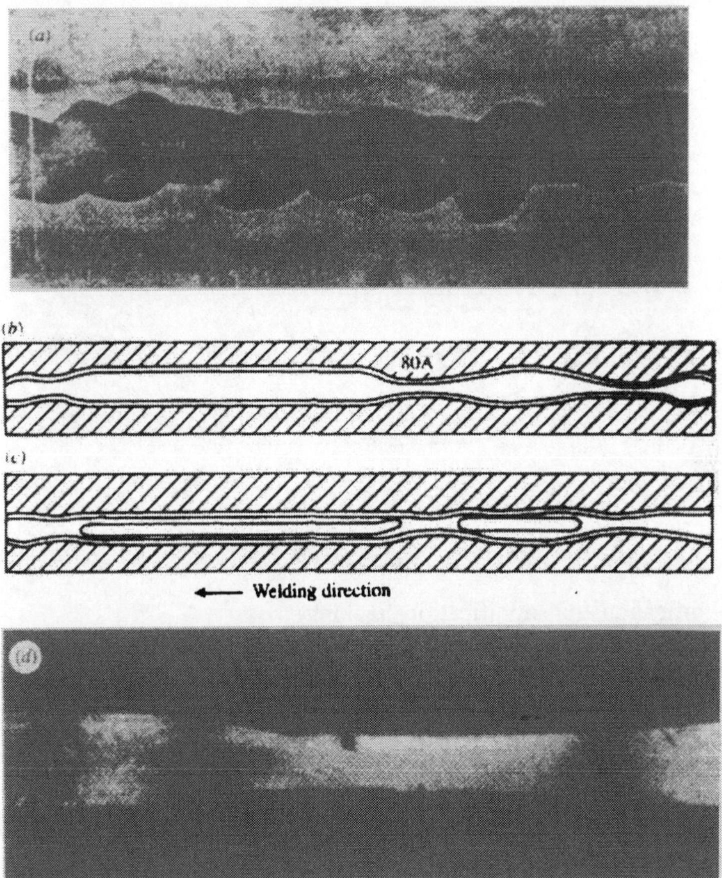

Figure 22 Examples of arc wander: (b) and (c) show the front and back faces of a full penetration weld; (d) is a magnified section of (b) showing presence of slag spots.

3.1.9 Porosity

Poor weld penetration is often accompanied by porosity. Kou and Wang [42] proposed that the direction of the flow in the weld pool could be responsible for the presence of pores in the weld. When the flow is radially inward the weld pool motion will assist the escape of bubbles away from the solidification front (Figure 23a). In contrast, when the flow is radially outward the bubbles will tend to be swept towards the solidification front (Figure 23b). It is obvious that the presence of a solid-slag film at the rear of the pool will tend to create a stagnant region which will assist the entrapment of gas bubbles by the solidification front, thus it is not surprising that porosity problems are encountered when welding steels with high Ca levels.

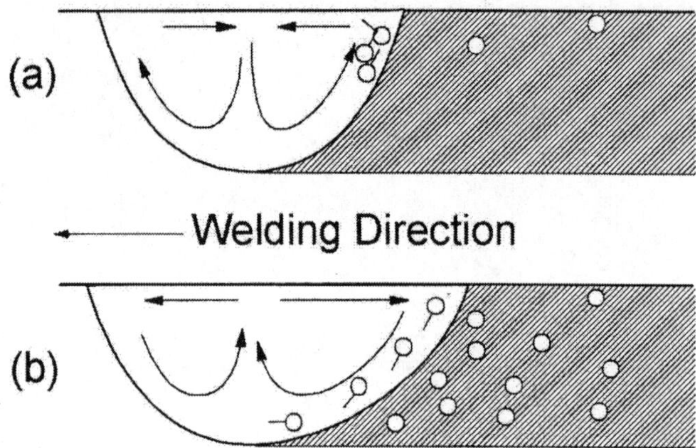

Figure 23 Influence of weld pool motion on porosity: (a) inward surface flow; (b) outward surface flow.

3.1.10 Solidification and solidification cracking

Several workers have proposed that Marangoni convection plays a part in the fluid movements of the interdendritic liquid or in bubbles entrapped at the interface during solidification [43]. However, Marangoni flow will tend to be suppressed where the liquid at the interface has a high viscosity and on this basis it is difficult to see how Marangoni flow could affect inter-dendritic fluid flow.

Hot cracking of welds results from the presence of low melting liquid during the solidification which allow boundaries to separate when subject to shrinkage stresses. The formation of dendrites during solidification hinders the feeding of liquid metal and hot tearing occurs when the rate of shrinkage is greater than the rate of feeding (of liquid metal). Holt et al [44] proposed that the direction of Marangoni flow affected the rate of feeding and hence hot cracking.

For the liquid metal in the dendritic region

(i) the temperature of the tip is higher than that at the root

(ii) the S content tends to accumulate in the liquid metal (and not the solid).

It is usually considered that Marangoni forces would be eliminated for a solid free surface. Thus, for the above mechanism to apply, it is necessary to have a bubble at the free surface. Although bubbles are generated near the solidification front the cause proposed is unlikely to be a major mechanism.

3.1.11 'Humping' and 'undercutting'

'Humping' is the formation of a raised section in the centre of the weld and 'undercutting' is the depression at the edge of the weld (Figure 24). 'Humping' and 'undercutting' were found to be prevalent for high sulphur (HS) casts and when using high travel speeds. One characteristic of Marangoni flow is that the surface is raised in regions where the liquid is being driven downwards and depressed where the flow is upwards (Figure 24). Thus for HS casts with a radially inward flow, it is obvious Marangoni flows can account for both humping and undercutting. However, on the basis of the flow patterns shown in Figure 24 it is difficult to account for the undercutting in LS casts.

Gratzke et al [45] rejected the thermocapillary mechanism and proposed that humping and undercutting were caused by Rayleigh instability, i.e. the break-up of a liquid cylinder by the action of surface and gravity forces. They concluded that (i) the (width/length) radio of the pool was the most important factor and (ii) the surface tension does not affect the onset of humping, only the kinetic behaviour which is a function of $(\rho/\gamma)^{1/2}$. The latter conclusion would seem to be inconsistent with the observation that it was prevalent in HS casts.

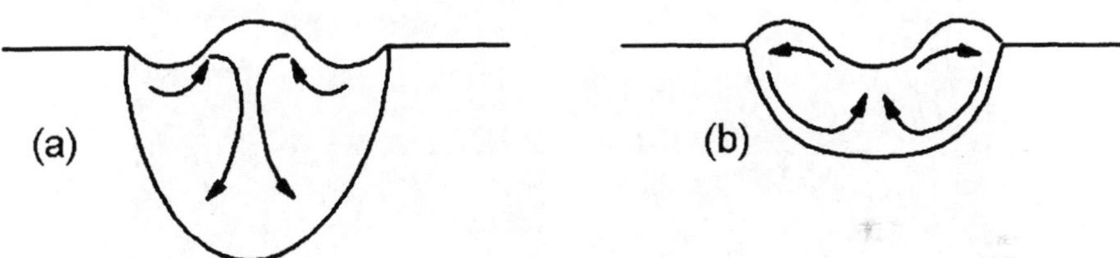

Figure 24 Predicted surface profiles of weld showing humping and undercutting: (a) inward surface flows with downward flow in centre; (b) outward surface flows with downward flow at periphery of weld pool.

3.1.12 Surface rippling

The surface of weld pools produced by low sulphur (LS) casts tend to be flat and placid in contrast to the surfaces of HS casts which tend to be turbulent and agitated. The solidified welds of both HS and LS alloys exhibit regularly spaced fine ripples, possibly caused by the oscillation frequency of the arc or laser. However, for the HS casts there is a series of deeper ripples of longer wavelength superimposed on the fine ripple background. In full penetration welds the coarse ripples were seen on both free surfaces and pulsation of the laser would not be expected to affect the back surface. If the rippling is associated with the surface properties of the steel melt, then it must be related to a low surface tension and a positive $(d\gamma/dT)$. In the weld pool there are also thermal gradients in the direction perpendicular to the surface. Although these gradients do not produce any substantial fluid flow in the weld pool they can produce thermocapillary instabilities. These instabilities arise when a metal with a negative ($d\gamma/dT$) is heated from below or a metal with positive ($d\gamma/dT$) is heated from above and which give rise to capillary waves [46]. The thermocapillary forces acting parallel to the surface amplify these capillary waves and produce instabilities. It also explains whey they only occur in HS casts since they have positive($d\gamma/dT$) coefficients and welding is usually carried out by heating the upper surface.

3.2 GAS-METAL ARC WELDING (GMA)

This process is similar to GTA welding but a filler metal is used. Takasu and Toguri [47] have shown that there are four forces affecting the fluid motion (Figure 25) in the pool when the molten filler metal drop hits the molten pool, namely: (i) a stirring force due to the momentum of the drop; (ii) a buoyancy force related to the density difference between the drop and the pool; (iii) a 'curvature' force related to the surface tension normal to the surface; and (iv) the Marangoni force related to the difference in surface tension of the drop and pool. Takasu and Toguri showed that when (a) $\gamma_{drop} > \gamma_{pool}$ the droplet penetrated into the pool and (b) $\gamma_{drop} < \gamma_{pool}$ the drop will spread out over the surface.

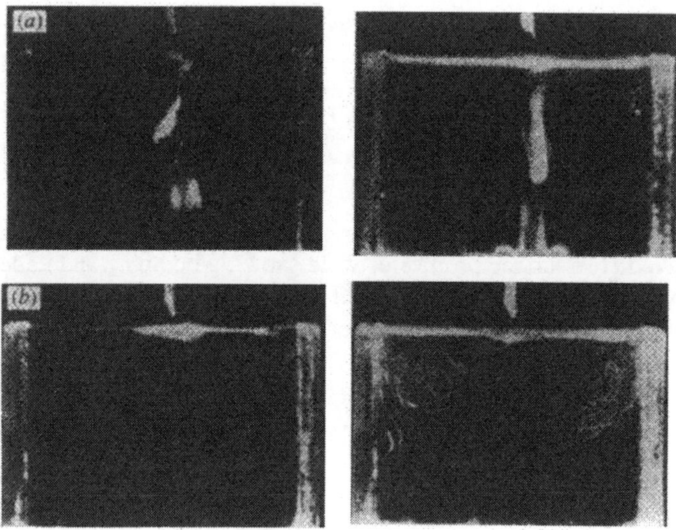

Figure 25 Behaviour of a droplet falling into a liquid pool, and the resultant fluid flow

3.3 CERAMIC-METAL JOINING

The brazing of a metal to a ceramic (say Si_3N_4) is achieved using a braze based on Sn or Cu/Ag alloy. However, the contact angle between Si_3N_4 and the braze is > 140° i.e. non-wetting. In high temperature systems wetting is usually associated with reaction. Consequently, in order to provide strongly wetting conditions ($\theta^o = 20^o$) it is common practice to add 1.5% Ti to the braze since this will react with the Si_3N_4 to produce silicides. This decrease in contact angle is accompanied by increases in both the work of adhesion (W_A) and metal-ceramic bond strength.

4 INTERFACIAL EFFECTS IN IRON- AND STEELMAKING

4.1 IRONMAKING

4.1.1 Bath smelting

Bath smelting has attracted considerable interest in recent years since it was realised that coke for the blast furnace may not be available at reasonable cost in the future because of environmental concerns. The HI smelt, AISI and DIOS processes all involve (i) the injection of coal, pre-reduced iron ore and oxygen and air and (ii) the production of a slag foam containing Fe-C droplets. The latter provide huge (surface area/mass) ratios [48] and consequently, high reaction rates which lead to high production rates. The factors affecting foam formation are given in a previous section on formation of foams.

The coal particles pyrolise producing volatiles and char and dissolve in the molten iron. The iron oxide dissolves in the slag and forms a foam and is reduced by:

(i) reaction of the iron oxide with C in iron droplets ($FeO + C = Fe + CO$)

(ii) gas/metal reactions ($C + CO_2 = 2CO$)

(iii) by reaction with the char

(iv) by reaction with C in the molten iron melt.

Belton and Fruehan showed that the rate of the reactions (i) and (ii) increased as the S content of the iron decreased (Figure 26) (34). This is usually attributed to the S atoms occupying most of the interfacial sites and thereby blocking the sites available for the reaction.

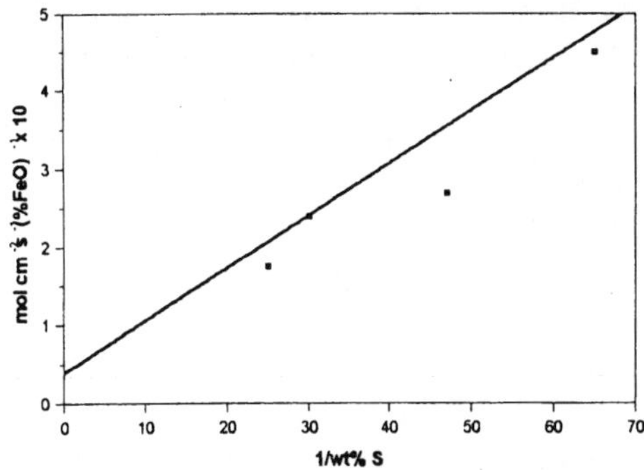

Figure 26 Rate of reaction as a function of reciprocal of the sulphur conent for the

reaction of Fe-C droplets with slag at 1400 °C.

The rate of dissolution of carbon in iron is also affected by the S content, the rate decreasing with increasing S content. The adsorption of S on the surface of the carbon has been proposed as the limiting step in the dissolution of carbon [49,50]. Belton and Fruehan pointed out there was no evidence to support the strong adsorption of S on carbons and suggested that the effect of S was due to the reduction of the work of adhesion W_A (bond strength) with increasing S content (34).

4.1.2 Iron carbide production

Iron carbides are being considered as an alternative iron source for charging into the Electric Arc Furnace. The process involves the reaction of iron oxides with methane in a fluidised bed. It has been found that H_2S in the gas has the beneficial effects of suppressing carbon precipitation [51]. It has been suggested [51] that sulphur prevented carbon precipitation by balancing two effects viz (i) reducing the reaction rate by surface blockage and (ii) promoting the formation of iron with a fibrous and porous structure, which provides increases in surface area and, consequently, reaction rate. Whatever mechanism obtains, the fact that sulphur is responsible for suppressing carbon formation strongly suggests that interfacial properties are involved.

4.2 STEELMAKING

4.2.1 Basic-oxygen steelmaking (BOS)

The creation of a stable slag foam containing metal droplets in top-blown BOS processes provides, sequentially, a high (surface area/mass) ratio of steel in slag, rapid decarburisation and dephosphorisation reactions and high productivity. Slag foaming is particularly vigorous when (i) the bubbles are small (< 2 mm diam) and (ii) the slags contain higher concentrations of surface active components such P_2O_5, CaF_2 (and to some extent SiO_2). Olette [52] and Kozakevitch [53] have proposed that 'slopping' probably occurs at the stage of the BOS process where the slag composition has a high solidification temperature (< 1650 °C) and produces a large number of solid particles. The factors affecting foaming are given in Section 2.6.

However, modelling studies [55] have indicated that, for the high gas velocities used in BOS, (i) "expanded foam" produced differs from the "classical slag foam" studied in the laboratory and (ii) "expanded foam" is dependent upon gas velocity and interfacial properties are relatively unimportant.

Kazakov [39] has pointed out that (i) dusts are formed by the evaporation of FeO from the slag and (ii) that when an electric current is applied to the reaction zone of the converter, the FeO droplets take on a charge. When liquid particles take on an electric charge their surface tension is reduced and this results in a reduction in dust formation. A 1 kW source was attached to the tuyeres and insulated from earth and metallic structures and voltage was maintained at < 40 V. Resulting dust formation was halved when the tuyere was connected to the negative pole.

4.2.2 Electric Arc Furnace (EAF)

Foaming slags are preferred to exposed arcs in EAF since they provide the following advantages:

(i) more stable arc conditions since it provides (a) a longer arc, (b) lower radiation losses to the furnace walls, resulting in higher electrical and thermal efficiency;

(ii) a cover from the atmosphere which results in lower nitrogen contents for the metal;

(iii) lower dust contents.

Stable foams are produced by (a) reducing the surface tension of the slag (usually by P_2O_5) and (b) by using a slag with a high viscosity (usually by ensuring the liquid slag contains solids) to increase drainage times. Additions of SiO_2 (which is mildly surface active and increases viscosity) have been used previously to improve foam stability but are not particularly beneficial. Probably a better way on ensuring a stable foam is to operate with either a high P_2O_5 concentration or a composition with a high solidification temperature providing solid particles to stabilise the foam.

4.3 LADLE REFINING

Ladle refining involves the following: (i) addition of the alloying elements, (ii) homogenisation of the steel, (iii) degassing, deoxidation and desulphurisation of the steel, (iv) removal of inclusions. Interfacial phenomena play a central role in the latter two functions.

4.3.1 Kinetics of refining reactions

The presence of surface active elements at the interface, such as sulphur, tends to block (or "poison") interfacial reactions. The classical case is the kinetics of nitrogenisation of iron where it has been shown that the rate is inversely proportional to the S content of the iron.

Gaskell and Saelim [56] reported that the rate of desulphurisation was greater than that calculated from the diffusion of S in Fe and attributed this enhanced rate to Marangoni turbulence at the interface. However, deoxidation and desulphurisation reactions will lead, sequentially, to rapid mass transfer, a low, dynamic interfacial tension, emulsification, and very rapid kinetics. However, emulsification will also lead to a higher inclusion content in the steel.

4.3.2 Inclusion removal

Inclusions can arise from varied sources, (i) through oxidation reactions (e.g. $2Al+3\underline{O} \rightarrow Al_2O_3$ or $3/2\ SiO_2 + 2\underline{Al} \rightarrow Al_2O_3 + 3/2\underline{Si}$), (ii) from slag emulsification and entrapment, (iii) from ladle 'glazes' (formed on the refractory during draining of the metal when the descending slag comes in contact with the refractory and glazes it; this glaze is detached into the metal flow when a new cast is poured into the ladle).

The thickness of the "glaze" and the infiltration into the pores of the capillary lining have been reported [57] to be functions of the surface tension and contact angle of the slag.

The removal of inclusions and gas bubbles from liquid steel is an important process in the ladle, tundish and continuing casting mould. The principles of inclusion removal are summarised in Section 2.4

4.4 NOZZLE CLOGGING

Deposition of Al_2O_3 or TiO_2 on the inner walls of the submerged entry nozzle (SEN), when casting Al or Ti killed steel grades can contribute to significant clogging which prevents a smooth steel flow into the continuous casting mould. Their most detrimental effect is the release of the clog which may lead to sticker breakouts in the mould or entrapment of large inclusions in the steel. It is not known for certain whether the clogs nucleate at the SEN walls or if pre-existing oxides agglomerate and stick at the walls but centrifugal forces will assist clogging. What is certain however, is that once the oxides reach the nozzle wall (or form there) there has to be enough interfacial attraction to hold them there. Calculations show that the attractive forces acting on a micron sized inclusion at a particle/refractory interface is about an order of magnitude greater than the drag force acting on the particle [58]. The clogs resume a coral like structure with no metal inside them. The ejection of molten steel from within the clog structure is governed by a balance between forces of static pressure and interfacial tension between steel and the clog. The critical distance, between particles constituting the clog, necessary to start the ejection of molten steel can be calculated with the following equation that simplifies the geometry to the space between three particles in a plane [58]:

$$-\gamma_m \cos(\theta)\,\frac{7.3}{rP} = 0.4 + \frac{4\delta}{r} + \left(\frac{\delta}{r}\right)^2 \tag{12}$$

where γ_m is the surface tension of molten steel, θ is the contact angle between molten steel and oxide, r is the particle radius, P is the static pressure of molten steel and δ is the critical distance between the particles necessary to start ejection of the molten steel. Boron nitride has been used to coat the inside of SENs to minimise clogging. Alternatively, a basic internal coating has been used to produce low-melting calcium aluminates which are liquid at the pouring temperature.

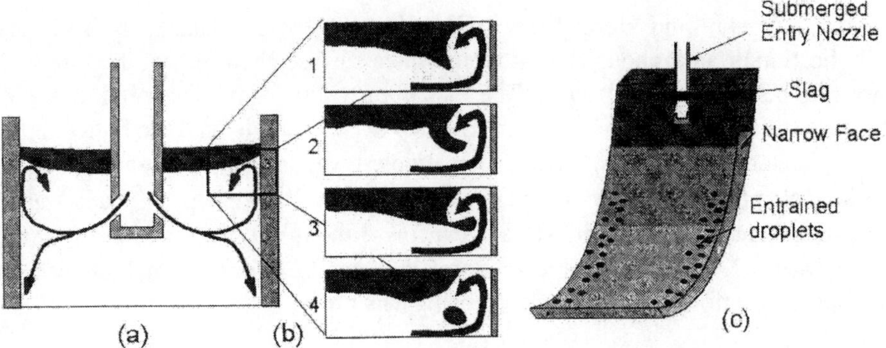

Figure 27 Schematic drawings illustrating (a) metal flow in the mould, (b) entrainment of slag droplets in the metal flow and (c) location of surface inclusions in the slab.

4.5 CONTINUOUS CASTING

In continuous casting various defects such as "blister", "pencil pipe" and "slivers" occur through the entrapment of gas bubbles, inclusions and mould flux on the newly-solidified steel meniscus. These defects are more likely to occur when the metal flow in the mould is excessively turbulent (Figure 27). However, in the case of slag entrapment the detachment of slag is more likely to occur when the metal/slag interfacial tension (γ_{ms}) is low [59]. Several mould powders have been formulated to provide high values of γ_m, these usually involve the removal of Na_2O and any NaF from the flux. However, γ_{ms} also involves the surface-tension of the metal (Equation 1) and since a high surface tension value would increase γ_{ms} thus a decrease in the sulphur content would be an equally effective way of minimising slag entrapment.

It has been reported that there was some lowering of the dynamic interfacial tension γ_{ms} when a molten, 321-stainless steel was in contact with a mould flux [4] due to mass transfer between the metal and slag phase and this would promote entrapment.

Modelling studies were carried out to determine the effect of wettability of the metal on the SEN refractory on the behaviour of gas bubbles in the mould [60]. Contact angle was found to have a significant effect with non-wetting conditions leading to the coalescence of bubbles, deep penetration of bubbles and a fluctuating flow [60].

4.6 STRIP CASTING

Strip casting involves the rapid solidification of molten steel between two rotating, cold rolls (Figure 28). It has been pointed out that it is essential to have good surface quality of the strip because (surface area/mass) ratio is at least 100 times greater than that in slab casting and this requires a better understanding of the interfacial heat transfer. Stresov and Herbertson [61] also noted the similarities with the rapid solidification of amorphous alloy ribbons where surface chemistry of the substrate is known to be an important factor. The "splat" formation of molten metal droplets on stainless steel substrates [62] has been studied experimentally and the flattening ratio (ζ = diameter of solidified disc/diameter droplet) was found to be functions of both the Reynolds number and Weber number, $We = \rho_L D v^2/\gamma$ where ρ, D, v and γ are the density, diameter, velocity and surface tension of the metal droplet. They found $\zeta = C.We^j$ where $C \approx 0.54$ and j varied between 0.29 and 0.36 for the metals studied, thus the flattening ratio increases with decreasing tension.

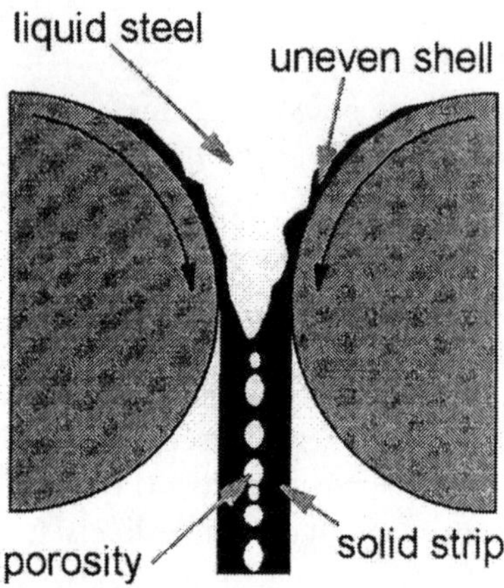

Figure 28 Schematic representation of strip casting process.

Recent work [63] has indicated that when a hot droplet is pressed against a cold substrate the liquid does not wet the substrate since thermocapillary (Marangoni) flows are established in the drop and induce a flow of air into the contact region which prevents wetting of the substrate (Figure 29).

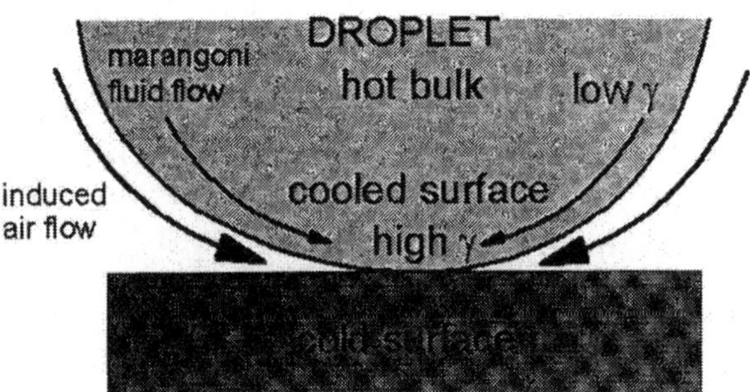

Figure 29 Schematic representation of the flow of air produced when a hot drop is brought into contact with a cold substrate

Obviously, in strip casting the metal stream would have much greater momentum than the drops in these experiments. However, thermocapillary forces could still be created and induce airflow. Inspection of Figure 29 shows that these conditions only apply to low S steels (< 40 ppm, at 1700 °C and ca 10 ppm S at 1500 °C). For steels with higher S content the thermocapillary flow (and the induced air flow) would be reversed and thus wetting would occur. On the basis of this argument the heat transfer would be expected to be higher for high S casts than for low S casts. This is the case in practice.

5 INTERFACIAL EFFECTS IN NON-FERROUS METAL PRODUCTION

6.1 ALUMINIUM PRODUCTION

6.1.1 Effect of electrocapillarity on efficiency of electrolysis

The Hall-Heroult process is used to produce aluminium (Figure 30). It uses an electrolyte of $3NaF.AlF_3$ (cryolite) + CaF_2 + AlF_3 and about 3-8% Al_2O_3 is dissolved in the cryolite, The Al_2O_3 is the source of the aluminium. The process uses a carbon cathode and a consumable carbon anode.

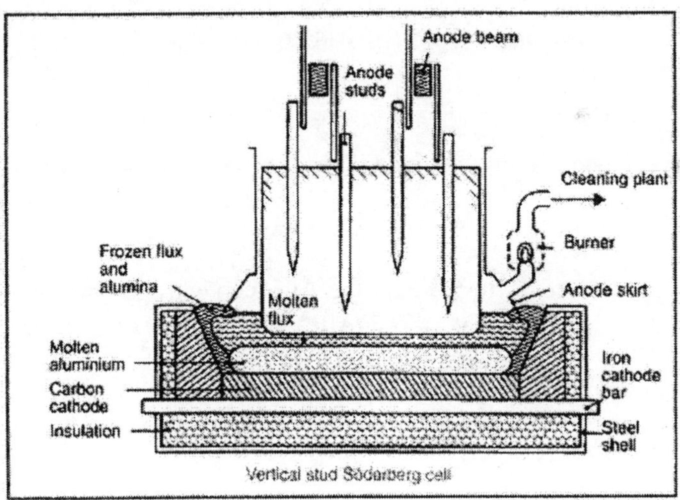

Figure 30 Schematic representation of the Hall-Heroult process

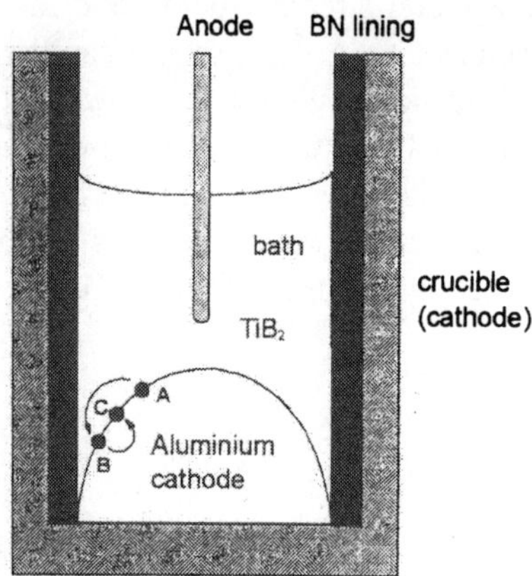

Figure 31 Schematic diagram showing the experiments carried out by Takasu and Toguri [47].

Reactions:

$$2Al_2O_3 \rightarrow Al^{3+} + 3AlO_2^- \quad (13)$$

$$\text{Cathode: } Al^{3+} + 3e- \rightarrow Al \quad (14)$$

$$\text{Anode: } 3AlO_2^- \rightarrow 3Al + 3O_2 + 3e- \quad (15)$$

$$3O_2 + 3C \rightarrow 3CO_2 \quad (16)$$

A back reaction can occur at the cathode and reduces the efficiency of the process.

$$Al \rightarrow Al^{3+} + 3e- \quad (17)$$

Thus the back reaction tends to increase the AlF_3 content and thus decreases γ_{ms}, (Figure 32a), since it is known that as (NaF/AlF_3) ratio increases γ_{ms} increases.

Takasu and Toguri [47] used the apparatus shown in Figure and used TiB_2 particles to trace the movement of the aluminium.

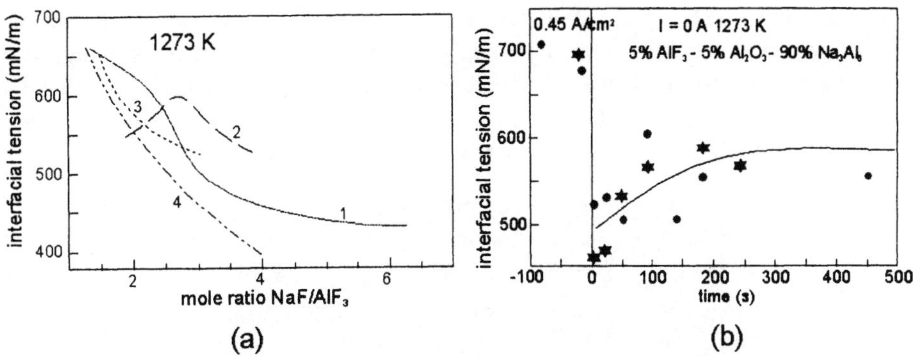

Figure 32 Interfacial tension as a function of (a) (NaF/AlF3) ratio and (b) time [47].

It was found that when a positive potential (Al cathode) was applied the apex of the Al was observed to increase in height. When the potential was reversed i.e. a negative potential (Al anode) then the apex was found to decrease.

When the potential was interrupted then the interfacial tension, γ_{ms}, was observed to decrease and then subsequently showed a gradual increase (Figure 32b).The current density is high in the centre and low at the walls. It was noted that when a positive potential was applied the particle moved from the centre to the wall. However, when a negative potential was applied the particle moved from the wall to the centre. This was due to the build-up of AlF_3 in the high current density region and NaF in the low density region. Since NaF increases the interfacial tension, γ_{ms} at wall will be high but will be low in the centre. Marangoni flow will thus be in the direction from the centre towards the wall for a positive potential.

For a negative potential, NaF will be depleted in the wall (low current density) region. Consequently, γ_{ms} will be low but will be high in the centre and thus the motion due to Marangoni forces will be in the direction of wall to centre. Thus the back reactions give rise to electrocapillarity forces causing movement of the liquid aluminium.

5.1.2 Aluminium grain-refining alloys

Although emulsification is desirable since it provides rapid reactions and high productivities, it has a downside since it can lead to entrapment of reaction products in the metal, which is very undesirable. One such case of unwanted emulsification occurs in the production of master alloys of Al containing TiB_2 and $TiAl_3$ which act as grain refiners in Al alloys. The master alloy is prepared by adding K_2TiF_6 and KBF_4 to the Al where it reacts to form TiB_2, $TiAl_3$ and a mixture of KF and AlF_3. If pure K_2TiF_6 and KBF_4 are used, emulsification of KF-AlF_3 melt occurs in the Al alloy (Figure 33) which is undesirable [64]. However, if the K_2TiB_6 and KBF_4 contain as little as 40 ppm Ca then there is no emulsification, i.e. no liquid fluoride drops are formed in the Al master alloy. The extremely low levels of Ca required to overcome emulsification indicate that this must be an interfacial phenomena. It has been suggested that the Group VII elements are highly surface active (although it is difficult to keep fluorine in the metal) and thus one possible explanation is that the Ca in the Al significantly reduces the soluble fluorine concentration (F%) in Al and thereby increases the interfacial tension between Al and the fluoride flux.

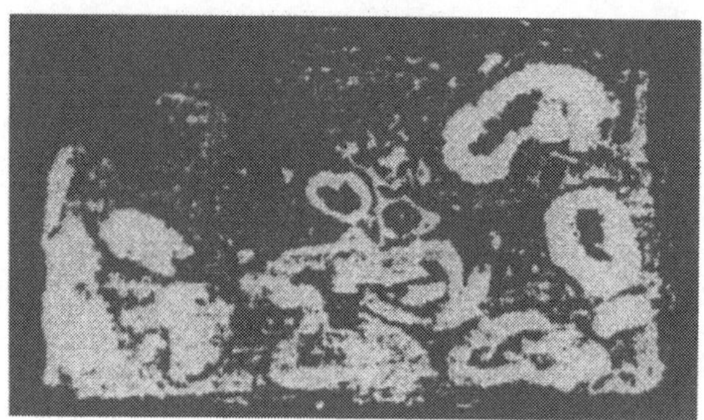

Figure 33 Photograph of a sectioned sample showing fluoride emulsion in Al master alloy containing TiB2 + TiAl3

5.2 METAL MATRIX COMPOSITES (MMCs)

MMCs are used to combine metal strength, increased stiffness and lightness and are widely used in sporting goods etc. MMCs consist of oxide or other fibres in a matrix of metal e.g. Al. The mechanical strength of bonds formed between metal and ceramic phases can be enhanced by improving the wettability of the metal phase i.e. increasing W_A. The contact angle of, say, Al on Al_2O_3 is $> 90^o$ giving poor bonding. Consequently, in order to improve the bonding Mg is added to the Al and this reacts with the alumina fibres to form spinel ($MgO.Al_2O_3$) and this reactivity results in better wetting and improved bonding of the fibre.

5.3 REFINING OF NICKEL-BASED SUPERALLOYS

In the electroslag refining (ESR) process an impure alloy electrode is heated in molten slag, it melts and the metal drops fall through a molten slag layer and forms a molten pool. The process provides fast kinetics for reactions like desulphurisation and the long residence time of the molten pool (before solidification) allows inclusions to float out. The observations of Takasu and Toguri [47] that the metal droplet does not penetrate far into the liquid when $\gamma_{pool} > \gamma_{drop}$ (Figure 25, Section 4.2) would assist the removal of inclusions. Since desulphurisation occurs during the descent of the drop it is probable that $\gamma_{pool} > \gamma_{drop}$ and thus the inclusions will remain close to the surface and can thus float out.

5.4 ELECTRON BEAM BUTTON MELTING (EBBM)

It is essential that alloys used in critical applications (e.g. aero engines) should have high-temperature strength. Non-metallic inclusions cause a marked deterioration in fatigue strength, the strength decreasing with increasing concentration, and size of the inclusions. Consequently, the alloys used in these applications must have great cleanness (i.e. as few inclusions as possible). Nickel-based alloys are used in these applications and great efforts are made to get these alloys as clean as possible. It is also important that these alloys should be checked for cleanness before use and conventional techniques (such as metallography) are not entirely successful for these superclean alloys. The electron beam button melting (EBBM) is an established technique for cleanness evaluation in these alloys. The principle of the method is shown in Figure 34. About 1 kg of the alloy is heated with electron beam and dripped into a water cooled mould. The beam is partially diverted to maintain a liquid pool of alloy to allow the inclusions to float to the surface of the pool, where they collect and form a *raft*. When all the alloy has been melted the metal button is allowed to solidify and the raft is subsequently examined in a scanning electron microscope to determine the number and the size of the inclusions.

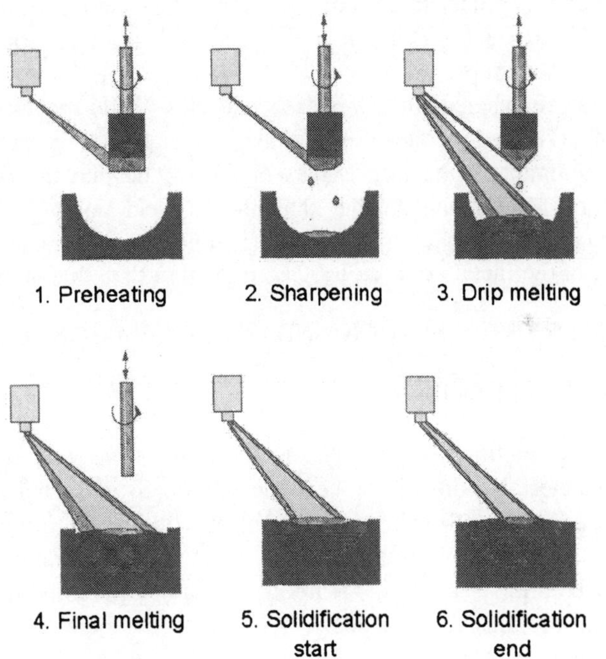

Figure 34 Electron beam button melting process

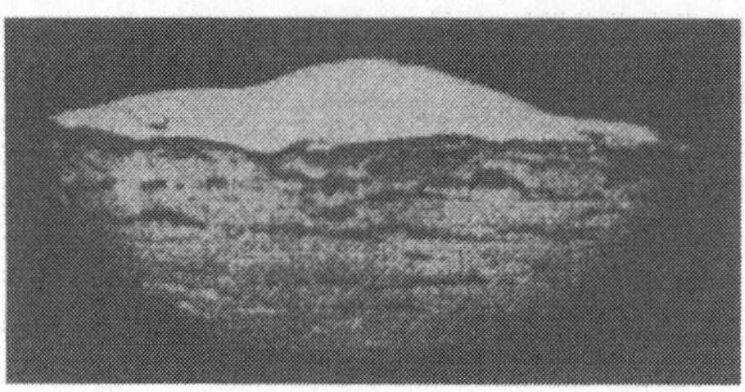

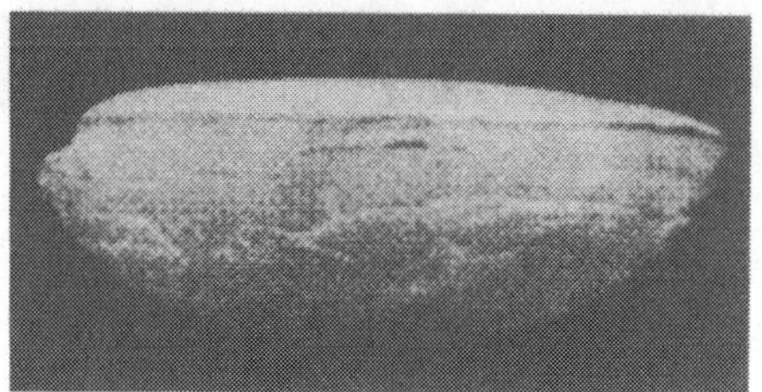

Figure 35 Comparison of the shape of the button and the cap of EBBM button on heats (VIM/VAR) low-S cast (left) and (VIM/ESR) high-S cast (right) of IN 718 under "identical" process conditions.

It was found that some 'clean' casts did not form a raft. Inspection of the buttons shown in Figure 35 indicated that the profile of the low S cast was flat whereas that of the high S cast was 'humped'. This is reminiscent of weld pool behaviour and suggests that thermocapillary forces are operative. Observations of the movement of particles on the surface high and low S casts [64] showed radially-inward and radially-outward movement, respectively, i.e. particles would congregate in high-S casts and form a raft whereas in low S casts particles would move outwards and be distributed around the outer edges of the molten metal pool. Thermocapillary forces appear to be dominant in the later stages of EBBM despite the fact that ($d\gamma/dT$) values (10 K mm^{-1}) are much smaller than in the weld pool (> 100 K mm^{-1}). It should be noted that the critical or 'crossover' composition is around 10 ppm S since the temperature of the metal pool (ca 1500 C) is smaller than that in the weld pool.

5.5 SINGLE CRYSTAL SILICON AND OTHER SEMICONDUCTORS

5.5.1 Floating Zone Method

The floating zone method (Figure 36) is used to refine impure Si and other semiconductor materials. In this process the impure Si is heated in an induction furnace and a seed of pure Si is brought in contact with it. The principle underlying the method is that most impurities decrease the liquidus temperature, consequently, pure material will solidify at a higher temperature leaving the impurities in the liquid phase. Thus the molten silicon tends to solidify on the pure Si seed leaving the impurities behind in the molten pool as the impure Si is gradually passed through the furnace. The products are therefore purer silicon and a 'heel' of silicon with a high concentration of impurities. Obviously if there is vigorous fluid flow in the molten metal the impurities will be churned up into the metal. Thus convection should be minimized. Experiments were carried out in microgravity to minimize the effects of buoyancy-driven-flows but it was discovered that there was little improvement because of the thermocapillary convection. It can be seen from Figure 37 that the temperature at the centre of the molten pool is higher than at the solid/liquid interfaces. The surface tension decreases with temperature so the surface tension is higher at the liquid/solid interface, thus flow will occur, along the surfaces from the centre to the edges (Figure 37b). This results in the formation of four vortices, as shown, which produces fluid motion in the molten pool. However, the transfer of hot liquid from the centre to the edge reduces the temperature gradient and thereby causes a reduction in thermocapillary flow. Consequently, this gives rises to an oscillating flow and vortex formation.

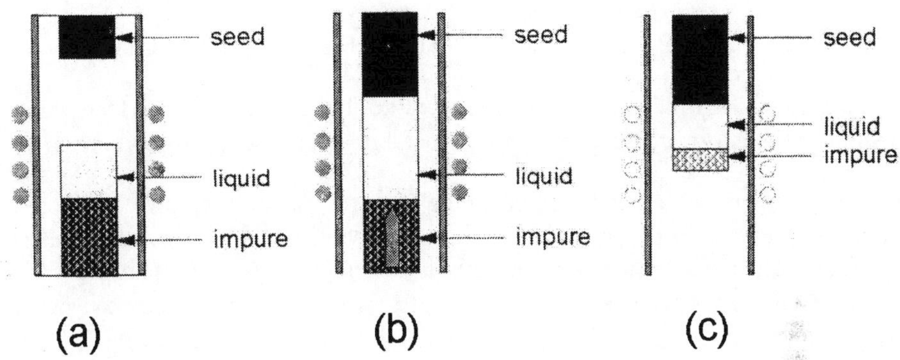

Figure 36 Schematic diagram showing the principles of the floating zone method.

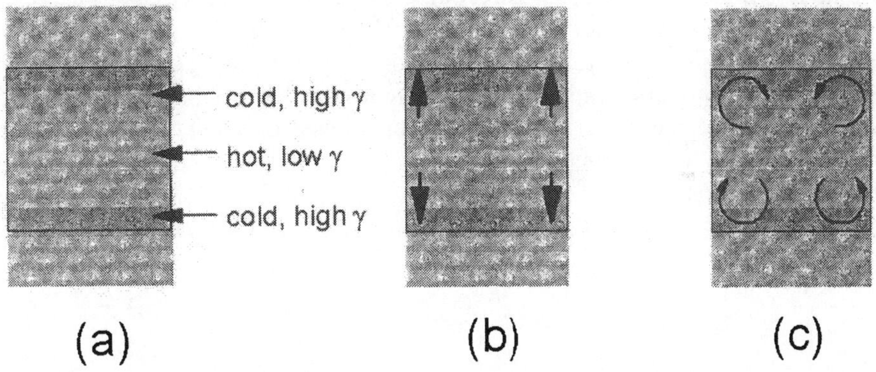

Figure 37 Schematic diagram demonstrating how thermocapillary forces lead to the formation of vortices

One way of minimizing thermocapillary forces is to create a surface with a very high viscosity (a solid surface can be considered to have an infinite viscosity). Thus if the surface of the melt was allowed to oxidize to form an oxide skin on the surface, this would effectively eliminate the Marangoni forces and produce a very quiescent molten pool [65].

5.5.2 Czochralski method

The Czochralski method is used for more than 90% of the silicon single crystals produced and Marangoni flows have a deleterious effect on the purity of the single crystal produced. Hibiya et al [66] pointed out that Marangoni flows do affect single crystal growth in Czochralski method. However, Hibiya et al [66] pointed out that the mechanism in the Czochralski method corresponds to that for a flat surface and arises from diffusocapillary forces arising from oxygen concentration differences between the SiO_2 crucible wall and the bulk. In contrast the Marangoni flows in the floating zone method arise from thermocapillary forces operating in a liquid column.

5.6 METAL POWDERS AND AMORPHOUS RIBBON PRODUCTION

Two applications which might be expected to be dependent upon the factors affecting jet break-up are the production of metal powders and metallic ribbons.

Metal powder production and spray-forming (Figure 38a) both involve the blowing of a metal stream with a jet of gas. The size distribution in both these processes is a key factor and it has been found that this is dependent inversely upon the Weber Number ($We = v2D\rho/\gamma$) where v = velocity of gas, D = nozzle diameter ρ = density and γ = surface tension. Thus the particle diameter has a direct dependence on the surface tension (Figure 8b) and an inverse dependency upon the S content of the alloy.

Amorphous ribbons are formed by directing a stream of liquid alloy onto a water-cooled drum. In the production of metallic ribbons by the melt spinning process some melts have been found to give serrated edges. It has been proposed that these serrations are due to instabilities which eventually lead to jet breakup. Serrated edges were found to occur [12] in metals with high levels of S and O, i.e. melts with low surface tension and Marangoni Numbers. This is consistent these are a result of the break-up of surface waves.

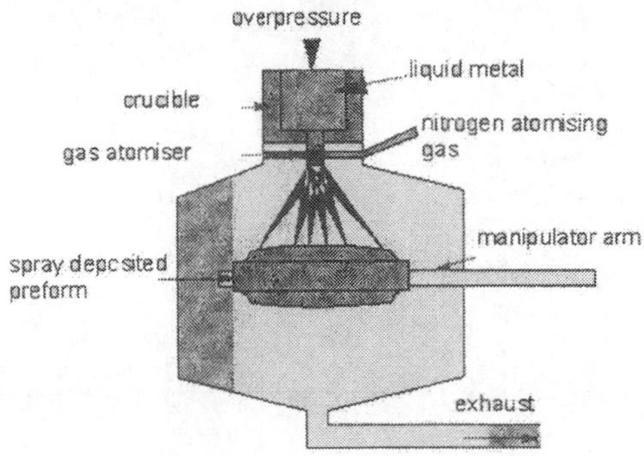

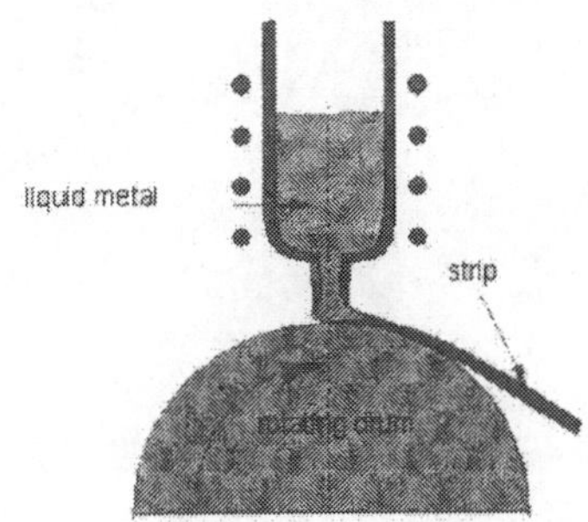

Figure 38 Schematic representations of (a) spray forming production and (b) metallic ribbon production

5.7 DIP COATING OF ZINC ALLOYS

In dip coating processes it is essential that the Zn-Al alloy used to coat the steel substrate should wet the substrate. The temperature of the steel substrate is important, for as can be seen from Figure 39 , the contact angle is >90° when the substrate is 300 °C but the reactive wetting occurs when the temperature of the substrate is heated to 550 °C [67].

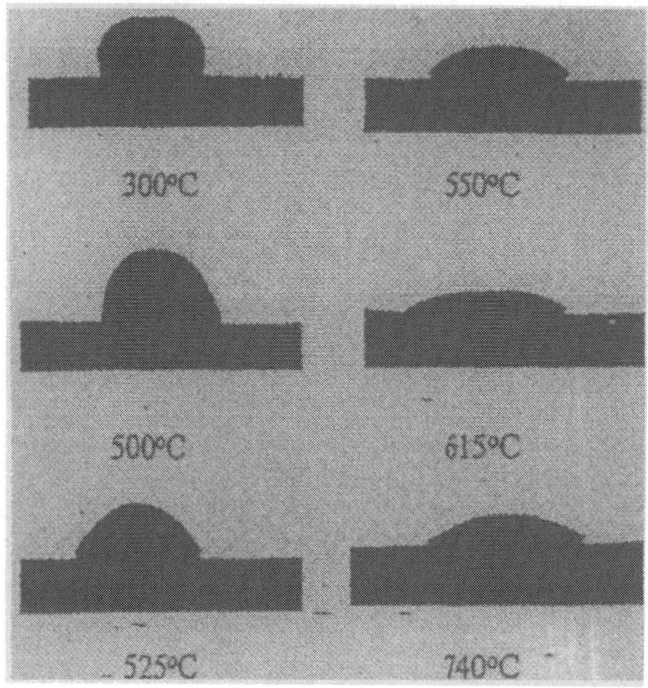

Figure 39 Influence of substrate preheat temperature on the quasi-equilibrium droplet shape [67]

6 SEPARATION OF PHASES

6.1 DISTILLATION COLUMNS

In distillation columns, a substance with a vapour pressure will transfer preferentially into the vapour phase. If the vaporising species is a surfactant, it will result in a local increase in surface tension, thus, it will produce diffusocapillary flows and pull liquid toward it. This, in turn will create better spreading and higher transfer coefficients and frequent replenishment of liquid (see Figure 40a). If, however, the vaporising species results in a decrease in the surface tension, the diffusocapillary flow will be away from the point of vaporisation and this could produce a 'dry spot' (Figure 40b).

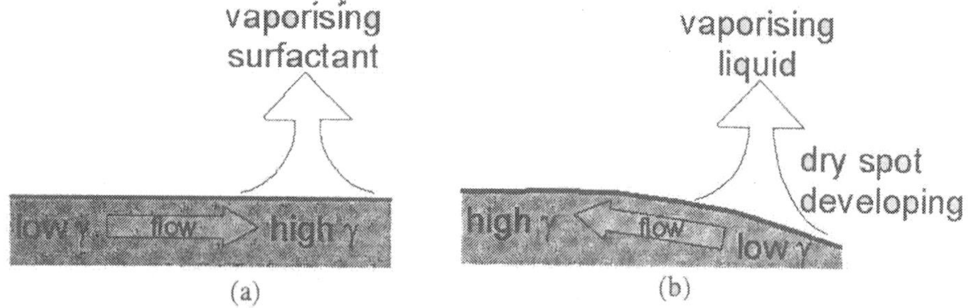

Figure 40 Schematic representation of (a) surfactant vaporisation and (b) vaporisation of substance which causes liquid to decrease in surface tension.

6.2 SEPARATION OF MINERAL FROM GANGUE

Frequently, mineral ores contain a very low concentration of the required metal e.g. 1% in the case of Cu. The froth flotation process used is to separate the required ore from the gangue by making use of differences in interfacial properties so that the ore attaches itself to a gas bubble whilst the gangue remains in the slurry of ore in water. For the ore to transfer from the water to a bubble it is necessary that the mineral be *hydrophobic*. However, most minerals are polar and are thus hydrophyllic. Consequently, collectors are added to the solution, these contain polar and non-polar parts, the polar part of the collector adheres to the polar part of the mineral leaving the covalent part of the collector poking out. Thus the mineral becomes hydrophobic (Figure 41). The mineral attaches itself to the bubble and is held in a foam. It is necessary that the foam be stable and this is promoted by a low surface tension. *Frothers* (such as pine oil) are added to the foam to lower the surface tension and stabilise the foam. Polar solutions have lower surface tensions. The *frother* also has covalent and polar parts, here the covalent part of the frother attaches itself to the covalent part of the treated mineral, leaving the polar part sticking out of the surface as this reduces the surface tension. This process allows the concentration to be increased to > 30%.

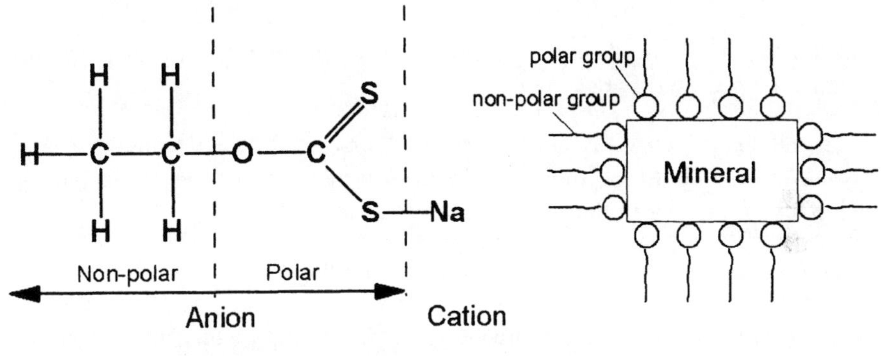

Figure 41 Schematic diagram showing the polar and non-polar parts of the collector, and its absorption onto the mineral

6.3 SEPARATION OF MINERALS BY

ELECTROCAPILLARY FORCES

When an electric field is applied to a system consisting of droplets of liquid phase B present in liquid A, electrocapillary forces can bring about the movement of these droplets. These forces can be used to recover trace metals and metal mattes from waste pyrometallurgical slags [47]. The driving force in thermocapillary flows is $(d\gamma/dT)$ i.e. the temperature dependence of the surface tension. Whereas in electrocapillarity the driving force is $(d\gamma/dE)$ where E is the electrical potential at constant chemical potential μ, and $(d\gamma/dE)_\mu$ is equal [47] to the surface excess charge density (q_E) at the droplet interface (Equation 18).

$$(d\gamma/dE)_\mu = -q_E \tag{18}$$

Figure 42demonstrates the mechanism responsible for the movement of a drop of phase B in liquid A (which acts an electrolyte) in an electric field. From Figure 43b it can be seen that there are potential drop (ΔE) due to (i) the bulk resistance of electrolyte A (ΔE_{A1} and ΔE_{A2}), (ii) the double layer (ΔE_{DL1} and ΔE_{DL2}) and (iii) the interior of the drop (ΔE_{B1} and ΔE_{B2}). Figure 43c shows a typical plot of interfacial tension (γ_{BA}) as a function of electrical potential, this is usually referred to as an electrocapillarity diagram. It can be seen that:

(i) at the centre of the drop there is a vertical line representing those points where the potential E=0;

(ii) the bell-shaped curve is asymmetric with respect to E=0 with the centre occurring a E<0 which implies by Equation 18 that there is a positive, q_E (surface excess charge density);

(iii) γ_{BA} is a maximum at a potential of zero charge (E_{PZC}).

Since Marangoni flow occurs in the direction of low γ to high γ the flow will occur (Figure 42c) along the surface of B towards regions corresponding to E_{PZC}. The build-up of pressure in the centre of the drop will result in a movement of the drop towards the left. Thus the net motion of the drop is a result of the asymmetry of the electrocapillarity diagram.

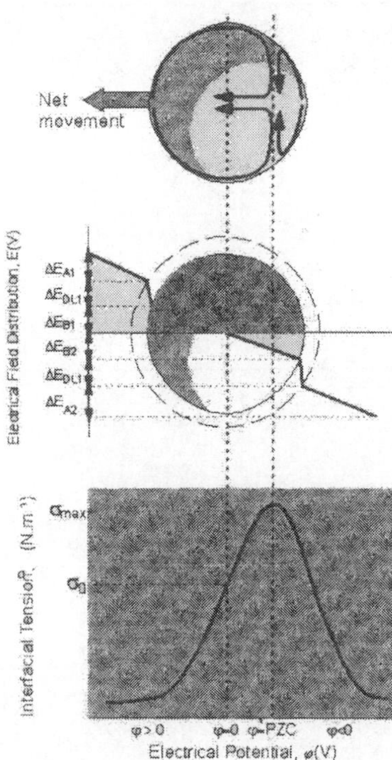

Figure 42 Spherical drop of phase B in phase A when no chemical (Faradaic) reactions occur at the electrode (a) Marangoni flow in the droplet and net movement of the drop (b) electrical field distribution and (c) electrocapillarity diagram.

More recently Takasu and Toguri [42] used electrocapillary forces to separate Cu_2S, FeS and Cu droplets in Fayalite (FeO, SiO_2) slag. A positive value in Figure 43 indicates that the Cu and Cu_2S migrate to the anode and a negative value indicates FeS will move to the cathode.

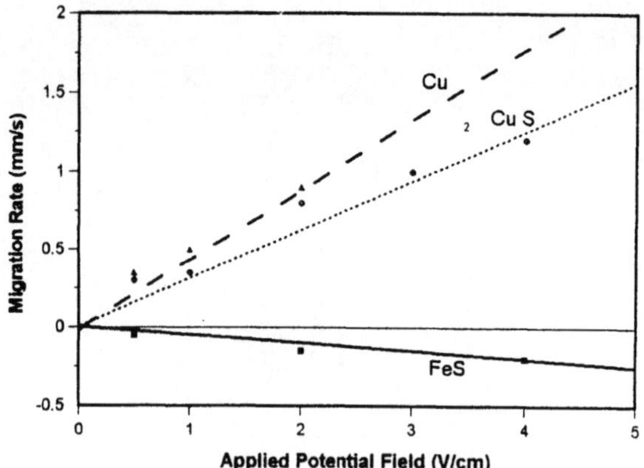

Figure 43 Effect of applied electric field on the migration rates of Cu, Cu2S (80% Cu) and FeS droplets on the surface of synthetic fayalite slag (70% FeO) at 1523K

6.4 REMOVAL OF INCLUSIONS FROM MOLTEN STEEL

The removal of an inclusion to the slag phase is promoted by a low value for the work of adhesion, W_A (see Section 2.4).

$$W_A = \gamma_{MG} (1+\cos\theta) \tag{19}$$

Thus inclusion removal is promoted by a low surface tension of metal (high concentration of S, or Se, Te) and a negative value for cos θ i.e. θ > 90° i.e. when the metal does not wet the inclusion.

For flotation the flotation coefficient Δ should be both positive and have a high value for inclusion removal. These conditions correspond to a high value of γ_{MG} or a low S concentration and once more a value of θ > 90°

$$\Delta = \gamma_{MG} (1-\cos\theta) \tag{20}$$

Liquid iron usually has a contact angle of θ > 90° for most oxides but this is not the case for the TiO_2 and may not be the case where the steel reacts with the inclusion (e.g. steel containing Al reacting with MgO inclusions to form Spinel $MgO.Al_2O_3$).

7 CORROSION, EROSION PROCESSES

7.1 HOT CORROSION OF TURBINE BLADES

This is an example of how relatively low levels of contamination occurring in service affect performance. Turbine blades are protected by an oxide layer (Al_2O_3 or Cr_2O_3) covering the surface. When operating at high temperatures in corrosive atmospheres, a thin layer of molten salt (such as cobalt and sodium sulphates) is formed on the blade and this dissolves the oxide layer especially in the high temperature regions. Thermocapillary convection occurs in the direction shown in Figure 44 i.e. from the hot to cold surface regions. When liquid saturated with Al_2O_3 in the high temperature region reaches the cold region, the Al_2O_3 becomes super-saturated and precipitates. The cold liquid returns by the circulatory flow to the hot region and dissolves more Al_2O_3 and the process is repeated. Thus there is a gradual, eventually catastrophic, dissolution of the protective oxide in the hotter region of the blade and precipitation of the oxide in the cooler regions.

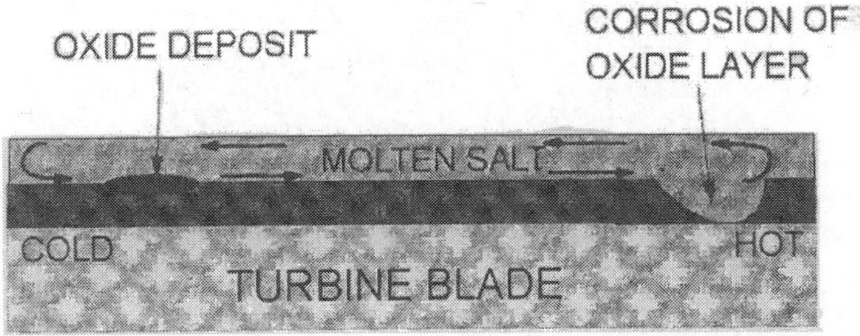

Figure 44 Schematic diagram showing the thermocapillary motion and circulation flows in the molten salt layer on the turbine blade [68].

7.2 REFRACTORY EROSION

The cost of refractories on a global scale is around 2 billion US $ per annum. Frequently, erosion of refractories occurs around the metal/slag interface or the gas/slag interface, which is known as "slag line" attack (Figure 45). This has frequently been attributed to Marangoni convection [69].

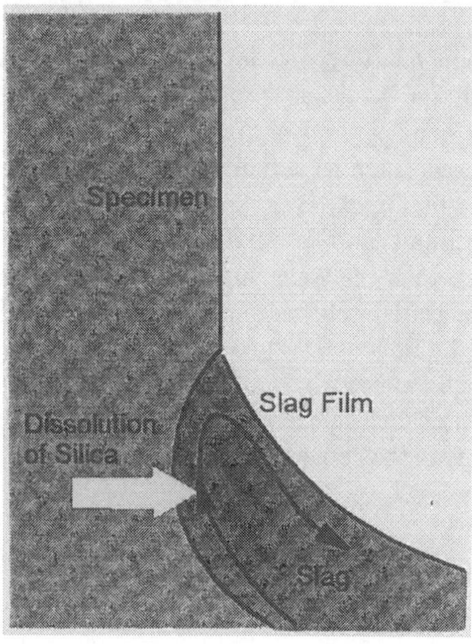

Figure 45 Marangoni convection of slag film in the local corrosion zone of a SiO_2(s)-(FeO-SiO_2) slag system showing typical "slag line" attack.

7.2.1 PbO – SiO_2 slag on SiO_2 refractory

The surface tension (γ) of SiO_2 is greater than that of PbO. The PbO-SiO_2 slag will attack the SiO_2 refractory and will dissolve the silica, thereby causing an increase in surface tension [69]. Consequently, the upper slag film (Figure 46), which has been in contact with the refractory for a longer time, will have a higher silica content and thus a higher surface tension than the lower slag film. Thus Marangoni flow will occur from a region of low surface tension to one of high surface tension (from lower to higher). Thus this will result in a slag film being pulled up the wall until it is balanced by the forces of gravity (in a similar way to wine tears).

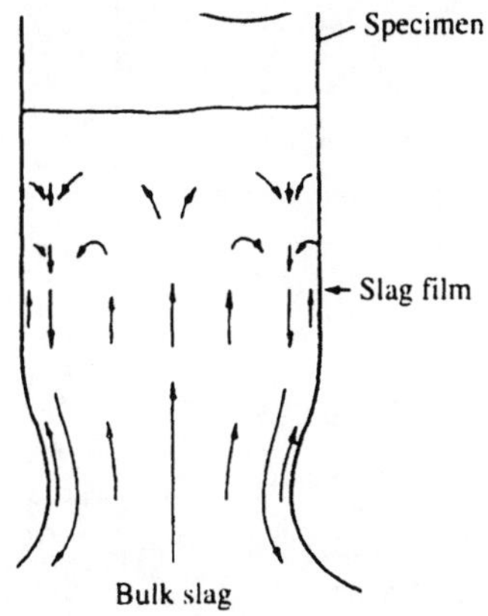

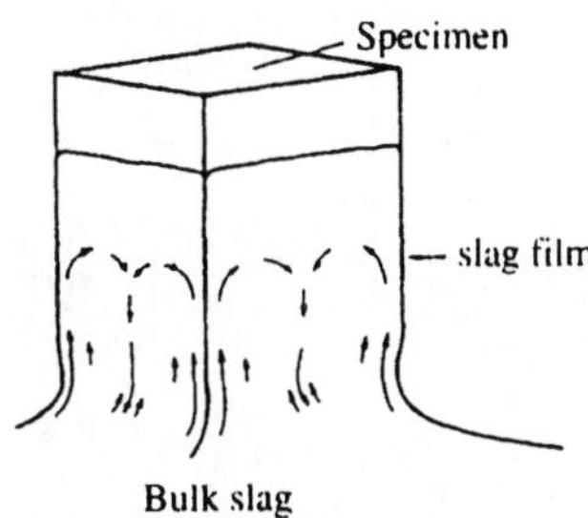

Figure 46 Typical flow patterns of a $PbO-SiO_2$ slag film on a cylinder, and prism of SiO_2.

7.2.2 . $FeO-SiO_2$ slag on SiO_2 refractory

The surface tension of silica is less than that of FeO. Thus, dissolution of silica will result in a decrease in surface tension of the slag. The local corrosion zone is narrow.

The slag was observed [69] to exhibit two different forms of motion:

(i) rotational motion around the specimen and

(ii) "up and down" motion which occurred during the developed stage of corrosion (Figure ure 47).

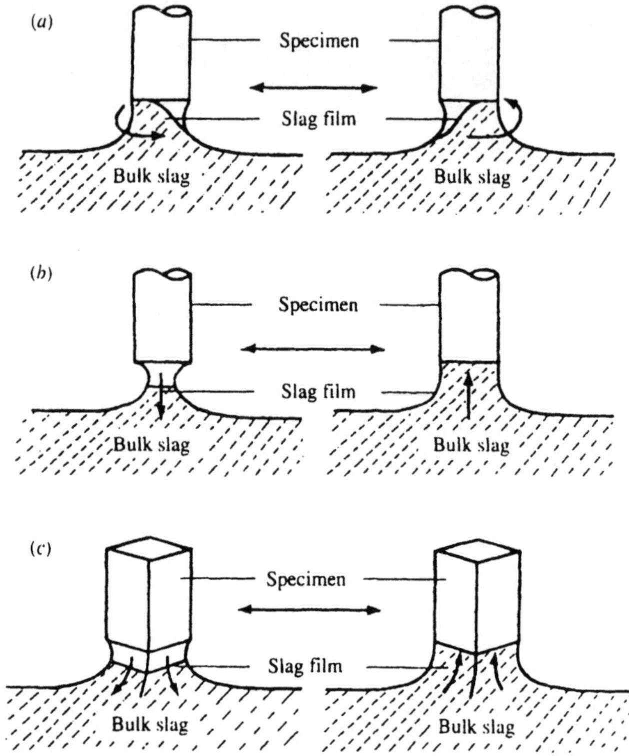

Figure 47 Slag film movements for rod and prism silica specimens dipped in a FeO-SiO_2 slag [69]: (a) rotational movement of slag film for the rod specimen; (b) up-and-down movement of slag film for the rod specimen; (c) up-and-down movement of slag film for the prism specimen

Since the silica content of the slag is greater at the wall than that in the bulk, Marangoni flow will sweep slag away from the refractory and into the bulk. This sets up a vortex and which enhances the erosion rate.

7.2.3 Corrosion of MgO-Carbon refractories

Modern refractories contain 10-20% graphite which improves:

(i) the thermal shock resistance of the refractory

(ii) the resistance to slag attack.

MgO is soluble in slag and carbon dissolves in steel. Wettability in high-temperature systems is usually associated with reactivity between the components. If the surface of the refractory is predominantly covered by MgO then this will favour the slag to come in contact with, and wet, the refractory. The MgO will then dissolve in the slag and this will cause an increase in surface tension and Marangoni flow will result in vortex formation and enhanced erosion around the interface.

When the MgO has dissolved then the surface will be predominantly carbon and the slag is non-wetting to slag, so the slag will withdraw and the surface will be covered by the steel which proceeds to dissolve the carbon until the surface is predominantly MgO. Whereupon, the steel withdraws and the slag covers the surface. Thus a new cycle begins. The mechanism is shown in Figure 48

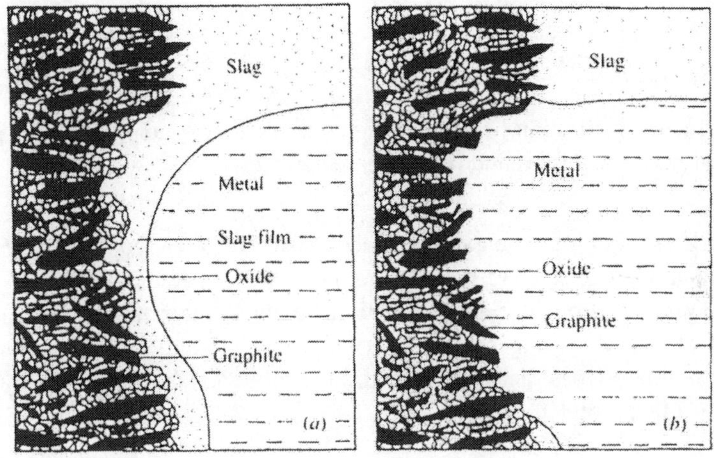

Figure 48 Schematic representation of the manner in which local corrosion of the immersion nozzle at the slag-metal interface proceeds [69].

7.3 EFFECT OF ELECTRICAL POTENTIAL

Kazakov et al [70] measured the vertical penetration of slag into the pores of a refractory when applying an electrical potential between the refractory and the melt, and found that the minimum penetration of slag into the refractory occurred when -1 volt was applied. Recent experiments [71] have confirmed these findings, as shown in Figure .

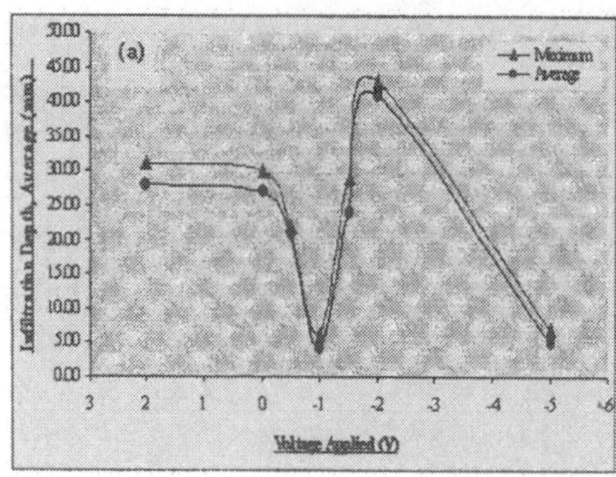
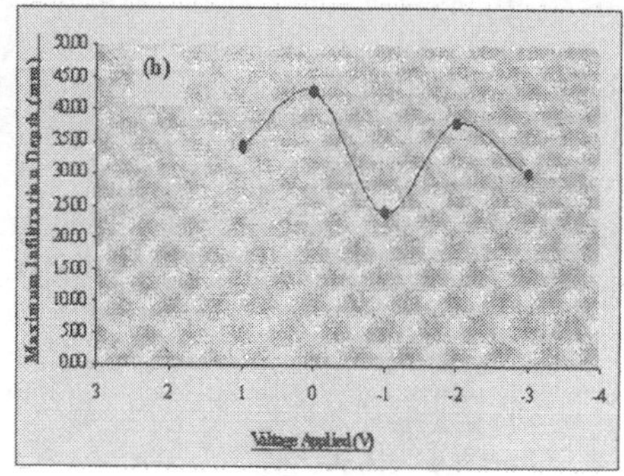

Figure 49 Inside infiltration height of slag as a function of applied voltage for (a) - cylindrical samples and (b) – square samples.

Kazakov suggested that an increased surface tension resulted from the change in the electrical potential across the refractory and melt, and was at a maximum at –1.0V, to explain the sudden drop in slag penetration height. Subsequently, Kazakov [55] carried out plant trials using the same principle and found that by applying a 2V electrical potential across the refractory and the molten slag, the refractory life was increased by a factor of 10.

The height (h) of vertical penetration of slag up a capillary derived using the Hagen-Poisseulle relationship is given by:

$$h = 2\gamma \cos\theta / (R \rho g) \tag{21}$$

Where R = the inner diameter zero, ρ = density of slag, η = viscosity of slag and g = gravitational constant, at equilibrium this reduces to:

Inspection of these equations shows that penetration increases with increasing surface tension, which is the exact opposite of the proposal due to Kazakov. However, recent work in which an electrical potential was applied to a sessile drop of slag (Figure 50) has shown that both the contact angle and surface tension both increase dramatically when an electric potential of -1 volt is applied [71].

Thus it is the increase in the contact angle which is responsible for the minimum in the penetration and not the increase in surface tension.

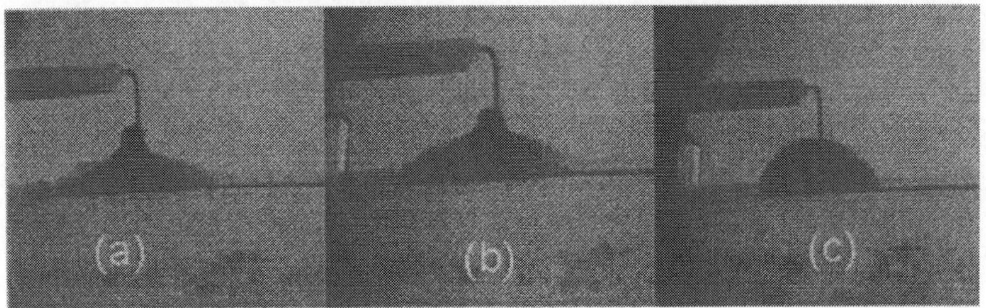

Figure 50 Variation in Slag droplet geometry as a function of Electric Potential after 2 minutes at (a) 0 v, (b) –1 v (c) –2 v [71].

REFERENCES

1. B J Keene: Intl. Materials Reviews 33 (1988) 1/38.

2. R F Brooks and K C Mills: High Temp-High Pressure 25 (1993) 657/664.

3. K C Mills and B J Keene: Intl. Materials Rev. 35 (1990) 185/216.

4. Y Chung: PhD Thesis "Surface and Interfacial phenomena of liquid iron alloys" Dept. Mater. Sci. and Eng., Carnegie-Mellon Univ., PA, April (1999).

5. A W Cramb and I Jimbo: Steel Research 60 91989) 157/165.

6. F D Richardson "Physical Chemistry of Melts in Metallurgy" 446/453

7. R Minto and W G Davenport: Trans. Inst. Min. Metall. 81C (1972) C36.

8. P Kozakevitch and L D Lucas: Rev. Metall. 65(9) (1968) 589.

9. B J Keene: "Contact angle and work of adhesion between Ni-based melts and non-metallic solids" NPL Report DMM(A)23 March (1991).

10. P Kozakevitch: Symp. Phys. Chem. of Steelmaking (1957) publ. MIT press Boston, p 134.

11. A Jakobsson, M Nasu, J Mangwiro, K C Mills and S Seetharaman: Phil. Trans. Roy. Soc. (London) A356 (1998) 991.

12. P V Riboud and L D Lucas: "Influence of mass transfer upon surface phenomena in iron- and steelmaking", Can. Met. Quart. 20 (1981) 199/208/

13. H Gaye, L D Lucas, M Olette and P V Riboud: Can. Met. Quart. 23 (1984) 179/191.

14. Y Kawai, N Shinozaki and K Mori: Can. Met. Quart. 21 (1982) 385/391.

15. C F Cooper and J Kitchener: JISI 193 (1959) 48/58.

16. J H Swisher and C L McCabe: Trans. Met. Soc. AIME 230 1669/1675.

17. K Ito and R J Fruehan: Metall. Trans. 20B (1989) 509/521.

18. C W Nexhip, S Sun and S Jahanshahi: Phil. Trans. Roy. Soc. (London) 336 (1998) 995/996.

19. C W Nexhip: "Fundamentals of foaming in molten slag systems" Dept. Chem. Eng., Melbourne Univ., Australia, Nov (1997).

20. Lord Rayleigh: Proc. Royal Soc. (London) 29 (1879) 71.

21. J Berg: Can. Met. Quart. 21 (1982) 121/136.

22. S S Glickstein and W Yeniscavich: Weld Res. Council Bull. 226 (1997) May.

23. W F Savage, E F Nippes and G M Goodwin: Welding J. 56 (1977) 126s/132s.

24. J R Roper and D L Olsen: Welding J. 57 (1978) 104s/107s.

25. C A Heiple and J R Roper: Welding J. 61 (1982) 975s.

26. J L Robinson and T G Gooch: Proc. ASTM and Intl. Conf. on Trends in Welding Res., Gatlinburg, TN, USA, 14-18/5/1989.

27. C A Heiple, J R Roper, R T Stagner and R J Aden: Welding J. 62 (1982) 72s.

28. K M Gupt, N I Yavoisky, A F Vishkaryov and S A Bliznakov: Met. Eng. Ind. Inst. Technol., Bombay (1972/3) 39/45.

29. K M Gupt, N I Yavoisky, A F Vishkaryov and S A Bliznakov: Trans. Ind. Inst. Metals 29 (1976) 286/291.

30. E F Friedman: Welding J. 57 (1978) 161s/166s.

31. K Ishizaki: J. Jap. Weld. Soc. 34 (1965) 146.

32. B J Keene, K C Mills, J W Bryant and E D Hondros, Can. Metall. Quart. 21 (1982) 393.

33. J F Lancaster: "The stability of meridional liquid flow induced by a gradient of surface tension or by electromagnetic forces" Int. Inst. Welding, Doc 212-682-87 (1987).

34. G M Oreper and J Szekely: J. Fluid Mech. 147 (1984) 55.

35. S Kou and D K Sun: Metall. Trans. A, A16 (1985) 203.

36. K C Mills, B J Keene, R F Brooks and A Olusanya: "The surface tensions of 304 and 316 stainless steels and their effect on weld penetration". Proc. Centenary Conf. Metallurgy Dept., Univ. Strathclyde, Glasgow, June (1984) paper R.

37. J L Robinson and T G Gooch: Proc. ASTM Intl. Conf. on Trends in Welding Res., Gatlinburg, TN, USA, 14-18/5/1989.

38. A A Shirali and K C Mills; Welding J. 72 (1993) 347s/352s.

39. B Pollard: Welding J. 67 (1988) 202s/213s.

40. P Burgardt and C A Heiple: Welding J. 65 (1986) 150s/155s.

41. M J Tinkler, I Grant, G Mizuno and C Gluck: Proc. Intl. Conf. on the effects of residual impurities and microalloying elements on weldability and weld properties, held London 1983, publ. The Welding Inst., Abington, Paper 27.

42. S Kou and Y H Wang: Welding J. 65 (1986) 633.

43. B N Antar, F G Collins and G H Fichtel: Intl. J.Heat Mass Transfer 23 (1980) 191/201.

44. M Holt, D L Olson and C E Cross: Scripta Met., 26 (1992) 1119/1121.

45. V Gratzke, P D Kapadia, J Dowden, J Kroos and G Simon: J. Phys. D25 (1992) 1640/7.

46. V A Nemchinsky: Intl. J. Heat Mass Transfer 40 (1997) 881/891.

47. T Takasu and J M Toguri: Marangoni and Interfacial Phenomena in Materials Processing, edited E D Hondros, M McLean and K C Mills, publ. Inst. Materials, London (1998) 153/166.

48. G R Belton and R J Fruehan: Proc. of ET Turkdogan Symp. publ. ISI, Warrendale (1994) 3/22.

49. V Sapiajwalla, I F Taylor and J K Wright: Proc. 52nd Ironmaking Conf., publ. AIME, Warrendale, PA (1993) 355/365.

50. Y Shinego, M Tokuda and M Ohtani: Trans. J. Inst. Metals 26 (1985) 33/43.

51. S Hagashi and Y Iguchi: ISIJ Intl. 37 (1997) 345/348.

52. M Olette: ISIJ Intl. 33 (1993) 1113/1124.

53. P Kozakevitch: J. Metals (1969) 57/68.

54. H Gou, G A Irons and W-K Lu: Proc. ET Turkdogan Symp., publ. ISI Warrendale, PA (1994) 83/91.

55. A A Kazakov: Russian Metall. 6 (1997) 25/29.

56. D R Gaskell and A Saelim: Phil. Trans. Roy. Soc. (London) 336 (1998) 995/996.

57. R Sebring and M Franken: Proc. 24.

58. K Umera, M Takahashi, S Koyama and M Nitta: ISIJ Intl. 32 (1992) 150/156.

59. S Feldbauer and A W Cramb: Proc. 78th Steelmaking Conf., Nashville, TN (1994).

60. Z Wang, K Mukai and D Izu: ISIJ Intl. 39 (1999) 959/966.

61. L Strezov and J Herbertson: ISIJ Intl. 38 (1998) 959/966.

62. S Amada, K Tomoyasu and M Haruyama: Surf. Coating Technol. 96 (1997) 176/183.

63. R Monti, R Savino and S Tempesta: Eur. J. Mech., B, Fluids, 17 (1998) 51/77.

64. P D Lee, P N Quested and M McLean: Marangoni and Interfacial Phenomena in Materials Processing, edited E D Hondros, M McLean and K C Mills, publ. Inst. Materials, London, 213/229.

65. D Schwake, A Scharmann, F Preissler and R Veder: J. Cryst. Growth 43 (1978) 305.

66. T Hibiya et al: Marangoni and Interfacial Phenomena in Materials Processing, edited E D Hondros, M McLean and K C Mills, publ. Inst. Materials, London, (1988) 85/96.

67. N Ebrill et al: Proc. Belton Symp. held Sydney, January 2000, publ. ISS, Warrendale, PA, pp

68. S K Lau and S C Singh: J. Mater. Sci. 18 (1983) 3743/3748.

69. K Mukai: Marangoni and Interfacial Processing edited E D Hondros, M McLean and K C Mills, publ. Inst. of Materials, London (1998) 201/212.

70. A A Kazakov.

71. S Riaz, S Teo, A Karam, K C Mills: unpublished results, Imperial College, London (2000).

Zeitfracht Medien GmbH
Ferdinand-Jühlke-Straße 7
99095 Erfurt, Deutschland
produktsicherheit@kolibri360.de